Undergraduate Lecture Notes in Physics

Series Editors

Neil Ashby, University of Colorado, Boulder, CO, USA

William Brantley, Department of Physics, Furman University, Greenville, SC, USA

Michael Fowler, Department of Physics, University of Virginia, Charlottesville, VA, USA

Morten Hjorth-Jensen, Department of Physics, University of Oslo, Oslo, Norway

Michael Inglis, Department of Physical Sciences, SUNY Suffolk County Community College, Selden, NY, USA

Barry Luokkala⊙, Department of Physics, Carnegie Mellon University, Pittsburgh, PA, USA

Undergraduate Lecture Notes in Physics (ULNP) publishes authoritative texts covering topics throughout pure and applied physics. Each title in the series is suitable as a basis for undergraduate instruction, typically containing practice problems, worked examples, chapter summaries, and suggestions for further reading.

ULNP titles must provide at least one of the following:

- An exceptionally clear and concise treatment of a standard undergraduate subject.
- A solid undergraduate-level introduction to a graduate, advanced, or non-standard subject.
- A novel perspective or an unusual approach to teaching a subject.

ULNP especially encourages new, original, and idiosyncratic approaches to physics teaching at the undergraduate level.

The purpose of ULNP is to provide intriguing, absorbing books that will continue to be the reader's preferred reference throughout their academic career.

More information about this series at https://link.springer.com/bookseries/8917

Jean Bricmont

Making Sense of Statistical Mechanics

 Springer

Jean Bricmont
UCLouvain
Louvain-la-Neuve, Belgium

ISSN 2192-4791   ISSN 2192-4805 (electronic)
Undergraduate Lecture Notes in Physics
ISBN 978-3-030-91793-7   ISBN 978-3-030-91794-4 (eBook)
https://doi.org/10.1007/978-3-030-91794-4

This Springer imprint is published by the registered company Springer Nature Switzerland AG
The registered company address is: Gewerbestrasse 11, 6330 Cham, Switzerland

Contents

Chapter 1
Introduction

Statistical mechanics can be quite disconcerting for a student. One the one hand, she or he is told that one must use probabilistic notions because of the large number of variables involved, but the meaning of the word "probability" is often not made explicit, beyond its mathematical definition. Does probability mean a judgment about the likelihood of some event? But then, what do our judgments have to do with things, like heat flowing from hot to cold, that occur independently of human beings' beliefs? Or should one understand probabilities as the frequencies of occurrence of certain events? But which events exactly? And then, why does statistical mechanics derive laws like those of thermodynamics, whose frequency of occurrence is equal to one, since they are, or at least seem to be, deterministic?

Soon after being told to use probabilities, the student is instructed to employ the method of ensembles and to compute physical quantities by averaging them over the set of microscopic configurations (atoms or molecules), with certain probability distributions whose choice depend on whether the system is isolated (the micro-canonical ensemble), can exchange energy but not particles with its environment (the canonical ensemble) or can exchange both (the grand-canonical ensemble). But the relation between these ensembles is not always clear and, more fundamentally, it is unclear why we would want or need to compute those averages. After all, what we observe in any given physical situation described by statistical mechanics are, in a sense, events that occur only once: we may see many times heat flowing from hot to cold, but we can be quite sure that the microscopic configurations of the system under consideration vary between different such events. So, what is exactly the role of averages?

We are sometimes told that the reason to take averages is related to the law of large numbers in probability theory, but unless we have a clear notion of what probabilities are, this justification remains somewhat obscure. In order to justify these averages, we are also referred to ergodic theory, which is part of the theory of dynamical systems, but that approach meets several difficulties that will be discussed in this book.

© Springer Nature Switzerland AG 2022
J. Bricmont, *Making Sense of Statistical Mechanics*, Undergraduate Lecture
Notes in Physics, https://doi.org/10.1007/978-3-030-91794-4_1

One way to answer all those questions is to turn to the theory of information and to consider statistical mechanics as a branch of that theory: isn't the whole of science about information anyway? If so, let us use some principle like the one of maximum entropy to choose the appropriate probability distribution with respect to which we take averages of physical quantities. If the predictions based on that method are verified, then there is no further question to be raised. This approach is linked to a philosophical attitude with respect to the goals of science (which are, for that attitude, to describe or to predict phenomena rather than to explain them), and we will criticize both that attitude and the information theoretic approach precisely because of their lack of explanatory power.

All of the above deals with the foundation or justification of equilibrium statistical mechanics, where at least the mathematical formalism is entirely clear and can be put on a rigorous mathematical basis. But the explanation of why systems approach an equilibrium state, starting from a non equilibrium one, or transit from one equilibrium state to another is often even more obscure.

The student is sometimes directed again to dynamical concepts such as ergodicity, mixing or "chaos" or else, to irreversible equations like Boltzmann's equation. The former approach meets the same problems as for the justification of the equilibrium formalism, while the latter has met enormous resistance (in its inception) and a good deal of confusion ever since.

The validity of Boltzmann's equation depends crucially on an hypothesis: the one of molecular chaos or the *Stosszahlansatz* in German (using German words always looks more serious). But what is the status of that assumption? Is it purely mechanical? Does it follow from the fact that this equation holds for dilute gases? Or is is statistical? If it is mechanical, then how can one derive an irreversible equation, describing convergence to equilibrium, from mechanical laws that are well-known to be reversible in time?

But if the assumption of molecular chaos is statistical, what is it precisely? Actually, it is not even clear, when one goes through the literature, what is exactly the f function (as it is usually called), whose behavior is governed by Boltzmann's equation. Does it refer to an average, taken again over some ensemble, or does it refer to individual systems?

Finally, there is the perplexing concept of entropy: what is it exactly, what does it measure and what does it mean? We will encounter five notions of entropy: the thermodynamic one of Carnot and Clausius, the one of Boltzmann and the one of Gibbs in statistical mechanics, Shannon's entropy in the theory of information and the Kolmogorov-Sinai entropy in the theory of dynamical systems. It is important to distinguish the mathematical similarity between the formulas expressing those notions from their real world significance.

It would preposterous to claim that this book will clarify all those issues and even more to claim any originality in doing so. But attempts at clarification is the goal of this book; it will rely on the Boltzmannian (sometimes called neo-Boltzmannian) approach to statistical mechanics, based on the work of Joel Lebowitz, Sheldon Goldstein and their collaborators. Besides, we hope to convince the reader that many issues in the "foundations" of statistical mechanics are unclear in large parts of the literature, even in the writings of great physicists and mathematicians.

This book is not mathematically sophisticated; there is some measure theory in Appendix 2.A (but that the reader may skip in large parts), which is used to some extent in Chap. 4, but the rest of the book uses basic calculus and sometimes linear algebra. Most of the time, we avoid the technical complications associated with interacting systems and limit ourselves to non-interacting ones, since our goal is conceptual rather than mathematical (we will give references to extensions to interacting systems).

This book is aimed at undergraduate or graduate students who fail to get a clear picture of statistical mechanics from their courses (my situation when I was a student) or more advanced physicists who are in search of such a clearer picture and, also, at philosophers of science.

We have chosen not to include a discussion of quantum statistical mechanics; partly because the book would be much longer, partly because there are additional difficulties caused by the necessity of a proper understanding of the quantum formalism and also because, once the statistical notions used in classical statistical mechanics are understood and the meaning of the quantum formalism is clarified, the extension to the quantum situation is not that difficult.[1]

We will not discuss either genuine non equilibrium phenomena, namely for "open" systems, interacting with their environment; the main reason being that this subject is too vast and not very well understood.

On the other hand, when we discuss probabilities (in Chap. 2), dynamical systems (in Chap. 4) and the theory of information (in Chap. 7), we will go well beyond what is needed in the rest of the book.

Here is an outline of the book's content:

- In Chap. 2, we discuss the meaning of the concept of probability, about which there are at least two opposing views: the subjective or Bayesian one and the objective or frequentist view. Our main goal is to clarify the status of the law of large numbers and its relation with probabilistic explanations.
- In Chap. 3, we give the basics of classical mechanics needed in statistical mechanics.
- In Chap. 4, we give an elementary but detailed discussion of dynamical systems; there are two reasons for that: one reason is that those systems can be analyzed by statistical methods similar to those used in statistical mechanics and the second reason is that it is sometimes thought (erroneously in our view) that the "chaotic" properties of some dynamical systems are relevant to statistical mechanics, in particular in order to understand the approach to equilibrium.
- Chapter 5 is a standard review of equilibrium thermodynamics, which is included mainly for reasons of completeness.
- In Chap. 6, we introduce the Boltzmannian approach to equilibrium statistical mechanics, define Boltzmann's entropy, as well as the one of Gibbs, define the

---

[1] See e.g. the lecture note by Tumulka [314] for such an extension.

various "ensembles" (or probability distributions), explain in what sense they are equivalent and discuss their meaning.

- Chapter 7 presents the information theoretic approach, based of the notion of Shannon's entropy and its maximization. Although we are critical of that approach, we try to give it a fair hearing, which is required because of its popularity.
- Chapter 8 is the central chapter of this book, where Boltzmann's ideas about approach to equilibrium are explained in detail. We will answer various objections to that approach, discuss alternative misleading "solutions," see precisely what is the status of Boltzmann's equation and illustrate those ideas in several simple models. We will also consider the difference between Boltzmann and Gibbs entropies in the context of approach to equilibrium, and discuss the (sometimes alleged) subjectivity of entropy and irreversibility.
- Chapter 9 discusses the theory of phase transitions, but is not more than a guided tour through that theory, a subject that would deserve a whole book to be treated adequately. We include it here mainly to show that the equilibrium formalism can lead to highly non trivial conclusions.
- In the final Chap. 10, we make some remarks about the notion of reductionism in science.

There is, at the end of each chapter, a summary of its content; the reader may wish to start with that summary. We have also added exercises, whose solutions are given or sketched in Chap. 11. Even if the reader cannot solve certain exercises, reading their solution may be enlightening. We indicate by a * the most difficult exercises.

Acknowledgements I have benefited from discussions with so many people over the subjects treated in this book that I could not possibly list all of them. Let me say that I have been inspired specially by discussions and work with Joel Lebowitz and Sheldon Goldstein and by their lectures and writings. I have also greatly benefitted from lectures by Roderich Tumulka [314] and Nino N. Zanghì [333] and also from my collaboration with Antti Kupiainen, Koji Kuroda, Christian Maes, Charles-Édouard Pfister and Joseph Slawny while working on phase transitions.

I thank Valia Allori, Michel Ghins, Dustin Lazarovici, John Norton and Roderich Tumulka for comments and discussions on parts on this book. I thank Luc Haine for his collaboration in writing lecture notes that were used in parts of Chap. 4. I also thank Antoine Bricmont for his help with the pictures.

Finally I thank my editor, Angela Lahee, for her encouragement and her patience.

Chapter 2
Probability

*Probability is the most important concept in modern science,
especially as nobody has the slightest notion what it means.*

Bertrand Russell (From a 1929 Lecture; quoted by Bell [21, p. 587])

2.1   Introduction

The goal of this chapter is to discuss the meaning of the word "probability" as it is used in the natural sciences. Probability theory is of course an important branch of mathematics (formalized by Kolmogorov [196] and others), which itself is part of measure theory, and that we will summarize in Appendix 2.A.

But that branch of mathematics does not tell us what "probability" means when it is used in physics or other sciences. There are broadly speaking, two schools of thought on this issue: the so-called subjective or Bayesian approach and the objective or frequentist one. After a brief introduction to the problem, we will discuss the first approach in Sect. 2.2 and the second in Sect. 2.4 after a mathematical interlude in Sect. 2.3. In Sect. 2.4, we will also discuss the propensity approach of Popper which tries to be a sort of third way beyond Bayesianism and frequentism. Our main goal however will be to discuss the role of probabilistic reasonings in scientific explanations and that will be done in Sect. 2.5. It will rely on what is known as Cournot's principle or typicality.

But let us start with a very simple problem: a family has two children, one of them is a boy; what is the probability that it has two boys? There are two wrong answers to that question: one is $\frac{1}{4}$, because that is the probability for a family with two children of having two boys (assuming that the probability of having a boy or a girl are equal and that the sex of different children are independent of one another) and neglecting the information that we know that one of the children is a boy. Another wrong answer (the "sexist" one) is $\frac{1}{2}$, because it assumes that the boy is the first child.

© Springer Nature Switzerland AG 2022
J. Bricmont, *Making Sense of Statistical Mechanics*, Undergraduate Lecture Notes in Physics, https://doi.org/10.1007/978-3-030-91794-4_2

Of course, the correct answer is $\frac{1}{3}$, since there are three "possibilities" for families with two children, one of them being a boy: Boy-Boy, Boy-Girl, Girl-Boy; and one "favorable" one, Boy-Boy.

This exemplifies the standard definition of the probability of an event E:

$$P(E) = \frac{\text{number of favorable events}}{\text{number of possible events}} \qquad (2.1.1)$$

where "favorable" means events where E is realized. Of course, here there is only one such event, but if one asks the probability that the number on the face that is up when a die is thrown is even, then there are three favorable events (2, 4, 6) and six possible ones and so the probability is $\frac{1}{2}$.

In the reasoning above, one wrong answer, $\frac{1}{4}$, occurs because one neglects some information, (that one child is a boy) and the other $\frac{1}{2}$ occurs because one thinks one has some information that is not given (that the boy is the first child).

This illustrates the principle that, in calculating probabilities, one must use all the information available but not additional information that one does not have.

In the classical approach to probabilities, as formulated among others by Laplace, in order to compute the denominator in (2.1.1), one must count the number of cases about which we are "equally uncertain":

> The theory of chances consists in reducing all events of the same kind to a certain number of equally possible cases, that is to say, to cases whose existence we are equally uncertain of, and in determining the number of cases favorable to the event whose probability is sought. The ratio of this number to that of all possible cases is the measure of this probability, which is thus only a fraction whose numerator is the number of favorable cases, and whose denominator is the number of all possible cases.

> Pierre-Simon Laplace [209]

It is well known that Laplace started his essay on probability by imagining a "demon" or "intelligence" possessing exact knowledge of everything in the world:

> Given for one instant an intelligence which could comprehend all the forces by which nature is animated and the respective situation of the beings who compose it—an intelligence sufficiently vast to submit these data to analysis—it would embrace in the same formula the movements of the greatest bodies of the universe and those of the lightest atom; for it, nothing would be uncertain and the future, as the past, would be present to its eyes.

> Pierre Simon Laplace, [209], p. 4.

But Laplace immediately added that *we* shall "always remain infinitely removed" from this imaginary "intelligence" and its ideal knowledge of the "respective situation of the beings who compose" the natural world, that is, in modern language, of the precise initial conditions of all the particles. He distinguished clearly between what nature does and the knowledge we have of it. For Laplace, probability is nothing but a method that allows us to reason in situations of partial ignorance (i.e. our situation, but not the one of the demon). The meaning of Laplace's quote is completely misrepresented if one imagines that *he* hoped that one could arrive someday at a perfect knowledge and a universal predictability, for the aim of his essay was precisely to explain how to proceed in the absence of such a perfect knowledge.[1]

For Laplace the notion of "possible cases" is related partly to our knowledge and partly to our ignorance: in the example of the family with two children, we know that one of them is a boy but we do not know if he is the first or the second child.

However, one may still ask what does the word "possible" mean. If we take a given family with two children, either it has two boys or it does not. What does it mean to say, for that particular family, supposing that it has two boys, that it is "possible" for that family not to have two boys?

A frequent answer to that question goes through the law of large numbers (which we will just state informally here and discuss in detail in Sect. 2.3): suppose we take a large number (say, a thousand) of families with two children, one of which is a boy, then, "on average" there will be one third of those families with two boys and two thirds with at least one girl. But if each particular family with two children either has two boys or does not, the same is true for the thousand families: the fraction of them that has two boys is whatever it is; in what sense could it "possibly" be different from what it is?

One answer is that it is "very probable" for the fraction of families with two boys to be "approximately" $\frac{1}{3}$. But now we use the word probable or very probable to define the word "possible" that enters into the definition of probable. There seems to be some circularity here.

Another way to give a meaning to probabilities is related to gambling: suppose I am asked to give one dollar in order to guess whether a given family with two children, one of which is a boy, has two boys. And I gain x dollars if I guess right, but I lose my dollar if not. Obviously, I should always guess that the family does not have two boys, since this is the more probable situation. But what is the value of x that makes the game worthwhile for me to play? Since I will win in $\frac{2}{3}$ of the cases and lose in $\frac{1}{3}$ of them, my average gain is:

$$\frac{2}{3}x - \frac{1}{3}$$

which is strictly positive if $x > \frac{1}{2}$.

Then, the word probability acquires a meaning as a rule of practical rationality: it is *rational* for me to pay one dollar to play that game if the amount x that I gain if I guess correctly is larger than a half dollar.

But suppose we play that game and lose systematically: almost all families encountered in the game with two children, one of which is a boy, have two boys! What should one say then?

One possible answer is that that set of families is "special" or is not "random". But what does *that* mean? How to define random without appealing to the notion of probability, which is what we are just trying to define?

---

[1] See Sect. 4.8 for a further discussion of Laplace's views.

Another possibility could be that there is an unknown law of nature that makes the occurence of two boys in a family with one boy very probable or, in other words, that the sexes of different children are not independent of each other.

So, is probability only related to our knowledge and to our ignorance? But then, what does it have to do with the physical world? Does it acquire a meaning through many repetitions of the same event? Or is its meaning coming only from rules of practical rationality? Do events need to be random in order to apply probability to them? And if so, what does random mean?

And, last but not least, if one applies probability to physics, which is the point of this book, which meaning of probability are we going to use? Is the fact that heat flows from hot to cold (which can be considered as a law of statistical physics) something that depends on our knowledge or our information? Is that related to gambling? Do we have to repeat many times an experiment where a hot and a cold reservoir are brought into contact to see in which way heat flows?

All this seems absurd (at least I hope it does seem so to the reader)! In the next sections, we will consider different notions of probability and then try to see what their connection to physics is, without pretending do go in depth into the "philosophy of probability", a subject that would require several volumes to be treated adequately (see [166] for an overview).

2.2 "Subjective" Versus "Objective" Probabilities

As we said, there are, broadly speaking, two different meanings given to the word 'probability' in the natural sciences. The first notion is the so-called "objective" or "frequentist" one, namely the view of probability as something like a "theoretical frequency": if one says that the probability of the event E under condition X, Y, Z equals p, one means that, if one reproduces the 'same' conditions X, Y, Z "sufficiently often", the event E will appear with frequency p. Of course, since the world is constantly changing, it is not clear what reproducing the 'same' conditions means exactly; besides, the expression "sufficiently often" is vague and this is the source of much criticism of that notion of probability.[2] But, putting those objections aside for a moment, probabilistic statements are, according to the "frequentist" view, factual statements that can in principle be confirmed or refuted by observations or experiments. We will come back to the discussion of the frequentist view in Sect. 2.4 below, but now we will turn to the other meaning of the word 'probability', the "subjective" or Bayesian one.

---

[2] The notion of 'event' is also ambiguous. Suppose that we toss a thousand coins. Can we regard this as a single event? Should one identify probabilities of the number of heads and tails in that 'event' by repeating a thousand times the tossing of a thousand coins? And what about considering that latter repetition as a single event?

In this approach, probabilities refer to a form of reasoning and not to a factual statement. Assigning a probability to an event expresses a judgment on the likelihood of that single event, based on the information available at that moment. Note that, here, one is not interested in what happens when one reproduces many times the 'same' event, as in the objective approach, but in the probability of a single event. This is of course very important in practice: when I wonder whether I need to take my umbrella because it may rain, or whether the stock market will crash next week, I am not mainly interested in the frequencies with which such events occur but with what will happen here and now; of course, these frequencies may be part of the information that is used in arriving at a judgment on the probability of a single event, but, typically, they are not the only information available.

One may even ask probabilistic questions, like "what is the probability of life or of intelligent life in the universe or in our galaxy?" or "what is the probability that the value of a given physical constants lies in a given interval" that do not make sense from a frequentist point of view.[3] Yet, people do try to answer these questions; those answers may not be better than educated guesses, but these examples show that our intuitive notion of probability is not restricted to theoretical frequencies.

Note that, to add to the confusion, one has to make a distinction between the "objective" or "rational" Bayesian approach and the "subjective" Bayesian approach. In the latter approach, which was championed among others by the Italian mathematician de Finetti [93, 94], probabilities can be assigned more or less arbitrarily, provided one follows the rules of probability, like

$$P(A \cup B) = P(A) + P(B), \tag{2.2.1}$$

whenever $A \cap B = \emptyset$.

If the equality in (2.2.1) is replaced by a strict inequality, then a gambler putting bets according to this modified rule could be victim of a "Dutch book", meaning that there is a strategy that would make him lose with certainty whatever the outcomes of the game are.[4] For example, suppose that you violate the additivity rule by assigning probabilities so that:

$$P(A \cup B) < P(A) + P(B),$$

for some A and B with $A \cap B = \emptyset$.

Then a cunning bettor could buy from you a bet on $A \cup B$ for $P(A \cup B)$ units (meaning that he pays you an amount $P(A \cup B)$ units of money and wins a fixed amount of money M if the event $A \cup B$ occurs), and sell you bets on A and B individually for $P(A)$ and $P(B)$ units of money respectively (meaning that you pay him an amount $P(A)$ and $P(B)$ units of money and that he will pay you the same amount of money M if the event A occurs or if the event B occurs). He pockets an initial profit of $P(A) + P(B) - P(A \cup B)$, and retains it whatever happens (since

---

[3] See Cox [90] for more discussion of such examples.

[4] The origin of the expression "Dutch book" is not clear. 'Book' here refers to booking, see [322].

he will loose and win an amount of money M if either A or B occurs: if A occurs, he looses M, but he also wins M since then, the event $A \cup B$ has occurred, and the same is true if B occurs).[5]

The Dutch book argument shows that rationality requires your probabilities to obey the probability calculus.[6]

By contrast, the "rational" Bayesian approach, championed among others by the physicist Jaynes [180, 183] considers probabilities as being subjective in the sense that they consist in judgments (made by us) and not facts "out there" in the world, but also insists that assignments of probabilities to events should not be made arbitrarily (i.e. following nothing more than rules like (2.2.1) in order to avoid "Dutch book" problems) but should also be constrained rationally by incorporating all the information that we have about the situation, but not information that we do not have.

How does a rational Bayesian assign subjective probabilities to an event? This relied originally on what the economist John Maynard Keynes (who started his career as a theoretician of probability) called the "indifference principle" ([190, Chap. 4]), and Laplace called the "principle of insufficient reason", and that we will explain now.

2.2.1 The Indifference Principle

This principle says: first, list a series of possibilities for a "random" event, about which *we know nothing*, namely that we have no reason to think that one of them is more likely to occur than another one (so that "we are equally ignorant" with respect to all those possibilities). Then, assign to each of them an equal probability. If there are N possibilities, we have:

$$P(i) = \frac{1}{N}, \tag{2.2.2}$$

$\forall i = 1, \ldots, N.$

---

[5] If the reader prefers a concrete example, let us consider the throw of a die and let A be the event "the number on the face that is up is even" and let B be the event "the number on the face that is up is odd" and let M be one dollar. Then, if you believe that the probability that the number on the face that is up is even *or* the number on the face that is up is odd is strictly less than the probability that the number on the face that is up is even *plus* the probability that the number on the face that is up is odd (which would be silly of course, but goal here is to show that violating the equality (2.2.1) is silly), then you can convince yourself by following the argument here that you will be victim of a Dutch bet.

[6] There is also the converse theorem that, if your subjective probabilities conform to the probability calculus, then no Dutch book can be made against you [189]. For a further discussion of Dutch book arguments, see [322].

This "principle" is just another expression of our equal ignorance.[7]

There are many problems with this definition and several objections have been raised against it. First of all, when are we in this situation of indifference? In games of chance where there is a symmetry between the different outcomes of the random event (tossing of a coin, throwing of a die, roulette wheels etc.) it is easy to apply the indifference principle. But for more complicated situations, it is not obvious how to proceed.

Some people object that we use our ignorance to gain some information about that random event: at first, we do not know anything about it and from that we deduce that all those events are equally probable. But, from a subjectivist view of probabilities, not knowing anything about a series of possibilities and saying that all those possibilities have equal probabilities are equivalent statements, since, in that view, a probability statement is not a statement about the world but about our state of knowledge.

In more complicated situations, where there is no symmetry between the different possibilities one uses the *maximum entropy principle*. Namely one assigns to each probability distribution $\mathbf{p} = (p_i)_{i=1}^{N}$ over N objects its Shannon entropy, given by:

$$S(\mathbf{p}) = -\sum_{i=1}^{N} p_i \log p_i.$$

One then chooses the probability distribution that has the maximum entropy, among those that satisfy certain constraints that incorporate the information that we have about the system.

The rationale behind this principle, as for the indifference principle, is not to introduce bias in our judgments, namely information that we do not have (like people who believe in lucky numbers). And one can argue that maximizing the Shannon entropy is indeed the best way to formalize that notion. We will discuss in detail this idea and its justification in Sect. 7.2, see also Shannon [291] and Jaynes [180], [183, Sect. 11.3].

---

[7] Since there is no uniform probability distribution on $\mathbb{N}$, we cannot extend formula (2.2.2) to events with a countable infinity of possible outcomes. But one can extend it to subsets of $\mathbb{R}^N$ with finite Lebesgue measure (see Appendix 2.A). If $\Omega \subset \mathbb{R}^N$ has $\mu_{\text{Leb}}(\Omega) < \infty$, then the extension of (2.2.2) is, for $A \subset \Omega$,

$$P(A) = \frac{\mu_{\text{Leb}}(A)}{\mu_{\text{Leb}}(\Omega)}.$$

2.2.2 Cox' "Axioms" and Theorem

As an aside, let us mention also that, in 1946, Cox [90], inspired by previous ideas
of Keynes [190], gave a foundation to the "subjective" approach to probability based
on reasonings about the plausibility of *propositions* (see Jaynes [183] for an extensive discussion of this approach). Instead of assigning probabilities to events, as in
elementary probabilities, or to sets, as in the mathematical version (see Appendix
2.A), Cox gives a numerical value to the *plausibility* $\mathcal{P}(p \mid q)$ of a proposition p
given that another proposition q is true.[8]

Then, Cox imposes some rules of rationality on those plausibility assignments
and derive from them, for a given proposition r, the sum rule:

$$\mathcal{P}(p \text{ or } q \mid r) = \mathcal{P}(p \mid r) + \mathcal{P}(q \mid r) - \mathcal{P}(p \text{ and } q \mid r), \tag{2.2.3}$$

and the product rule:

$$\mathcal{P}(p \text{ and } q \mid r) = \mathcal{P}(p \mid q \text{ and } r)\mathcal{P}(q \mid r) = \mathcal{P}(q \mid p \text{ and } r)\mathcal{P}(p \mid r). \tag{2.2.4}$$

If we replace the propositions by sets (e.g. sets of events that render the propositions true), (2.2.3), expressed in terms of probabilities of sets, means:

$$P(A \cup B) = P(A) + P(B) - P(A \cap B), \tag{2.2.5}$$

which reduces to (2.2.1) when p and q are incompatible, namely when the sets A and
B of events for which those propositions are true are disjoint; besides (2.2.5) follows
by applying (2.2.1) to the disjoint union $A \cup B = (A \setminus B) \cup (B \setminus A) \cup (A \cap B)$ and
using $A = (A \setminus B) \cup (A \cap B)$ and $B = (B \setminus A) \cup (A \cap B)$.

Equation (2.2.4), expressed in terms of probabilities of sets, means:

$$P(A \cap B \mid C) = P(A \mid B \cap C)P(B \mid C) = P(B \mid A \cap C)P(A \mid C),$$

where by definition,

$$P(A \mid B) = \frac{P(A \cap B)}{P(B)} \tag{2.2.6}$$

is the *conditional probability* of event A given event B.

---

[8] Both Cox and Jaynes argue that this notion of plausibility is more general than the one of probability
of events. Of course, if the propositions asserts something about properties of a set of events, its
plausibility will coincide with probability of the set of events that renders the proposition true, but
one can think of more general propositions.

From (2.2.6), we get Bayes' formula[9]:

$$P(A \mid B) = \frac{P(B \mid A)P(A)}{P(B)},$$

a formula which has the distinctive feature of being both trivial and all-important. Indeed, it is the basis on which rational Bayesians change their probabilities when new information arises.

2.2.3 Bayesian Updating

Suppose that we have a certain number of hypotheses $H_1, H_2, \ldots, H_n$ and that we have assigned probabilities $P(H_i)$ to each of them, probabilities that exhaust all possibilities and are mutually exclusive:

$$\sum_{i=1}^{n} P(H_i) = 1.$$

Those probabilities are called the *prior probabilities.*

Now, we collect new data (D) and we want to know how to change our assignments of probabilities to those various hypotheses. We will write $P(H_i \mid D)$ for the (updated) probability of hypothesis H_i, given D.

We assume that we know enough about the system to compute the probabilities of the data, for each hypothesis: $P(D \mid H_i)$, $i = 1, \ldots, n$. Those probabilities are called the *likelihoods.*

Then we simply use Bayes' formula:

$$P(H_i \mid D) = \frac{P(D \mid H_i)P(H_i)}{P(D)}, \tag{2.2.7}$$

where $P(D) = \sum_{i=1}^{n} P(D \mid H_i)P(H_i)$; this implies that the new probabilities still add up to one:

$$\sum_{i=1}^{n} P(H_i \mid D) = 1.$$

The probabilities $P(H_i \mid D)$ are called the *posterior probabilities.*

To illustrate this method, consider the well known and apparently paradoxical example of "false positives" in medical testing: assume that the prevalence of a disease in the general population is 0.5% and assume that a test for that disease gives a 99% true positive rate (99% of the people who have the disease are tested positive)

---

[9] For a historical discussion of the origins of that formula, which was known before Bayes' work and was developed by Laplace, see Jaynes [183, Sect. 4.6.1].

and with a false positive rate of 2% (2% of the people who do not have the disease are tested positive). If a random person tests positive, what is the probability that this person has the disease?

In formulas, let $n = 2$, let H_1 be the hypothesis that a person has the disease and let H_2 be the hypothesis that this person does not have the disease. Likewise, let the data D be the positive test. We are asked to compute the posterior probability $P(H_1 \mid D)$.

We are given the likelihoods $P(D \mid H_1) = 0.99$, $P(D \mid H_2) = 0.02$, and the prior probabilities $P(H_1) = 0.005$ and thus $P(H_2) = 0.995$; we compute: $P(D) = P(D \mid H_1)P(H_1) + P(D \mid H_2)P(H_2) = 0.99 \cdot 0.005 + 0.02 \cdot 0.995$. From (2.2.7), we get the posterior probability:

$$P(H_1 \mid D) = \frac{P(D \mid H_1)P(H_1)}{P(D)} = \frac{0.99 \cdot 0.005}{0.99 \cdot 0.005 + 0.02 \cdot 0.995} = 0.1992 \approx 20\%,$$

which is surprising at first sight since it means that being tested positive implies that you have only about 20% chances of having the disease. The reason for that is that the false positive rate (2%) is much higher than the prevalence of the disease in the population (0.5%).

To practice Bayesian updating, the reader may want to solve exercise 2.15.

Also as an exercise, the reader may consider a coin and various hypotheses about the possible bias of the coin: let [0, 1] be divided into n intervals: $I_j = [\frac{j}{n}, \frac{j+1}{n}[$, $j = 0, \ldots, n - 1$ and let H_j be the hypothesis that the frequency with which the coin lands heads lies in the interval I_j. Then, assume that, after N tosses of the coin, we have approximately $\frac{N}{2}$ heads and consider that as our data D. Then, compute from (2.2.7) $P(H_j \mid D)$ for $j = 0, \ldots, n - 1$, and conclude that (for n even) $P(H_{\frac{n}{2}-1} \mid D) + P(H_{\frac{n}{2}} \mid D) \to 1$ and $P(H_j \mid D) \to 0$ as $n \to \infty$ for all $j \neq \frac{n}{2} - 1$, $j \neq \frac{n}{2}$ as $n \to \infty$.

So, even if we started with a wrong hypothesis as our prior probability (say $P(H_0) \approx 1$), if the data are those of a fair coin, the Bayesian updating will lead to the correct hypothesis if the data are large enough. So, in that sense the choice of the prior distribution does not matter at least in the long run (see also Jaynes [183, Chap. 4]).

Of course, once one has obtained a posterior probability, and new data arrive, one can repeat the procedure, taking as prior probability the previous posterior probability and applying (2.2.7) to obtain new posterior probabilities and this can be iterated as much as one wants when new data come in.

It should be emphasized that, although all these probabilities are "subjective", in the sense that they depend on the amount of data that we have (i.e. on our "information"), the procedure to obtain posterior probabilities, based on (2.2.7), is perfectly independent of the possession of those data by the agents. In other words, two agents possessing the same data will assign, if they are rational Bayesians, the same subjective posterior probabilities to the various hypotheses under consideration.

2.2.4 Objections to the "Subjective" Approach

Let us now consider frequent objections to the "subjective" approach.

1. **Subjectivism.** Some people think that a Bayesian view of probabilities presupposes some form of subjectivism, meant as a doctrine in philosophy or philosophy of science that regards what we call knowledge as basically produced by "subjects" independently of any connection to the "outside world". But there is no logical link here: a subjectivist about probabilities may very well claim that there are objective facts in the world, that the laws governing it are also objective, and consider probabilities as being a tool used in situations where our knowledge of those facts and those laws is incomplete. In fact, one could argue that, if there is any connection between Bayesianism and philosophical subjectivism, it goes in the opposite direction; a Bayesian should naturally think that one and only one among the 'possible' states is actually realized, but that there is a difference between what really happens in the world and what we know about it. On the contrary, the philosophical subjectivist position often starts by confusing the world and our knowledge of it (for example, much of loose talk about everything being 'information' often ignores the fact that 'information' is ultimately information about something which itself is not information).

   Moreover, there is nothing arbitrary or subjective in the assignment of rational "subjective" probabilities. What is subjective here is simply the fact that there are no true or real probabilities "out there" in the world. But the choice of probabilities obeys rules (maximizing Shannon's entropy and doing Bayesian updating) that do not depend on any individual's whims, although it does depend on his or her information.

2. **(Ir)relevance to physics.** One may think that the Bayesian approach is useful in games of chance or in various practical problems of forecasting (as in insurance) but not for physics. Our answer in Sect. 2.5 will be based on the law of large numbers (discussed in Sect. 2.3).

3. **Ambiguities in the assignment of probabilities.** It is often difficult to assign unambiguously a (subjective) probability to an event. It is easy, of course, for coin tossing or similar experiments where there are finitely many possible outcomes, which, moreover, are related by symmetry. In general, one may use maximum entropy principles, but then, one may encounter various problems: how to choose the right set of variables, how to assign an a priori distribution on those, corresponding to maximal ignorance, and how to incorporate the "knowledge that we have".

   A paradigmatic example of such problems is "Bertrand's paradox", invented by the 19th century French mathematician Joseph Bertrand [27].

Fig. 2.1 The red chords are longer than a side of the triangle, because the chosen point is nearer the center of the circle than the point where the side of the triangle intersects the radius. The blue chords are shorter than a side of the triangle. The original uploader was Timecop at English Wikipedia. Transferred from en.wikipedia to Commons., CC BY-SA 3.0, https:// commons.wikimedia.org/w/ index.php?curid=2141740

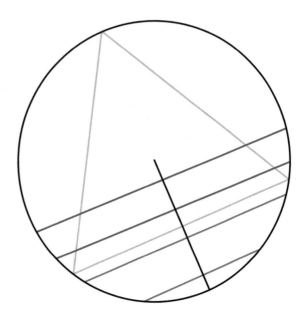

2.2.5 Bertrand's Paradox

Consider a circle and a set of straws that are thrown "at random" onto that circle. Assuming that the straw crosses that circle, its two points of intersection with the circle will define a chord. What is the probability that this chord is longer than the side of an equilateral triangle inscribed in that circle?

This was considered by Bertrand as an example of an ill-posed problem, because one obtains opposite answers depending on how one defines "at random". Here are several possibilities, where, in each case, 'random' means that we choose a uniform distribution, but on different variables:

1. One could draw a radius of the circle perpendicular to one of the sides of the equilateral triangle (the intersection of that radius with the side of the equilateral triangle will be the midpoint of that radius). Now choose "at random" a point on that radius and draw the chord having that point as its midpoint. Then, the chord is longer than a side of the triangle if the chosen point is nearer the center of the circle than the point where the side of the triangle intersects the radius, see Fig. 2.1. Since that intersection is the midpoint of the radius, the probability that this chord is longer than the side of an equilateral triangle inscribed in that circle is $\frac{1}{2}$.

2. One could choose at random the angle (comprised between 0 and 180 degrees) between the chord and the tangent of the circle at one of its intersections. The chord will be longer than a side of the triangle if that angle is greater than 60 degrees and less than 120 degrees, see Fig. 2.2. So the probability is $\frac{1}{3}$.

Fig. 2.2 The red chords have an endpoint that lies on the arc between the endpoints of the triangle side opposite the first point, so they are longer than a side of the triangle. The blue chords are shorter. The original uploader was Timecop at English Wikipedia. Transferred from en.wikipedia to Commons., CC BY-SA 3.0, https://commons.wikimedia.org/w/index.php?curid=2141731

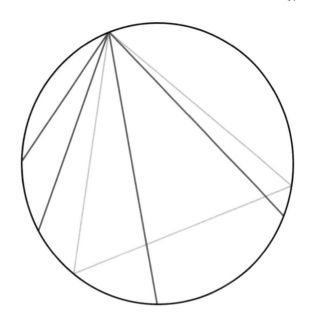

3. Finally, choose a point anywhere inside the circle and construct a chord with the chosen point as its midpoint. The chord is longer than a side of the inscribed triangle if the chosen point falls within a concentric circle of radius $\frac{1}{2}$ the radius of the larger circle see Fig. 2.3. The area of the smaller circle is one fourth the area of the larger circle, therefore the probability that a random chord is longer than a side of the inscribed triangle is $\frac{1}{4}$.

Jaynes, in an amusingly titled paper "the well-posed problem" [179], shows that actually it is the fact that we do not know the precise way in which the straws are thrown that determines a unique solution to that problem. Since the position of the chord is determined by the radial coordinates (r, θ) of its center, we must determine the probability density $f(r, \theta)$ of that center. First of all, since the formulation of the problem does not specify the position of the thrower relative to the circle, that density should be rotationally invariant: $f(r, \theta) = f(r)$. But Jaynes observes that there are other, less obvious symmetries: the result $f(r)$ should not depend on the diameter of the circle nor on its location (at least for small variations of those parameters); from that he deduces[10] that $f(r) \approx \frac{1}{r}$, which corresponds to solution 1 i.e. to a probability equal to $\frac{1}{2}$.

What is clever in that solution is that it uses all the information that we do have (the symmetries) but not those that we do not have (the exact values of the radius and the location of the circle).

Jaynes even checked that prediction with a colleague by tossing broom straws on a circle drawn on the floor. His results were in agreement with solution 1, but he

---

[10] See Jaynes [179], [183, Sect. 12.4.4] for the details of the calculation.

Fig. 2.3 Choose a point anywhere within the circle and construct a chord with the chosen point as its midpoint. The red chords are longer than a side of the inscribed triangle because the chosen point falls within a concentric circle of radius half of the radius of the larger circle. The blue chords are shorter than a side of the inscribed triangle. The original uploader was Timecop at English Wikipedia. Transferred from en.wikipedia to Commons., CC BY-SA 3.0, https:// commons.wikimedia.org/w/ index.php?curid=2141745

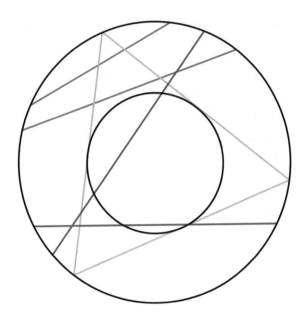

admits modestly that "experimental results would be more convincing if reported by others."

But it is interesting to contrast Jaynes' attitude with the one of Richard von Mises, the great advocate of the frequency interpretation, who regarded problems such as Bertrand's as not belonging to the field of probability at all.[11] But then, as Jaynes observes, "if we were to adopt von Mises' philosophy of probability strictly and consistently, the range of legitimate physical applications of probability theory would be reduced almost to the vanishing point."[12]

But there are other problems similar to the one above and that seem genuinely ill-defined: for example, consider a vessel containing both water and wine, with a ratio $x = \frac{\text{amount of wine}}{\text{amount of water}}$ lying between $\frac{1}{3}$ and 3. What is the probability that x is larger than 2? The answer depends on whether one puts a uniform distribution on the ratio x in which case the probability equals $\frac{\int_2^3 dx}{\int_{\frac{1}{3}}^3 dx} = \frac{3}{8}$ or on the ratio $y = \frac{\text{amount of water}}{\text{amount of wine}} = \frac{1}{x}$ in which case the answer is $\frac{\int_{\frac{1}{3}}^{\frac{1}{2}} dy}{\int_{\frac{1}{3}}^3 dy} = \frac{1}{16}$. For further discussion of this paradox and of proposed solutions, see Deakin [91].

---

[11] The Austrian mathematician and engineer Richard von Mises was the brother of the economist Ludwig von Mises; as an aside, one may note that John Maynard Keynes was both an opponent of Ludwig von Mises and a great defender of the Bayesian approach to probability [190].

[12] In [179], Jaynes quotes various reactions to Bertrand's paradox, from Bertrand himself, Borel, Poincaré, Gnedenko, von Mises, among others.

Many people consider such problems as objections to the subjectivist approach; but that misses the point of what this approach tries to achieve: finding the best thing to do in a bad situation, namely one in which we have a large degree of ignorance. Of course, the greater the ignorance, the worse the situation. But what is the alternative? If we want to move on (instead of giving up) in situations of partial ignorance, a Bayesian approach still seems to be the most rational procedure.

2.3 The Law of Large Numbers

2.3.1 A Simple Example

A way to establish a connection between the Bayesian and the frequentist views on probability relies on the law of large numbers: the calculus of probabilities—viewed now as part of deductive reasoning—leads one to ascribe subjective probabilities close to one for certain events that are precisely those that the objective approach deals with, namely the frequencies of some events, when the 'same' experiment is repeated many times. So, rather than opposing the two views, one should carefully distinguish between them, but regard the objective one as, in a sense, derived from the subjective one (i.e. when the law of large numbers leads to subjective probabilities sufficiently close to one). Let us state the law of large numbers, using a terminology that will be useful when we turn to statistical mechanics later.

Consider first the simple example of coin flipping. Let 0 denote 'head' and 1, 'tail'. The 'space' of results of any single flip, $\{0, 1\}$, will be called the 'individual phase space' while the space of all possible results of N flips, $\{0, 1\}^N$, will be called the 'total phase space'.

The variables N_0, N_1 that count the number of heads (0) or tails (1) will be called *macroscopic*, in anticipation for their later use in statistical mechanics.

Here we introduce a distinction which will be essential throughout this book between the *macroscopic variables*, or the *macrostate*, and the *microstate*. The microstate, for N flips, is the sequence of results for all the flips, while the macrostate simply specifies the values of N_0 and N_1.

Now, define a sequence of sets of microstates $\mathcal{T}_N \subset \{0, 1\}^N$ to be *typical* relative to a given sequence of probability measures P_N on $\{0, 1\}^N$, if

$$P_N(\mathcal{T}_N) \to 1. \tag{2.3.1}$$

as $N \to \infty$. If the typical sets $\mathcal{T}_N$ are defined by a property, we will also call that property typical.[13]

Let $G_N(\epsilon)$ be the set of microstates such that

---

[13] This use of the word typical is not always the usual one, which refers to the probability of a given set, not a sequence of sets, to be close to 1. But to give a precise meaning to the expression "close to 1" we need to consider sequences of sets.

$$|\frac{N_0}{N} - \frac{1}{2}| \le \epsilon \tag{2.3.2}$$

Here the letter G stand for "good", because we will use the same expression later in the context of statistical mechanics.

Then, (a weak form of) the law of large numbers states that $\forall \epsilon > 0$

$$P_N(G_N(\epsilon)) \to 1 \tag{2.3.3}$$

as $N \to \infty$, where P_N the product measure on $\{0, 1\}^N$ that assigns independent probabilities $\frac{1}{2}$ to each outcome of each flip. This is the measure that one would assign on the basis of the indifference principle: give an equal probability to all possible sequences of results. In words, (2.3.3) says that the set of sequences $G_N(\epsilon)$ is typical in the sense of definition (2.3.1), $\forall \epsilon > 0$.

It is called *the law of large numbers* for obvious reasons: it deals with the law-like behavior of large numbers of variables.

A more intuitive way to say the same thing is that, if we simply count the number of microstates that belong to $G_N(\epsilon)$, we find that they form a fraction of the total number of microstates close to 1, for N large.

To prove that, it is enough to use the binomial formula that gives the number of microstates with a given N_0:

$$\frac{N!}{N_0! N_1!} = \frac{N!}{N_0!(N - N_0)!}, \tag{2.3.4}$$

and, using Stirling's formula ($\ln N! \approx N \ln N - N$, see Appendix 6.A.2) one sees that (2.3.4) reaches its maximum for $N_0 = \frac{1}{2}$ and is sharply peaked around that maximum. This is left as an exercise for the reader and, in any case, we will state a more general result in the next section.

Formula (2.3.4) allows us to introduce one of the most frequently used probability distribution: the binomial one. Suppose we have N independent events (meaning that their joint probabilities are products of individual ones, see (2.A.21)) with two possible outcomes, denoted 0 and 1, with respective probabilities p and $1 - p$, the probability of N_0 results 0 is:

$$\frac{N!}{N_0! N_1!} p^{N_0}(1 - p)^{N_1} == \frac{N!}{N_0!(N - N_0)!} p^{N_0}(1 - p)^{N - N_0}. \tag{2.3.5}$$

2.3.2 A More General Result

This subsection is somewhat more mathematical than the rest of this chapter and we refer to the Appendix 2.A for the notations, definitions and results mentioned here.

Let $\boldsymbol{\Omega}$, $\boldsymbol{\Sigma}$ be a product space $\times_{i=1}^{\infty} \Omega_i$ and $\boldsymbol{\mu}$ be a product measure $\times_{i=1}^{\infty} \mu_i$ on $\boldsymbol{\Omega}$, $\boldsymbol{\Sigma}$ (where all Ω_i's are copies of a given Ω and all μ_i's are copies of a given measure μ on Ω). Let $f_i : \Omega_i \to \mathbb{R}$ be a sequence of identical random variables (all f_i's are copies of a given $f : \Omega \to \mathbb{R}$) and form the sum $S_N(\mathbf{x}) : \boldsymbol{\Omega} \to \mathbb{R}$:

$$S_N(\mathbf{x}) = \frac{1}{N} \sum_{i=1}^{N} f_i(x_i) \tag{2.3.6}$$

Assume for simplicity that the function f is bounded. Then, if $G_N(\epsilon)$ denotes the set of microstates $\mathbf{x}$ such that

$$|S_N(\mathbf{x}) - \mathbb{E}(f)| \leq \epsilon, \tag{2.3.7}$$

with $\mathbb{E}(f) = \int_\Omega f(x) d\mu(x)$ the expectation value of f, we have $\forall \epsilon > 0$

$$\mu(G_N(\epsilon)) \to 1, \tag{2.3.8}$$

as $N \to \infty$. This is proven in Appendix 2.B, with explicit bounds on $\mu(G_N(\epsilon))$.

Formula (2.3.8) is called the weak law of large numbers because there is also a "strong" formulation of the law of large numbers:

$$\mu(\{\mathbf{x} \mid \lim_{N \to \infty} |S_N(\mathbf{x}) - \mathbb{E}(f)| = 0\}) = 1, \tag{2.3.9}$$

or, in words, the convergence of $\lim_{N \to \infty} |S_N(\mathbf{x}) - \mathbb{E}(f)|$ to 0 holds μ almost everywhere.

Here is another version of the law of large numbers: let $(A_1, A_2, \ldots, A_k)$ be a partition of $\mathbb{R}$. Given a sequence $(x_1, \ldots, x_N) \in \Omega^N$ define *the histogram* $(n_\alpha)_{\alpha=1}^k$ of the random variables $(f_1, \ldots, f_N)$ by:

$$n_\alpha(\mathbf{x}) = \frac{|\{i \in \{1, \ldots, N\} \mid f_i(x_i) \in A_\alpha\}|}{N} \tag{2.3.10}$$

where $|E|$ is the cardinality of the set E. The numbers $n_\alpha(\mathbf{x})$ give the fractions of x_i's, $i = 1, \ldots, N$, for which the random variables $f_i(x_i) \in A_\alpha$.

Let $P_\alpha = \mu(\{x \in \Omega \mid f(x) \in A_\alpha\})$ be the probability that the random variable f takes values in A_α.

Let $G'_N(\epsilon)$ denote the set of microstates $\mathbf{x}$ such that the fractions in the histogram are close to the corresponding probabilities:

$$|n_\alpha(\mathbf{x}) - P_\alpha| \leq \epsilon, \tag{2.3.11}$$

$\forall \alpha = 1, \ldots, k$.

We have again, $\forall \epsilon > 0$:

$$\mu(G'_N(\epsilon)) \to 1 \tag{2.3.12}$$

as $N \to \infty$.

The strong formulation of that result is:

$$\mu(\{\mathbf{x} \mid \lim_{N \to \infty} |n_\alpha(\mathbf{x}) - P_\alpha| = 0\}) = 1, \tag{2.3.13}$$

$\forall \alpha = 1, \ldots, k$.

To give a simple example of what we just stated, consider a finite set of real numbers $E = \{a_\alpha\}_{\alpha=1}^k$ and a random variable x whose probability distribution is: $P(x = a_\alpha) = p_\alpha, \alpha = 1, \ldots, k$, with $p_\alpha \geq 0, \sum_{\alpha=1}^k p_\alpha = 1$.

Then, take N independent variables $x_i, i = 1, \ldots, N$, with the same distribution as x, and define the frequency $n_\alpha(\mathbf{x})$ with which the variables $\mathbf{x} = (x_1, \ldots, x_N)$ take the value a_α by

$$n_\alpha(\mathbf{x}) = \frac{|\{i = 1, \ldots, N | x_i = a_\alpha\}|}{N} = \frac{\sum_{i=1}^N \delta(x_i = a_\alpha)}{N}, \tag{2.3.14}$$

where δ is the Kronecker delta.

The set of frequencies $(n_\alpha(\mathbf{x}))_{\alpha=1}^k$ is called the *empirical distribution* of the variables $x_i, i = 1, \ldots, N$ (it coincides with the histogram defined above if each element A_α of the partition of $\mathbb{R}$ contains exactly one $a_\alpha, \alpha = 1, \ldots, k$).

Let μ_N be the joint probability distribution of the independent variables $x_i, i = 1, \ldots, N$:

$$\mu_N(x_1 = a_1, \ldots, x_N = a_N) = \prod_{i=1}^N p_{\alpha_i}. \tag{2.3.15}$$

Then, the *law of large numbers* says that, $\forall \epsilon > 0$,

$$\lim_{N \to \infty} \mu_N(|n_\alpha(\mathbf{x}) - p_\alpha| \geq \epsilon) = 0,$$

$\forall \alpha = 1, \ldots, k$.

One way to rewrite (2.3.9) and (2.3.13) is:

$$S_N(\mathbf{x}) \approx \mathbb{E}(f)$$
$$n_\alpha(\mathbf{x}) \approx P_\alpha,$$

for N large, i.e. functions that a priori depend on $\mathbf{x}$ are actually constant in $\mathbf{x}$ with large probability and their constant value equals their average value. This is the simplest summary of the laws of large numbers and is related to what we will call the *fundamental formula* (6.6.21) of equilibrium statistical mechanics in Sect. 6.6.

2.3.3 Corrections to the Law of Large Numbers

A informal way to state (2.3.9) is

$$\sum_{i=1}^{N} f_i(x_i) \approx N\mathbb{E}(f)$$

which holds for typical configurations when $N \to \infty$.

One may ask: what is the correction to that approximation? It turns out that this correction is of order $\sqrt{N}$:

$$\sum_{i=1}^{N} f_i(x_i) \approx N\mathbb{E}(f) + \sqrt{N}X \tag{2.3.16}$$

where X is a Gaussian random variable. The precise formulation of (2.3.16) is:

Theorem 2.1 *The central limit theorem.*
Let $X_N = \frac{\sum_{i=1}^{N} f_i(x_i) - N\mathbb{E}(f)}{\sqrt{N}}$, with f_i as in (2.3.6).
Then, $\forall a, b \in \mathbb{R}, a < b$,

$$\lim_{N\to\infty} \mu(a \leq X_N \leq b) = \frac{1}{\sqrt{2\pi\sigma}} \int_a^b \exp\left(-\frac{x^2}{2\sigma^2}\right) \tag{2.3.17}$$

where $\sigma^2 = \mathbb{E}(f^2) - \mathbb{E}(f)^2$.

2.4 The Law of Large Numbers and the Frequentist Interpretation

A frequentist might want to use the law of large numbers, specially in its strong form, (2.3.9) or (2.3.13), in order to *define* the concept of probability. That might answer the objection that, if one defines that concept as a frequency of results in the repetition of a large number of the "same" experiment, the notion of "large" is imprecise. But, if we take the limit $N \to \infty$, then "large" becomes precise.

There are two obvious objections to that answer: first of all, it is true that many idealizations are made in physics by taking appropriate limits, of infinite time intervals or infinite spatial extension,[14] but it is difficult to see how a concept can be defined *only* through such a limit. Since the limit is never reached in nature and if the concept of probability makes no sense for finite sequences of experiments, then it cannot be used in the natural sciences.

---

[14] See Chap. 9 for such an idealization of infinite spatial extension in the theory of phase transitions.

The next objection is that statements like (2.3.9) or (2.3.13) are *probabilistic statements* even if they refer to events having probability one. But it is circular to define a concept by using a formula that involves that very concept.

The mathematician and defender of the frequentist interpretation, Richard von Mises, avoids referring to the law of large numbers and defines probabilities as limits of frequencies of particular attributes (like falling heads for a coin or landing on a 5 for a die) within what he calls a "collective." [324]

A collective is defined as an unlimited sequence of observations so that:

1. The limits of frequencies of particular attributes within the collective exist.
2. These limits are not affected by place selection, which means that the same limit would be obtained if we choose a subsequence of the original sequence of the collective according to some rule, for example the subsequence of events indexed by even numbers or by prime numbers or by numbers that are squares of integers. Of course, the rule must be specified independently of the results of the sequence of observations.

   This is a way of guaranteeing that the collective is "random". Consider the sequence $0, 1, 0, 1, 0, 1, 0, 1, \ldots$. The limits of frequencies of 0's and 1's is obviously $\frac{1}{2}$, but it would not be so if we chose the subsequence of events indexed by even numbers or by odd numbers. And that is a way of characterizing the sequence as nonrandom.[15]

The problem with von Mises' approach is that the application of probability to the physical world is severely restricted if not empty, if one follows that approach. Indeed, we never encounter literally infinite sequences of observations in the real world and the idea that probability applies only to such infinite collectives makes it either not applicable at all or only under conditions where we can 'imagine' such collectives.

But such 'imagination' implies a form of judgment (for example, how long a "random" sequence has to be to be considered as part of an infinite collective?) and if we introduce judgments in our definition of probability, why is that so different from the "subjective" theory of probability?

Another attempt to give an "objective" meaning to probabilities is due to the philosopher Karl Popper, who initially defended the frequentist interpretation (see [267, Chap. 8]), but later introduced the idea of "propensity" to give a meaning to the probability of finite sequences of events or even of single events.

Propensity is supposed to be a property of empirical setups that cause definite frequencies to occur. For example, the combination of the design of a coin and the way it is tossed explains or causes the approximately $\frac{1}{2}$ frequencies of heads and tails. The same holds for die-throwing or the shuffling of a deck of cards.[16]

---

[15] As an aside, von Mises shows that in random sequences (as in the game of roulette for example), there is no strategy (like betting on red if the previous result is black and vice versa or any more complicated rule) that will allow you to win, in the long run. That is apparently contrary to the beliefs of some casino players.

[16] Actually, a similar idea was introduced earlier by the American philosopher Charles Sanders Pierce: "I am, then, to define the meaning of the statement that the probability, that if a die be

Like the frequency interpretation, the propensity interpretation considers probabilities as objective properties of entities in the real world. Probability is thought of as a physical propensity, or disposition, or tendency of a given type of physical situation to yield an outcome of a certain kind, or to yield in the long run a relative frequency of such an outcome.

Obviously, there is some truth in this idea. If the result of coin tossing could be perfectly controlled and predicted,[17] then we could obtain any sequence of heads and tails that we want and we would not consider the phenomenon as being random, in the same way that we do not consider eclipses as random.

But for every irregular and unpredictable phenomenon, there are specific mechanisms that make it so. The dynamics of those unpredictable systems is always such that the result depends sensitively on the initial conditions.[18] If one tosses the coin with a slightly larger velocity or make it spin slightly more, it will fall heads rather than tails.[19] And the same is true for all the other familiar random phenomena: throwing of dice, the roulette, shuffling cards etc.

This is correct but to try to subsume all these particular physical mechanisms under a grand idea of "propensity" is, as observed by Elliott Sober, like attributing the fact that opium causes sleep to its somniferous property, as in Molière's parody of the doctors of his time (see [301, p. 64]).

Popper makes an analogy between his notion of propensity and the notion of force in classical mechanics; for him, empirical statistics are a way to verify statements about propensities just as the motions of bodies are a way to verify statements about the nature of forces. The problem with this analogy is that forces are given by specific formulas from which the motions of bodies are computed (at least in some cases); but there is nothing analogous to that in the notion of propensity: there is no formula for the propensity of a coin that will imply the frequency with which it will fall either heads or tails; the only "formulas" here are the observed frequencies, but since these are what the notion of propensity is supposed to explain, the analogy with mechanical forces is completely misleading.

Popper is obsessed with the idea of objectivity and thus rejects Bayesianism on the ground that it is "subjective". But Bayesians incorporate the information about coins (for example, that it is a fair coin, or not), in order to determine their "subjective" probabilities. And of course, the properties of the coin and the way it is tossed explain why the results are what they are. There is nothing subjective about that and

---

thrown from a dice box it will turn up a number divisible by three, is one-third. The statement means that the die has a certain 'would-be'; and to say that the die has a 'would-be' is to say that it has a property, quite analogous to any habit that a man might have." ([253, p. 79–80], quoted in [166]).

[17] For an example of machines that are able to perform such a control, so that the coin always lands heads, see e.g. Diaconis, Holmes and Montgomery [99].

[18] In Chap. 4, we will study a more sophisticated class of dynamical systems that exhibit a sensitive dependence on the initial conditions and that notion will be defined precisely.

[19] See [99, 187] for models of the dynamics of tossed coins, which show that the regions in the space of initial conditions that leads to coins falling heads or tails are very narrow and thus that the result of coin tossing changes under a small change of those initial conditions.

the behavior of the coin is not affected by our knowledge, but our probabilities are, because probabilities are tools that we use in situations of incomplete knowledge.

Popper wrote: "It is clearly absurd to believe that pennies fall or molecules collide in a random fashion *because we do not know* the initial conditions, and that they would do otherwise if some demon were to give their secret away to us: it is not only impossible, it is absurd to explain objective statistical frequencies by subjective ignorance." ([265, p. 106]).

Obviously, the world does what it does, whether we know about it or not. So, indeed, if "some demon" were to provide us with a detailed knowledge of the state of falling pennies, nothing would change in their future evolution. But who ever thought otherwise? Popper assumes that Bayesians must hold a view (that knowledge of some random events would affect their behavior) that no Bayesian actually has. As we will explain in the next section, our subjective probabilities do play a role in our *explanations* of random events, but not in their behavior.

To be charitable to Popper, or maybe to others who may understand him that way, one may interpret his use of the word "to know" as meaning "to be able to control." One may imagine situations where one can *control* more variables, hence to "know" more about the system, but not simply in the sense that a demon gives us that knowledge. For example, if we could control whether pennies fall heads or tails, as Diaconis, Holmes and Montgomery manage to do ([99]), then the statistics of the results could change, but not because of any abstract "knowledge" that we may have, but because of our ability to control the behavior of the pennies.

2.5   Explanations and Probabilistic Explanations

One way to connect probabilities to the physical world is via the so-called *Cournot's principle* which says that, if the probability of an event A is very small, given some set of conditions C, then one can be practically certain that the event A will not occur on a single realization of those conditions.[20]

Of course, the event and its probability have to be specified before doing the experiment where that event could occur. Otherwise, if one tosses one thousand

---

[20] Here is how Cournot himself stated that principle: "*The physically impossible event is therefore the one that has infinitely small probability*; and only this remark gives a consistance, an objective and phenomenal value to the theory of mathematical probability." [88, p. 78]. See Shafer and Vovk [289, 290] and Goldstein [151] for a discussion of the history of that principle and the fact that it was endorsed by many famous mathematicians and probabilists. Jacob Bernoulli, long before Cournot stated that: "Because it is only rarely possible to obtain full certainty, necessity and custom demand that what is merely morally certain be taken as certain" [26]. According to Paul Levy, Cournot's principle is the only connection between probability and the empirical world. He calls it "the principle of the very unlikely event" [230]. Jacques Hadamard refers to "the principle of the negligible event" [165]. Andrei Kolmogorov, in [196, Chap. 1] uses Cournot's principle to connects the mathematical formalism with the real world. Similarly Émile Borel writes that "The principle that an event with very small probability will not happen is the only law of chance" [39].

coins, we will obtain a definite sequence of heads and tails and that sequence does occur, although its a priori probability is very small: $\frac{1}{2^{1000}}$.

Besides, the probability assigned to A must be properly chosen: if one were to assign probabilities $(\frac{1}{3}, \frac{2}{3})$ to heads and tails and toss a thousand coins, the event A defined by (2.3.2) would have a very small probability (exercise: estimate that probability), although it has a probability close to 1 if one assigns the usual probabilities $(\frac{1}{2}, \frac{1}{2})$ to heads and tails.

Another way to state Cournot's principle is that atypical events never occur.[21] However, in reality, atypical events do occur: a series of coin tossing could give significant deviations from the $(\frac{1}{2}, \frac{1}{2})$ frequencies. But that would mean that one has to revise one's probabilities (and this is the basis of Bayesian updating: adjust your probabilities in light of the data).

But Cournot's principle helps us to understand what is a *probabilistic explanation* of "random" physical phenomena.

A first form of scientific explanation is given by *laws*. If state A produces state B, according to deterministic laws, then the occurence of B can be explained by the occurrence of A and the existence of those laws.[22] If A is prepared in the laboratory, this kind of explanation is rather satisfactory, since the initial state A is produced by us.

But if B is some natural phenomena, like today's weather and A is some meteorological condition yesterday, then A itself has to be explained, and that leads potentially to an "infinite" regress, going back in principle to the beginning of the universe. In practice, nobody goes back that far, and A is simply taken as "given", which means that our explanations are in practice limited when we go backwards in time.

It is worth noting that there is something "anthropomorphic" even in this type of explanation: for example if A is something very special, one will try to explain A as being caused by anterior events that are not so special. Otherwise our explanation of B in terms of A will look unsatisfactory. Both the situations A and B and the laws are perfectly objective but the notion of explanation is "subjective" in the sense that it depends on what we, humans, regard as a valid explanation.

Consider now a situation where probabilities are involved, starting with the simplest example, coin tossing, and trying to use that example to build up our intuition about what constitutes a valid explanation.

---

[21] At least as long as the number of events is finite. By contrast, to take an example of Borel, if apes type at random on a typewriter for a long enough time, they will produce Shakespeare's work or any other finite collection of symbols. Indeed, if an event has probability $p > 0$, then the probability that it does *not* occur in N trials is $(1 - p)^N$, which goes to 0 as $N \to \infty$, so that the event is certain to occur in that limit, no matter how small p is; the probability that apes typing at random on a typewriter produce Shakespeare's work is certainly strictly positive even if it extremely small. This estimate also shows that we need $N \approx \frac{1}{p}$ for that occurrence to become probable.

[22] This is the main idea behind the deductive nomological model, according to which scientific explanations are deductive arguments with laws as one of the premises, see Hempel [169], [170] and [330]. For a detailed discussion of reductionism, see Chibarro, Rondoni, and Vulpiani [78].

First observe that, if we toss a coin many times and we find approximately half heads and half tails, we do not feel that there is anything special to be explained. If, however, the result deviates strongly from that average, we will look for an explanation (e.g. by saying that the coin is biased).

This leads to the following suggestion: suppose that we want to explain some phenomenon when our partial knowledge of the past is such that this phenomenon could not have been predicted with certainty (for coin tossing, the past would be the exact initial conditions of the coins when they are flipped). We will say that our knowledge, although partial, is *sufficient* to explain that phenomenon if we would have predicted it using probabilistic reasoning (which is thus "subjective" since it is a reasoning) and the information we had about the past. That notion of 'explanation' incorporates, of course, as a special case, the notion of explanation based on deterministic laws. Also, it fits our intuition concerning the coin-tossing situation discussed above: being ignorant of any properties of the coin leads us to predict a proportion of heads or tails approximately equal to one-half. Hence, such a result is not surprising or, in other words, does not "need to be explained", while a deviation from it requires an explanation.

We will rely on that notion of explanation when we discuss the explanatory status of statistical mechanical laws. We will also see in Sect. 7.8.2 examples where deviations from those laws force us to seek new explanations and, sometimes, to revise some of our deepest physical assumptions.

2.6 Final Remarks

The opposition between the frequentist and Bayesian approaches to probability theory can be viewed, at least in some versions of that opposition, as part of a larger opposition between a certain version of empiricism and a certain version of rationalism. By this we mean that Bayesianism relies on the notion of rational (inductive) inference, which by definition, goes beyond mere analysis of data. The link to rationalism is that it trusts human reason of being able to make rational judgments that are not limited to "observations". By contrast, frequentism is related to a form of skepticism with respect to the reliability of such judgments, in part because their answers can be ambiguous, as exemplified by Bertrand's paradoxes.

Therefore, the frequentist will say, let's limit the theory of probability to frequencies or to "data" that can be observed and forget about those uncertain reasonings. And that reaction has definitely an empiricist flavor. We have already explained our objections to that approach in Sect. 2.4. We will simply add here the remark that this move away from rationalism and towards some form of empiricism occurred simultaneously in different fields in the beginning of the twentieth century and was a somewhat understandable reaction to the "crises in the sciences" caused by the replacement of classical mechanics, that had been the bedrock of science for centuries, both by the theories of relativity and by quantum mechanics.

Here are some examples, besides frequentism, of such moves:

Logical positivism in the philosophy of science: in the Vienna Circle, there was a strong emphasis on observations or sense-data as being the only sort of things one can meaningfully speak about or "verify". On the other hand, the logical positivists were (rightly) reacting to the metaphysical traditions in philosophy and they were also interested in inductive logic.

Formalism in the philosophy of mathematics: while realists in the philosophy of mathematics (often called Platonists) think that mathematics studies something real (numbers or sets), formalists argued that mathematicians are just deducing theorems from axioms according to given rules, but nothing more and that the axioms and the rules do not attempt to capture some "hidden" reality.

Behaviorism in the philosophy of mind: instead of postulating some "hidden" mental structures, behaviorists insisted that on should only study the links between stimuli and reactions.

The Copenhagen interpretation of quantum mechanics: the goal of science, for Bohr, Heisenberg and their followers, as opposed to the "realists" like Einstein and Schrödinger, is not to discover the properties of some microscopic reality but only to predict "results of measurements."

Of course, for each of these positions, there are pros and cons and various nuances of these positions and we do not intend to discuss them in detail. But what is common to them is a sort of modesty with respect to science and knowledge: let's focus on what we know for certain: empirical frequencies, sense-data, formal manipulations, input-output reactions, or "measurements". But, in doing so, one abandons the explanatory character of science.

In general, modesty is praiseworthy; the problem here is that, if it goes too far, it tends to make science devoid of meaning and therefore, of interest.

2.7 Summary

The main opposition in the philosophy of probability is between the "subjective" and "objective" approaches. From the subjective or Bayesian point of view, probabilistic propositions are judgments while from the objectivist point of view, those propositions are statements of fact.

But there can be different objective approaches. First of all, if the laws of nature are intrinsically random (as they may be in quantum mechanics[23]), then, objective probabilistic statements are simply part of those laws. But, since we consider here only classical deterministic systems, this possibility is excluded by definition.

Objective probabilistic statements may also refer to frequencies of occurrence of certain events when "similar" experiments are repeated. This is the frequentist definition of probability. Our objections to the frequentist view is that, if one considers

---

[23] See e.g. [22, 30, 31, 57, 113, 114, 152, 242, 313] for expositions of the de Broglie-Bohm theory, which offers a deterministic alternative to the standard version of quantum mechanics.

literally infinite sequences, it is not clear what relationship this approach has with the physical world; and if one limits oneself to "long enough" sequences, one reintroduces a sort of judgment (what is long enough?) in the definition of probabilities.

The frequentist view is sometimes thought to be supported by the law of large numbers (Sect. 2.4) that says that empirical frequencies of events tend to the probabilities of those events, when the number of experiments or observations tends to infinity. But the law of large numbers is itself a probabilistic statement and it would be circular to base one's definition of the concept of probability on a probabilistic statement.

There an alternative objectivist approach, due to Popper and, before him, to Pearce, that tries to understand random events as due to something analogous to a force, and called propensity. But, as we explained in Sect. 2.4, it is difficult to give a non circular definition of this mysterious property of propensity.

Turning to subjective probabilities, one must distinguish between two different definitions: one associated mostly with de Finetti that considers probabilistic statements as expressions of degrees of belief that are only constrained by rules such as (2.2.1) so as to avoid being victim of a "Dutch book" namely a bet that one is certain to loose no matter what happens in the game.

Another version of the "subjective" approach to probabilities, is the "objective Bayesian" championed by E. T. Jaynes in the contemporary period, but which is quite close to the classical vision of Laplace and other founders of probability theory. Of course, this terminology, mixing the words subjective and objective is quite confusing. But objective Bayesians consider probabilistic statements as expressions of judgments (hence "subjective") but constrained by rules of rationality, which means trying to incorporate in those judgments all the information that we have but not more (hence "objective"). This is based on the indifference principle or the maximum entropy principle.

In Sect. 2.2.5, we answered various objections to that approach; in particular, there is nothing subjectivist or idealist in the philosophical sense in the objective Bayesian approach.

Finally, we explained why "typical" probabilistic statements, namely those applying to an overwhelming majority of cases (often through some application of the law of large numbers), play a crucial role in our explanations of phenomena that do not obey strict deterministic laws.

2.8 Exercises

2.1. What is the probability of getting at least 4 heads in 6 tosses of a coin?

2.2. N people are in a room and compare their birthdays. Compute the probability that two people have the same birthday as a function of N (assume that all days are equally probable birthdays and that there are 365 days in a year). Estimate how large must N be for that probability to exceed $\frac{1}{2}$.

2.3. Consider the (Gaussian) measure on $\mathbb{R}$:

$$d\mu_g(x) = \frac{1}{\sqrt{2\pi}\sigma} \exp\left(-\frac{x^2}{2\sigma^2}\right) dx$$

1. Check that $d\mu_g$ is a probability measure: $\int_{\mathbb{R}} \exp(-\frac{x^2}{2\sigma^2})dx = \sqrt{2\pi}\sigma$. Hint: change variables and compute $\int_{\mathbb{R}} \exp(-\frac{x^2}{2})dx$ by squaring it and doing a two dimensional integral.

2. Let $\mathbb{E}_g$ denote the expectation value with respect to μ_g. Check that

$$\mathbb{E}_g(\exp(\pm ikx)) = \exp\left(-\frac{\sigma^2 k^2}{2}\right).$$

Hint: complete the square in $-\frac{x^2}{2\sigma^2} \pm ikx$ and either take for granted the fact that

$$\int_{\mathbb{R}} \exp\left(-\frac{(x \mp ik\sigma^2)^2}{2\sigma^2}\right) dx = \sqrt{2\pi}\sigma,$$

or prove that formula by doing a change of variables and performing a suitable contour integration in $\mathbb{C}$.

3. Expanding both sides of $\mathbb{E}_g(\exp(ikx)) = \exp(-\frac{\sigma^2 k^2}{2})$ in a Taylor series in k, check that:

$$\mathbb{E}_g(x^{2m}) = \frac{(2m)!}{m!2^m}\sigma^{2m} \quad \forall m \in \mathbb{N}$$

(and $\mathbb{E}_g(x^{2m+1}) = 0, \forall m \in \mathbb{N}$).

2.4. Extend exercise 2.3 to n variables. Define the (Gaussian) measure on $\mathbb{R}^n$ by:

$$d\mu_g(\mathbf{x}) == \left[\frac{\det A}{(2\pi)^n}\right]^{\frac{1}{2}} \exp\left[-\frac{\mathbf{x} \cdot A\mathbf{x}}{2}\right] d\mathbf{x}$$

where $\mathbf{x} \cdot \mathbf{y}$ is the scalar product in $\mathbb{R}^n$, and A is a $n \times n$ symmetric matrix which is strictly positive definite: $\exists \epsilon > 0$ such that $\forall \mathbf{x} \in \mathbb{R}^n$, $\mathbf{x} \cdot A\mathbf{x} \geq \epsilon \|\mathbf{x}\|^2$ (this will guarantee the convergence of the integrals considered below).

1. Check that $d\mu_g$ is a probability measure: $\int_{\mathbb{R}^n} \exp[-\frac{\mathbf{x} \cdot A\mathbf{x}}{2}] d\mathbf{x} = [\frac{(2\pi)^n}{\det A}]^{\frac{1}{2}}$. Hint: diagonalize the matrix A, change variables and then use the computation done in exercise 2.3. Remember that $\det A = \prod_{i=1}^{n} \lambda_i$, where the λ_i's are the eigenvalues of A.

2. Let $\mathbb{E}_g$ denote the expectation value with respect to μ_g. Check that $\mathbb{E}_g(\exp(i(\mathbf{k}, \mathbf{x})) = \exp(-\frac{\mathbf{k} \cdot A^{-1} \mathbf{k}}{2})$, with A^{-1} the matrix inverse of A, by again completing the square in the exponential in the integral on the left hand side, as in exercise 2.3, or by diagonalizing A and changing variables.

3. Expanding both sides of $\mathbb{E}_g(\exp(i\mathbf{k} \cdot \mathbf{x}) = \exp(-\frac{(\mathbf{k} \cdot A^{-1} \mathbf{k})}{2})$ in a multivariable Taylor series in $\mathbf{k}$, check that:

$$\mathbb{E}_g(x^{2m}) = \frac{(2m)!}{m! 2^m} (A^{-1})_{mm} \quad \forall m \in \mathbb{N} \tag{2.8.1}$$

(and $\mathbb{E}_g(x^{2m+1}) = 0$, $\forall m \in \mathbb{N}$) and

$$\mathbb{E}_g(x_1 \dots x_{2m}) = \sum_{\Pi_2} \prod_{(i,j) \in \Pi} (A^{-1})_{ij} = \sum_{\Pi_2} \prod_{(i,j) \in \Pi} \mathbb{E}_g(x_i x_j) \tag{2.8.2}$$

where the sum runs over all partitions Π_2 of $\{1, \dots 2m\}$ into pairs, and where some of the variables x_i may coincide.

4. Check that (2.8.1) is a special case of (2.8.2).

2.5. Let X be a random variable taking values in $\mathbb{N}$, with distribution:

$$P(X = k) = \frac{\lambda^k}{k!} \exp(-\lambda) \quad \forall k \in \mathbb{N}$$

Check that P is a probability distribution and compute the mean and the variance of X. X is called a *Poisson random variable* with parameter λ.

2.6. Consider a sequence X_n of binomial random variables (see (2.3.5)) with parameter $p = p_n$, with $p_n = \frac{\lambda}{n}$. Show that the distribution of X_n converges, as $n \to \infty$, to the one of a Poisson random variable with parameter λ.

2.7. We have a sequence on N independent experiments with a probability p to observe a given outcome A for each of them. Let X be the random variable counting the number of times that the outcome A is observed.
What is the probability distribution of X? Compute the mean and the variance of X.

2.8. Suppose that, whenever we have N people they are all equally likely to be most unlucky (let's say at a certain game). You have a particularly unlucky event at that game and you meet one person after another to see if you are more unlucky than them. What is the probability distribution of the event that the first person more unlucky than you is the Nth one? What is the expected value of that variable (which is also the expected time it takes before you meet someone more unlucky than you).

2.9. *St Petersburg paradox*. Peter and Paul play the following game. Peter gives a amount N of money to Paul in order to play the game. Then they flip a coin. If it falls heads, Paul gives one euro to Peter and the game stops. If it falls tails, one flips the coin again and if it falls heads Paul gives two euro to Peter and the game stops, otherwise one repeats the operation doubling what Paul gives to Peter at each step.

So, if the coin falls heads for the first time at the nth flip of the coin, Paul gives to Peter 2^{n-1} euros.

What is the expected gain $\mathbb{E}$ of Peter?

Is it reasonable to equate the amount of money N given by Peter in order to play the game to the expected gain, i.e. to let $N = \mathbb{E}$? Would you do it if you were Peter?

2.10. If one assigns probabilities $(\frac{1}{3}, \frac{2}{3})$ to heads and tails estimate the probability that

$$|\frac{N_0}{N} - \frac{1}{2}| \leq \epsilon$$

where N_0 is the number of heads, for fixed $\epsilon > 0$, small, as $N \to \infty$.

2.11. Consider formula (2.3.5) with $p = \frac{1}{2}$ giving the probability of N_0 heads after N tosses of a coin. Write $N_0 = \frac{N}{2} + \sqrt{N}n$. Show, using Stirling's formula, that the distribution of n approaches a Gaussian one as $N \to \infty$.

2.12. An urn contains balls of L different colors and let p_i be the fraction of the balls of color $i = 1, \ldots, L$. We draw balls from the urn, note their color and put them back in the urn.

What is the probability of having, after N draws, a sequence $(n_i)_{i=1}^{L}$ of balls of color $i = 1, \ldots, L$? This probability distribution is called the *multinomial distribution* and generalizes the binomial distribution (2.3.5).

2.13. A drunken sailor moves on $\mathbb{Z}$ and makes a step of one unit to the right or to the left at random and independently of previous steps at each time $n \in \mathbb{N}$. Let $x(T)$ by the position of the sailor at time T.

Compute the expected value $\mathbb{E}(x(T)^2)$.

To what does the probability distribution of $\frac{x(T)}{\sqrt{T}}$ tend when $T \to \infty$?

2.14. Counterintuitive probabilities.

1. Bertrand's box paradox.
   There are three boxes:

   - a box containing two gold coins,
   - a box containing two silver coins,
   - a box containing one gold coin and one silver coin.

Choose a box at random and pick one coin from that box also at random. If that happens to be a gold coin, what is the probability of the next coin drawn from the same box is also a gold coin?

2. **The Monty Hall problem** You are on a game show, and you are given the choice of three doors: Behind one door is a treasure; behind the others, nothing. If you open the door with the treasure behind it, you get the treasure. First, you pick a door, without opening it, and the host, who knows what is behind the doors, opens another door, which has nothing behind it. He then says to you, "There are two doors left; which of the two remaining doors will you open? The one you chose first or the other one?" Which choice is to your advantage?

2.15. Bayesian updating. There are three types of coins which have different probabilities of landing heads when tossed:

1 Type A coins are fair, with probability 0.5 of heads.
2 Type B coins are bent and have probability 0.6 of heads.
3 Type C coins are bent and have probability 0.9 of heads.

Suppose I have a drawer containing 5 coins: 2 of type A, 2 of type B, and 1 of type C. I reach into the drawer and pick a coin at random. Without showing you the coin I flip it once and get heads. What is the probability that it is type A, B or C?

The remaining exercises concern the material in the appendix.

2.16. Check that the set $\mathbb{Q}$ of rational numbers is a Borel set but is neither open nor closed.

2.17. Check that the set of intervals in $\mathbb{R}$ or of rectangles in $\mathbb{R}^n$ (sets of the form $I_1 \times \cdots \times I_n$, where I_k are intervals in $\mathbb{R}$) are semi-algebras.

2.18. Check that the set of cylinder sets defined in (2.A.5) forms a semi-algebra.

2.19. Check that the indicator function of the rational numbers (say in $[0, 1]$) is measurable and its integral equals 0.

2.20. Prove (2.3.13).

Appendix 1

2.A Appendix A: Measure Theory

In this Appendix, we summarize basic facts about measures and integration that we use in this book.

2.A.1 Definition of a Measure

Given a set Ω, a measure μ is a map from subsets of Ω to the real numbers that will give the "size" of that set. The simplest example is when Ω is finite or countable, $\Omega = \{x_1, x_2, \ldots\}$ and we have a sequence of numbers $p_i \geq 0$, $i = 1, \ldots$; then, the measure μ is defined on subsets $A \subset \Omega$ by:

$$\mu(A) = \sum_{i; x_i \in A} p_i,$$

with $\mu(A) = \infty$ if the sum diverges.

However, it is in general not possible to define a "natural" measure on all subsets of uncountable sets. For a simple example, consider the set $[0, 1[$ with addition modulo 1. Define an equivalence relation $x \equiv y$ if $x - y \in \mathbb{Q}$. The set $[0, 1[$ is thus an uncountable union of equivalence classes, each of which is countable, since $\mathbb{Q}$ is countable and $[0, 1[$ is not. Let E be a set composed of one element taken from each equivalence class (we need the axiom of choice to prove that such a set exists but let us assume that), let q_n be an enumeration of the rational numbers in $[0, 1[$ and let $E_n = E + q_n$. The sets E_n's are two by two disjoint (by definition of equivalence classes: if $E_n \cap E_m \neq \emptyset$ for $n \neq m$ then there exists $x, y \in E$, with $x + q_n = y + q_m$, and that means that $x \equiv y$, which contradicts the definition of E) and $\cup_n E_n = [0, 1[$.

Now if we want the sets E_n to be measurable and if we want to define a translation invariant measure on $[0, 1[$ (with addition modulo 1) satisfying (2.A.1) (e.g. the Lebesgue measure defined after proposition 2.4), then we run into a contradiction since $\mu(E_n) = \mu(E)$, $\forall n$, by translation invariance, and $\mu(\cup_n E_n) = \mu([0, 1[) = 1$: the infinite sum of identical terms in (2.A.1) (which extends (2.2.1) to infinite sums) cannot equal 1.

A more sophisticated example of non-measurable sets, called the Banach-Tarski paradox, relies on the construction of a subtle partition of the unit ball in $\mathbb{R}^3$ into ten disjoint subsets $A_1, \ldots, A_{10}$ (this construction again uses the axiom of choice), such that there exist ten rotations $R_1, \ldots, R_{10}$ with the property that $R_1 A_1, \ldots, R_5 A_5$ form a partition of the unit ball and $R_6 A_6, \ldots, R_{10} A_{10}$ form also a partition of the unit ball. Thus, by partitioning adequately one unit ball and rotating without deformation the elements of the partition, one can construct two balls of the same size. This would be a paradox if the sets $A_1, \ldots, A_{10}$ were measurable, because the Lebesgue measure is invariant under rotations and then we would have, since the A_i's are disjoint:

$$1 = \mu_{\text{Leb}}(\cup_{i=1}^{10} A_i) = \sum_{i=1}^{10} \mu_{\text{Leb}}(A_i)$$

$$= \sum_{i=1}^{10} \mu_{\text{Leb}}(R_i A_i) = \mu_{\text{Leb}}(\cup_{i=1}^{5} R_i A_i) + \mu_{\text{Leb}}(\cup_{i=6}^{10} R_i A_i) = 2.$$

This proves that the sets $A_1, \ldots, A_{10}$ are not measurable.

To avoid such problems, measures are defined as maps on a restricted family of subsets. First, one defines a σ-algebra Σ of subsets of Ω as a family of subsets of Ω such that:

1. $\emptyset \in \Sigma; \Omega \in \Sigma$
2. If $A \in \Sigma$, $A^c \in \Sigma$.
3. If $A_i \in \Sigma$, $i \in \mathbb{N}$, $\cup_{i \in \mathbb{N}} A_i \in \Sigma$.

In other words, a σ-algebra is stable under complementation and countable unions (and therefore also countable intersections).

When the set Ω is a topological space, one often considers the Borel σ-algebra $\mathcal{B}(\Omega)$ which is the smallest σ-algebra containing all open sets (and, because of property 2, also all closed sets).[24] As an exercise, the reader may check that the set $\mathbb{Q}$ of rational numbers is a Borel set but is neither open nor closed.

Definition 2.2 Given a set Ω and a σ-algebra of subsets of Ω, a measure μ is a map $\mu : \Sigma \to \mathbb{R}^+$ satisfying:

1. $\mu(\emptyset) = 0$.
2. Given a family of sets $(A_i)_{i \in \mathbb{N}}$, $A_i \in \Sigma$, $\forall i$, that are two by two disjoint, $A_i \cap A_j = \emptyset$, $\forall i \neq j$,

$$\mu(\cup_{i \in \mathbb{N}} A_i) = \sum_i \mu(A_i) \tag{2.A.1}$$

The reader might wonder why one wants to define measures as maps on σ-algebras. The answer is that this provides a convenient family of sets to which one can assign a measure for the purpose of integration, defined below.

We will always assume that our measures are σ-finite, namely that one can write $\Omega = \cup_{n=1}^{\infty} A_n$, with $\mu(A_n) < \infty$, $\forall n$.

If one has $\mu(\Omega) < \infty$, the measure is *finite* and can always be normalized into a *probability measure*, namely such that $\mu(\Omega) = 1$.

A triple (Ω, Σ, μ) is called a *measure space*.

2.A.2 Constructions of Measures

The family of Borel sets, as well as other σ-algebras used here, is actually very large and it would be quite cumbersome to define explicitly the value of the map μ on every of those sets. Luckily, there exists extension theorems that guarantee that, if one defines μ on a much smaller class of sets, then it can be extended to the whole σ-algebra.

---

[24] Remember that the set of open sets is stable under arbitrary unions, i.e. not necessarily countable ones, and under finitely many intersections, while the set of closed sets is stable under arbitrary intersections and finitely many unions.

Definition 2.3 A *semi-algebra* S is a family of subsets of a set Ω such that:

1. $\emptyset \in S$.
2. $\forall A, B \in S$, we have $A \cap B \in S$ (the family is closed under pairwise intersections).
3. $\forall A, B \in S$, there exist disjoint sets $C_i \in S, i = 1, 2, \ldots, n$, such that $A \setminus B = \bigcup_{i=1}^{n} C_i$ (relative complements of elements of S can be written as finite disjoint unions of elements of S).

Proposition 2.4 *Extensions of measures: If a map defined on a semi-algebra of sets S satisfies the properties in Definition 2.2, and if Ω can be written as $\Omega = \cup_{i \in \mathbb{N}} A_i$, with $A_i \in S$, $\mu(A_i) < \infty$, $\forall i \in \mathbb{N}$, then that map can be extended in a unique way to a measure defined on the σ-algebra generated by S (i.e. the smallest σ-algebra containing S).*

For a proof, see e.g. Royden [278, Sect. 12.2].

It is easy to check that the set of intervals in $\mathbb{R}$ or of rectangles in $\mathbb{R}^n$ (sets of the form $I_1 \times \cdots \times I_n$, where I_k are intervals in $\mathbb{R}$) are semi-algebras (exercise). Thus, it is sufficient to define a measure on the intervals of $\mathbb{R}$ to have it defined on the Borel subsets of $\mathbb{R}$.

If we take the measure of an interval to be its length, or the measure of a rectangle in $\mathbb{R}^n$ to be its volume, one obtains by extension the *Lebesgue measure* μ_{Leb}, which is thus uniquely defined.[25] For the Lebesgue measure of a set E, we will write $\mu_{\text{Leb}}(E)$ or, when there is no ambiguity, $|E|$, which also denotes the cardinality of the set E for finite sets.

It is easy to see that any singleton has Lebesgue measure 0 and, by (2.A.1), any countable set (e.g. the set $\mathbb{Q}$ of rational numbers) has also Lebesgue measure 0.

It is often useful to consider discrete measures, also called delta functions (which are not ordinary functions defined on $\mathbb{R}$ but measures namely maps from $\mathcal{B}(\mathbb{R})$ or $\mathcal{B}(\mathbb{R}^n)$ to $\mathbb{R}^+$),[26] defined by:

$$\delta_x(A) = 1 \quad \text{iff} \quad A \ni x, \tag{2.A.2}$$

for $A \in \mathcal{B}(\mathbb{R})$, which implies $\delta_x(A) = 0$ if $A \not\ni x$. Given a sequence $\mathbf{x} = (x_i)_{i=1}^{\infty}$, $x_i \in \mathbb{R}$ and a sequence $(p_i)_{i=1}^{\infty}$, $p_i \geq 0$, one defines $\delta_{\mathbf{x}}$:

$$\delta_{\mathbf{x}}(A) = \sum_{i=1}^{\infty} p_i \delta_{x_i}(A). \tag{2.A.3}$$

Obviously those measures give non zero values to singletons.

---

[25] Actually, the Lebesgue measure is the unique measure on $\mathbb{R}^n$ that is both invariant under translations and normalized so that the measure of the unit cube in $\mathbb{R}^n$ equals 1, see e.g. Rudin [279, Theorem 2.20].

[26] The "delta functions" can also be introduced as distributions.

Note that $\delta(x - x_0)dx = \lim_{\lambda \to \infty} c_\lambda \exp(-\lambda(x - x_0)^2)dx$ in the sense that $\lim_{\lambda \to \infty} c_\lambda \int_{\mathbb{R}} f(x) \exp(-\lambda(x - x_0)^2)dx = f(x_0)$, for f continuous, for an appropriate c_λ such that $c_\lambda \int_{\mathbb{R}} \exp(-\lambda(x - x_0)^2)dx = 1$ (which means $c_\lambda = \sqrt{\frac{\lambda}{\pi}}$).

One can also extend such measures to subsets of $\mathbb{R}^n$ defined by $F(\mathbf{x}) = 0$:

$$\delta(F(\mathbf{x}))d\mathbf{x} = \lim_{\lambda \to \infty} c_\lambda \exp(-\lambda(F(\mathbf{x}))^2)d\mathbf{x}, \tag{2.A.4}$$

with c_λ such that $c_\lambda \int_{\mathbb{R}^n} \exp(-\lambda(F(\mathbf{x}))^2)d\mathbf{x} = 1$.

We will also use another class of measures defined through the extension theorem: the measures on product spaces. Let $\Omega = \times_{i \in \mathbb{Z}} \Omega_i$ be the infinite Cartesian product of copies Ω_i, $i \in \mathbb{Z}$, of a set Ω. If Σ is a σ-algebra of subsets of Ω, define a cylinder set in Ω as a set of the form

$$\times_{i < -N} \Omega_i \times A_{-N} \times A_{-N+1} \cdots \times A_N \times_{i > N} \Omega_i \tag{2.A.5}$$

where $A_i \in \Sigma$ for $-N \leq i \leq N$ (this definition is convenient, but one must keep in mind that some of the A_i's may be equal to Ω). Then, the set of cylinder sets forms a semi-algebra (exercise: check this) and the σ-algebra generated by those sets is called the *cylindrical σ-algebra*.

If μ is a probability measure on Ω, one defines the corresponding product measure on Ω by the following formula for the measure of cylinder sets:

$$\mu(\times_{i < -N} \Omega_i \times A_{-N} \times A_{-N+1} \cdots \times A_N \times_{i > N} \Omega_i) = \prod_{i = -N}^{i = N} \mu(A_i), \tag{2.A.6}$$

and one uses the extension theorem to extend it to the cylindrical σ-algebra.

The Lebesgue measure is related to a product measure in the following way: write the binary expansion of $x \in [0, 1[$, $x = \sum_{n=0}^{\infty} \frac{a_n}{2^{n+1}}$, with $a_n \in \{0, 1\}$, $n \in \mathbb{N}$ (with the convention that, when two binary expansions exists, we choose for instance the one that does not end with a string of 1's) and consider the space

$$\tilde{\Omega} = \{\mathbf{a} = (a_n)_{n \in \mathbb{N}}, a_n \in \{0, 1\}, \forall n \mid \forall N, \exists m > N, a_m = 0\}, \tag{2.A.7}$$

where the last condition comes from the fact that we have chosen binary expansions that do not end with a string of 1's. Then, the Lebesgue measure is equivalent to the product measure on $\tilde{\Omega}$ with its cylindrical σ-algebra that gives equal weights $(\frac{1}{2}, \frac{1}{2})$ to the symbols 0 and 1. This is also true if one write real numbers in any other basis: $x = \sum_{n=0}^{\infty} \frac{a_n}{p^{n+1}}$, with $a_n \in \{0, \ldots, p - 1\}$, $n \in \mathbb{N}$, for the corresponding product measure.

2.A.3 Integration

The idea of Riemann integration is to approximate the integral, that is, the area under a curve (for a positive valued function) by the sum of the areas of little vertical rectangles whose upper side lies just under that curve or just above it.

However that method of integration has two limitations: the set of functions that can be integrated with that method is restricted: for example, one cannot integrate à la Riemann the indicator function of the rational numbers, since the height of the only rectangles under the graph of that function is 0 and the height of the only rectangles above that graph is 1, although intuitively, since the set of rational numbers is of measure 0, that integral should exist and be also equal to 0. Moreover, if a sequence of Riemann integrable functions $F_n(x) \to F(x)$, $\forall x$, as $n \to \infty$, the conditions under which one can write the obviously desirable equation $\int F(x)dx = \lim_{n\to\infty} \int F_n(x)dx$ are not simple.

The idea of Lebesgue integration solves those problems. Consider for simplicity a bounded map $F : \mathbb{R} \to \mathbb{R}^+$ of bounded support (i.e. that vanishes outside a finite interval). Instead of dividing the domain of definition of that function into small intervals, as one does in the theory of Riemann integration, one divides the image of the function into small intervals and one seeks to approximate the integral by integrating functions of the form:

$$F_n = \sum_{m=0}^{\infty} \frac{m}{n} \mathbb{1}(A_m),\qquad\qquad (2.A.8)$$

with $A_m = F^{-1}([\frac{m}{n}, \frac{m+1}{n}[$ and where the sum $\sum_{m=0}^{\infty}$ is finite for any bounded F, since then $F^{-1}([\frac{m}{n}, \frac{m+1}{n}[) = \emptyset$ for m large enough. The integral of F_n is naturally defined as $\sum_{m=0}^{\infty} \frac{m}{n} \mu(A_m)$.

For that expression to make sense, it is enough for the measure of sets of the form $F^{-1}([\frac{m}{n}, \frac{m+1}{n}[$ to exists. It is convenient to introduce a more general notion.

Definition 2.5 Given a set Ω and a σ-algebra of subsets of Ω, a map $F : \Omega \to \mathbb{R}^n$ is measurable if

$$\forall A \in \mathcal{B}(\mathbb{R}^n), \quad F^{-1}(A) \in \Sigma,$$

where $\mathcal{B}(\mathbb{R}^n)$ are the Borel sets in $\mathbb{R}^n$.

One can then define the Lebesgue integral of F, $\int_{\mathbb{R}} F(x)dx$ as a limit of sums of the form (2.A.8) for $n \to \infty$ and one can show that any positive valued measurable bounded function of bounded support is Lebesgue-integrable.[27] In particular the indicator function of the rational numbers (say in $[0, 1]$) is measurable and its integral equals 0 (exercise: check that).

---

[27] We write dx for $x \in \mathbb{R}$ or $d\mathbf{x}$ for $\mathbf{x} \in \mathbb{R}^n$ instead of $d\mu_{\text{Leb}}(x)$.

This notion of integral extends to non positive valued functions, by writing any function as a difference of two positive valued functions $F = F_+ - F_-$ and defining $\int F(x)dx = \int F_+(x)dx - \int F_-(x)dx$. It can also be extended to unbounded functions or functions without bounded support by approximating those functions by measurable bounded functions of bounded support, when the approximating integrals converge.

A final extension (relative to the Riemann integral) is that one can define $\int_\Omega F(x)d\mu(x)$ as a limit of the form (2.A.8) for a function defined on an arbitrary measure space and not only on $\mathbb{R}$ or $\mathbb{R}^n$:

$$\int_\Omega F(x)d\mu(x) = \lim_{n \to \infty} \int_\Omega F_n(x)d\mu(x) = \lim_{n \to \infty} \sum_{m=1}^{\infty} \frac{m}{n}\mu(A_m), \qquad (2.A.9)$$

with A_m as in (2.A.8).

It is convenient to use the expression *almost everywhere* of *for almost all x* for some statement that is true except possibly on a set of measure 0.

On defines a *random variable* as any measurable function $f : \Omega \to \mathbb{R}$ or $\to \mathbb{R}^n$. One then defines the *expectation value* or average of a random variable:

$$\mathbb{E}(f) = \int_\Omega f(x)d\mu(x),$$

$\mathbb{E}(f)$ is often denoted $< f >$ in the physics literature. We will use both notations below.

One also defines the *variance* of a random variable:

$$\mathrm{Var}(f) \equiv \mathbb{E}[(f - \mathbb{E}(f))^2] = \mathbb{E}(f^2) - \mathbb{E}(f)^2 \geq 0 \qquad (2.A.10)$$

which equals 0 only if f is constant almost everywhere.

The main theorem allowing us to interchange a limit and an integral in the theory of Lebesgue integration is:

Proposition 2.6 (Dominated convergence theorem) *Let $F_n : \Omega \to \mathbb{R}$ be a sequence of functions converging to a function $F: F_n(x) \to F(x)$, for almost all x, as $n \to \infty$, and assume that there exists a function $G : \Omega \to \mathbb{R}$ with $\int_\Omega G(x)d\mu(x) < \infty$ and $\forall x \in \Omega \ \forall n, \ |F_n(x)| \leq G(x)$, then:*

$$\lim_{n \to \infty} \int_\Omega |F_n(x) - F(x)|d\mu(x) = 0$$

which implies $\lim_{n \to \infty} \int_\Omega F_n(x)d\mu(x) = \int_\Omega F(x)d\mu(x)$.

For a proof, see e.g. [278, Chap. 4] or [197, Chap. 5].

2.A.4 Approximation of Integrals

Since measurable functions can be quite irregular (think for example of the indicator function of the rational numbers), it is convenient to approximate them by regular functions. Let $F : \Omega \to \mathbb{R}$ be integrable: $\int_\Omega |F(x)| d\mu(x) < \infty$ and assume that Ω is a Borel subset of $\mathbb{R}^n$ for some n. Then, $\forall \epsilon > 0$, $\exists$ a continuous function $G : \Omega \to \mathbb{R}$ so that:

$$\int_\Omega |F(x) - G(x)| d\mu(x) \le \epsilon \tag{2.A.11}$$

The function G can chosen to be C^∞ and to vanish outside a bounded set. G can also be chosen as a function of the form $\sum_{m=1}^N c_m \mathbb{1}_{A_m}$ with A_m rectangles in Ω and $N < \infty$ (see e.g. [197, Chap. 7] for the proofs).

If $\Omega = \times_{i \in \mathbb{Z}} \Omega_i$ is a product space, μ a product measure on that space (see (2.A.6)) and $F : \Omega \to \mathbb{R}$ an integrable function, $\int_\Omega |F(\mathbf{x})| d\mu(\mathbf{x}) < \infty$, then, $\forall \epsilon > 0$, $\exists N < \infty$, $\exists G : \Omega \to \mathbb{R}$ which is a function only of the variables $(x_{-N}, \ldots, x_N)$, so that:

$$\int_\Omega |F(\mathbf{x}) - G(\mathbf{x})| d\mu(\mathbf{x}) < \epsilon \tag{2.A.12}$$

Bounds similar to (2.A.11) hold for square integrable F's, $\int_\Omega |F(x)|^2 d\mu(x) < \infty$, with $|F(x) - G(x)|$ replaced by $|F(x) - G(x)|^2$ and similarly for (2.A.12).

2.A.5 Invariant Measures

Consider a map $T : \Omega \to \Omega$ from Ω into itself. Later, specially in Chap. 4, we will think of T as a dynamical transformation on a space of physical states Ω. An important notion is the one of measures that are invariant under such transformations[28]:

Definition 2.7 Let (Ω, Σ, μ) be a measure space and $T : \Omega \to \Omega$ a map from Ω into itself. One says that μ is invariant under T or, equivalently, that the map T preserves the measure μ, if, $\forall A \in \Sigma$,

$$\mu(T^{-1}A) = \mu(A) \tag{2.A.13}$$

where

$$T^{-1}A = \{x | Tx \in A\}.$$

---

[28] In Sect. 4.1 we will give several examples of maps and of measures invariant under these maps.

Remark 2.8 If the map T is invertible, (2.A.13) is obviously equivalent to: $\forall A \in \Sigma$,

$$\mu(TA) = \mu(A) \tag{2.A.14}$$

The reader might wonder why one uses definition (2.A.13) and not (2.A.14). The logic is that, considering T to be a transformation acting on a set of states Ω, one wants to compare the measure of a subset of states A, with the measure of the set of initial conditions that are mapped onto that subset by the transformation T and that set is $T^{-1}A$. Equation (2.A.13) says that those two probabilities are equal and that expresses the fact that the map T preserves the measure μ.

Remark 2.9 Property (2.A.13) is equivalent to: $\forall F \in L^1(\Omega, d\mu)$,

$$\int_\Omega F(Tx)d\mu(x) = \int_\Omega F(x)d\mu(x) \tag{2.A.15}$$

For a function of the form $F(x) = \mathbb{1}_A(x)$, the equivalence of (2.A.13) and (2.A.15) is immediate. To prove that (2.A.13) implies (2.A.15) for more general functions $F \in L^1(\Omega)$, one uses (2.A.9). Since (2.A.15) holds for each term of the sum in the right hand side of (2.A.9), it holds also in the limit and that proves (2.A.15).

Remark 2.10 An important transformation that we will use often is the *shift* on a product space $\Omega = \times_{i \in \mathbb{Z}} \Omega_i$, equipped with the cylindrical σ-algebra, defined by, for $\mathbf{x} = (x_i)_{i \in \mathbb{Z}}$,

$$(T_{\text{shift}}\mathbf{x})_i = x_{i+1} \tag{2.A.16}$$

which leaves product measures μ defined by (2.A.6) invariant; the inverse of T is:

$$(T_{\text{shift}}^{-1}\mathbf{x})_i = x_{i-1}.$$

T_{shift} can also be defined on $\Omega^+ = \times_{i \in \mathbb{N}} \Omega_i$, but is not invertible.
   If Ω is a bounded subset of $\mathbb{R}^n$, a metric is defined on Ω by:

$$\text{dist}(\mathbf{x}, \mathbf{y}) = \sum_{i \in \mathbb{Z}} 2^{-|i|}|x_i - y_i|, \tag{2.A.17}$$

with $|\cdot|$ the Euclidean norm in $\mathbb{R}^n$. It is also defined on Ω^+ by the same formula with the sum in $\mathbb{N}$.
   The map T_{shift} is continuous for that metric:

$$\text{dist}(T_{\text{shift}}\mathbf{x}, T_{\text{shift}}\mathbf{y}) = \sum_{i \in \mathbb{Z}} 2^{-|i|}|x_{i+1} - y_{i+1}|$$
$$\leq 2 \sum_{i \in \mathbb{Z}} 2^{-|i+1|}|x_{i+1} - y_{i+1}| = 2 \sum_{i \in \mathbb{Z}} 2^{-|i|}|x_i - y_i| = 2\text{dist}(\mathbf{x}, \mathbf{y}). \tag{2.A.18}$$

2.A.6 Probability Densities, Marginal and Conditional Probabilities

A measure ν on (Ω, Σ) is *absolutely continuous* with respect to another measure μ on (Ω, Σ) if, $\forall A \in \Sigma$,

$$\mu(A) = 0 \to \nu(A) = 0.$$

In that situation, the *Radon-Nikodym theorem* implies that one can write, $\forall A \in \Sigma$,

$$\nu(A) = \int_A F(x) d\mu(x), \tag{2.A.19}$$

for a μ-integrable function $F : \Omega \to [0, \infty[$, see e.g. [278, Chap. 11] for a proof.

The function $F(x)$ is the *probability density* of ν relative to μ and one writes: $\frac{d\nu}{d\mu} = F$. We are often interested in situation where μ is the Lebesgue measure and then, one write (2.A.19) as: $\nu(A) = \int_A F(x) dx$.

Given a measure space (Ω, Σ, μ), the *marginal probability distribution* of a random variable $f : \Omega \to \mathbb{R}$ is the measure ν_f on $(\mathbb{R}, \mathcal{B}(\mathbb{R}))$ given by:

$$\nu_f(A) = \mu(f(x) \in A) = \mu(f^{-1}(A)). \tag{2.A.20}$$

For these ν_f and μ, $\frac{d\nu_f}{d\mu}$ is the probability density of the random variable f.

Two random variable f_1, f_2 are *independent* if, $\forall A, B \in \mathcal{B}(\mathbb{R})$,

$$\mu(f_1(x) \in A, f_2(x) \in B) = \mu(f_1(x) \in A)\mu(f_2(x) \in B). \tag{2.A.21}$$

This definition can be extended to any finite collection of random variables; an infinite collection of random variables is independent if any finite sub-collection of random variables is.

If $\mu(\Omega) = 1$, and if f_1, f_2 are independent, then:

$$\int_\Omega f_1(x) f_2(x) d\mu(x) = \int_\Omega f_1(x) d\mu(x) \int_\Omega f_2(x) d\mu(x). \tag{2.A.22}$$

We introduced in Sect. 2.2.2 the notion of *conditional probability* $P(A \mid B)$ of an event A, given some event B. For a discrete random variable X (taking a finite or countable number of values), one defines the *conditional probability distribution* of a random variable Y, given that $X = x$ by:

$$P(Y \in A \mid X = x) = \frac{P(Y \in A, X = x)}{P(X = x)} \tag{2.A.23}$$

for every measurable A and every x such that $P(X = x) \neq 0$.

If X is a continuous random variable, one cannot use (2.A.23), but if X and Y have a joint continuous distribution density $F_{X,Y}(x, y)$, $x, y \in \mathbb{R}$ one defines (up to subtleties about sets of measure 0) the *conditional probability density* of Y, given that $X = x$, by:

$$F(Y = y \mid X = x) = \frac{F_{X,Y}(x, y)}{\int F_{X,Y}(x, y')dy'}$$

for every x for which $\int F_{X,Y}(x, y')dy' \neq 0$.

If X and Y are independent, then it is easy to deduce from (2.A.21) that, for all x, the conditional probability density of Y equals the probability density of Y:
$F(Y = y \mid X = x) = F(Y = y)$.

2.A.7 Cantor Sets and Measures

As an aside, we mention a family of sets that are interesting from a topological and measure-theoretic viewpoint: the Cantor sets.

The original or "middle-third" Cantor set can be obtained by removing successively the "middle-third" intervals: let $C_0 = [0, 1]$, $C_1 = [0, 1] \setminus]\frac{1}{3}, \frac{2}{3}[= [0, \frac{1}{3}] \cup [\frac{2}{3}, 1]$, $C_2 = C_1 \setminus (]\frac{1}{9}, \frac{2}{9}[\cup]\frac{7}{9}, \frac{8}{9}[) = [0, \frac{1}{9}] \cup [\frac{2}{9}, \frac{1}{3}] \cup [\frac{2}{3}, \frac{7}{9}] \cup [\frac{8}{9}, 1]$, and so on. Then $C = \cap_{n=0}^{\infty} C_n$.

Actually, $C = \{x \mid x = \sum_{n=1}^{\infty} \frac{a_n}{3^n}, a_n \neq 1, \forall n\}$. To see this, observe that, if we write the ternary expansion of a number $x \in [0, 1]$, $x = \sum_{n=1}^{\infty} \frac{a_n}{3^n}$, then the numbers in $]\frac{1}{3}, \frac{2}{3}[$ correspond to $a_1 = 1$ those in $]\frac{1}{9}, \frac{2}{9}[\cup]\frac{7}{9}, \frac{8}{9}[$ correspond to $a_2 = 1$ and so on. So that the numbers in C are exactly those $x \in [0, 1]$ whose ternary expansion does not contain the symbol 1.

The Cantor set C has the following properties:

1. C is of zero measure since the sum of the lengths of the removed intervals is:
   $\sum_{n=1}^{\infty} \frac{2^{n-1}}{3^n} = 1$.
2. C is uncountable: the map $g : C \to [0, 1]$, $g(x) = \sum_{n=1}^{\infty} \frac{a_n}{2^{n+1}}$, where $x = \sum_{n=1}^{\infty} \frac{a_n}{3^n}$, $a_n \neq 1$, $\forall n$, is a surjection from C to $[0, 1]$, hence the cardinality of C must be at least the one of $[0, 1]$ (and also at most that cardinality, since $C \subset [0, 1]$).
3. C is a perfect set, which, by definition, means that it is closed (since it is the complement of a union of open intervals) and that each element of C is a limit of other elements of C: if $x = \sum_{n=1}^{\infty} \frac{a_n}{3^n}$, $a_n \neq 1$, $\forall n$, then $x = \lim_{k\to\infty} x_k$, where $x_k = \sum_{n=1}^{\infty} \frac{a_n^k}{3^n}$, with $a_n^k = a_n$, $\forall n \neq k$, and $a_k^k = 2 - a_k$.
4. C is nowhere dense which, for a closed set, means that its interior is empty: in any neighborhood of $x \in C$, there are points not in C. Given $x = \sum_{n=1}^{\infty} \frac{a_n}{3^n}$, $a_n \neq 1$,

$\forall n$ and $\epsilon > 0$, choose k so that $3^{-k} < \epsilon$, and let $y = \sum_{n=1}^{\infty} \frac{b_n}{3^n}$, with $b_n = a_n$, $\forall n \neq k$, and $b_k = 1$. Then, $y \notin C$ and $|x - y| \leq \epsilon$.

5. C is totally disconnected, i.e. its only connected components are singletons.

One can modify this construction by removing a fraction $\alpha < 1$ of the remaining set at each stage: the resulting set will still be of measure 0 since $\lim_{n \to \infty} (1 - \alpha)^n = 0$. A more interesting construction consists in removing at each step an interval of length α^n in the middle of each of the remaining intervals. Then we obtain a set C_α of size $1 - \sum_{n=1}^{\infty} 2^{n-1} \alpha^n = 1 - \frac{\alpha}{1-2\alpha} = \frac{1-3\alpha}{1-2\alpha}$, where we see that we need $0 < \alpha \leq \frac{1}{3}$ and the limit case $\alpha = \frac{1}{3}$ is the usual Cantor set. But all the C_α's for $\alpha < \frac{1}{3}$ have all the properties of C except of being of Lebesgue measure zero. Finally, let $\bar{C} = \cup_{k=1}^{\infty} C_{\frac{1}{k}}$. Then the Lebesgue measure of $\bar{C}$ is one (since $\mu_{\text{Leb}}(C_{\frac{1}{k}}) \geq 1 - \mathcal{O}(\frac{1}{k})$), but, since all $C_{\frac{1}{k}}$'s have an empty interior (hence, that their complement is open and dense in $[0, 1]$), the complement of $\bar{C}$, $[0, 1] \setminus \bar{C} = \cap_{k=1}^{\infty}([0, 1] \setminus C_{\frac{1}{k}})$ is also dense[29] in $[0, 1]$, by Baire's theorem.[30] $\bar{C}$ is an example of a set that is "large" from a measure-theoretic point of view, since it has measure 1, but is "small" from a topological point of view, since it is a countable union of nowhere dense sets.[31]

One can associate to the Cantor set a measure μ_C which is not absolutely continuous with respect to the Lebesgue measure but which is continuous, namely which gives zero measure to points, unlike the delta measures defined in (2.A.2), (2.A.3): take the product measure giving weights $(\frac{1}{2}, \frac{1}{2})$ to each symbol $a_n = 0, a_n = 2$, in the expansion $x = \sum_{n=1}^{\infty} \frac{a_n}{3^n}$ (and thus weight 0 to the symbol $a_n = 1$). Since $\mu_C(C) = 1$ while $\mu_{\text{Leb}}(C) = 0$, μ_C is not absolutely continuous with respect to the Lebesgue measure, but one easily checks that $\mu_C(\{x\}) = 0$, for any singleton $\{x\}$.

2.B Appendix B: Proofs of the Law of Large Numbers and of the Central Limit Theorem

Let us first prove (2.3.8). We need to estimate the probability of:

$$|\frac{1}{N} \sum_{i=1}^{N} f_i(x_i) - \mathbb{E}(f(x))| \geq \epsilon.$$

We may assume that $\mathbb{E}(f(x)) = 0$, by redefining f. We will use Markov's (or Chebyshev's) inequality: for a random variable F on (Ω, Σ, μ), and $a > 0$,

---

[29] A subset E of $\Omega \subset \mathbb{R}^n$ is dense in Ω if, $\forall x \in \Omega$, $\forall \epsilon > 0$, $\exists y \in E$ such that $|x - y| \leq \epsilon$.

[30] This theorem implies that the intersection of countable collection of open dense sets $(U_k)_{k \in \mathbb{N}}$ in $[0, 1]$, $\cap_{k \in \mathbb{N}} U_k$, is dense in $[0, 1]$. Take here $U_k = [0, 1] \setminus C_{\frac{1}{k}}$.

[31] In topology, such sets are called meager or of the first category, while sets such as $[0, 1] \setminus \bar{C}$ are called nonmeager or of the second category.

$$\mu(\mathbb{1}(F \geq a)) \leq \frac{\mathbb{E}(F^2)}{a^2}, \tag{2.B.1}$$

which follows trivially from $a\mathbb{1}(F \geq a) \leq F$.

Apply this to $F = |\frac{1}{N} \sum_{i=1}^{N} f_i(x_i)|$, $\mu = \mu$ as in (2.3.8) and $a = \epsilon$. We get:

$$\mu\left(|\frac{1}{N} \sum_{i=1}^{N} f_i(x_i)| \geq \epsilon\right) \leq \frac{\mathbb{E}((\sum_{i=1}^{N} f_i(x_i))^2)}{\epsilon^2 N^2}. \tag{2.B.2}$$

We have $\mathbb{E}((\sum_{i=1}^{N} f_i(x_i))^2) = \sum_{i,j=1}^{N} \mathbb{E}(f_i(x_i)f_j(x_j)) = \sum_{i=1}^{N} \mathbb{E}(f_i(x_i)^2) = N\mathbb{E}(f(x)^2)$, since $\mathbb{E}(f_i(x_i)f_j(x_j)) = \mathbb{E}(f_i(x_i))\mathbb{E}(f_j(x_j)) = 0$ for $i \neq j$ because the variables x_i, x_j are independent and, by assumption, $\mathbb{E}(f_i(x_i)) = 0, \forall i$.

Inserting this in the right hand side of (2.B.2), we get:

$$\mu\left(|\frac{1}{N} \sum_{i=1}^{N} f_i(x_i)| \geq \epsilon\right) \leq \frac{C}{\epsilon^2 N},$$

with $C = \mathbb{E}(f(x)^2)$ (we assumed f to be bounded), which implies

$$\mu(G_N(\epsilon)) \geq 1 - \frac{C}{\epsilon^2 N},$$

which proves (2.3.8).

To prove (2.3.12), we proceed similarly. Applying (2.B.1) to $|F_\alpha|$, for $\alpha = 1, \ldots, k$, where $F_\alpha = \frac{1}{N}(\sum_{i=1}^{N} \mathbb{1}(x_i \in A_\alpha) - \mathbb{E}(\sum_{i=1}^{N} \mathbb{1}(x_i \in A_\alpha))) = \frac{1}{N} \sum_{i=1}^{N} \mathbb{1}(x_i \in A_\alpha) - P_\alpha$, (since $\mathbb{E}(\mathbb{1}(x_i \in A_\alpha)) = P_\alpha$), we get:

$$\mu(|F_\alpha| \geq \epsilon) \leq \frac{\mathbb{E}(F_\alpha^2)}{\epsilon^2} \tag{2.B.3}$$

Using again the independence of the variables x_i, x_j for $i \neq j$, we get:

$$\mathbb{E}(F_\alpha^2) = \frac{1}{N^2} \left(\sum_{i=1}^{N} \text{Var}(\mathbb{1}(x_i \in A_\alpha))\right) = \frac{1}{N}(P_\alpha - P_\alpha^2) \tag{2.B.4}$$

Then, using (2.B.3) and (2.B.4):

$$\mu(\cup_{\alpha=1}^{k}(|F_\alpha| \geq \epsilon)) \leq \sum_{\alpha=1}^{k} \mu(|F_\alpha| \geq \epsilon) \leq \sum_{\alpha=1}^{k} \frac{P_\alpha(1 - P_\alpha)}{N\epsilon^2} \leq \frac{1}{N\epsilon^2}$$

since $P_\alpha \geq 0$ and $\sum_{\alpha=1}^{k} P_\alpha = 1$. This proves (2.3.12).

To prove the strong version of the law of large numbers, (2.3.9) we use two facts:

(1) The event $E_a = \{\mathbf{x} \mid \lim_{N\to\infty} S_N(\mathbf{x}) = a\}$, for $a \in \mathbb{R}$, is invariant under the shift map T_{shift}, defined in (2.A.16): $T_{\text{shift}}^{-1} E = E$. This is obvious, because if $\lim_{N\to\infty} S_N(\mathbf{x}) = a$ for some $\mathbf{x}$, the same will hold for $T_{\text{shift}}^{-1}(\mathbf{x})$.

(2) For every $E \in \Sigma$ and $\forall \epsilon > 0$, there exists $N < \infty$ and $A \in \Sigma$ but depending only the finite set of variables $(x_{-N}, \ldots, x_N)$ such that $\mu(E\triangle A) \leq \epsilon$, where $\triangle$ denotes the symmetric difference: $A\triangle B = A \cup B \setminus A \cap B$ (this approximation follows from (2.A.12) by taking $F = \mathbb{1}_E$ and $G = \mathbb{1}_A$, and using $\mu(E\triangle A) = \int |\mathbb{1}_E - \mathbb{1}_A| d\mu$).

From (2) we deduce that any event invariant under the shift T_{shift} must have measure equal to 0 or 1.[32] Indeed, let E satisfy $T_{\text{shift}}^{-1} E = E$ and let A be an event that depend on finitely many variables so that $\mu(E\triangle A) \leq \epsilon$; since $\mu(E\triangle A) = \int |\mathbb{1}_E - \mathbb{1}_A| d\mu$, we also have:

$$|\mu(E) - \mu(A)| = |\int (\mathbb{1}_E - \mathbb{1}_A) d\mu| \leq \epsilon. \tag{2.B.5}$$

Let $B = T^{-n}A$, where $n \geq 2N + 1$ so that B depends upon different variables from A. Since μ is a product measure, we have $\mu(A \cap B) = \mu(A)\mu(B) = \mu(A)^2$.

Since, both μ and E are shift-invariant, we have $\mu(E\triangle B) = \mu(T^{-n}E\triangle T^{-n}A) = \mu(E\triangle A) \leq \epsilon$. Since $E\triangle(A \cap B) \subset (E\triangle A) \cup (E\triangle B)$, we have $\mu(E\triangle(A \cap B)) \leq 2\epsilon$. Thus, since $|\mu(E) - \mu(A \cap B)| \leq \mu(E\triangle(A \cap B))$,

$$|\mu(E) - \mu(A \cap B)| = |\mu(E) - \mu(A)^2| \leq 2\epsilon. \tag{2.B.6}$$

Now, (2.B.5) implies that $|\mu(E)^2 - \mu(A)^2| = |\mu(E) - \mu(A)||\mu(E) + \mu(A)| \leq 2\epsilon$ and, since ϵ is arbitrary, combining this and (2.B.6), we get that $\mu(E) = \mu(E)^2$, which implies that $\mu(E)$ must equal 0 or 1.

So, using (1) above, we see that $\mu(E_a)$ equal 0 or 1 for all $a \in \mathbb{R}$. Since the average of $S_n(f)$ equals $\mathbb{E}(f)$, $\mu(E_a)$ equals 0 for $a \neq \mathbb{E}(f)$ and equals 1 for $a = \mathbb{E}(f)$.

The proof of (2.3.13) is similar and is left as an exercise.

We will only sketch the proof of the central limit theorem (2.3.17). Since we subtract the mean $\mathbb{E}(f)$ in the definition of X_N, we may assume it to be 0.

---

[32] This is similar to Kolmogorov's $0 - 1$ law that makes the same claim for events that are independent of any finite number of variables. Events like $E_a = \{\mathbf{x} \mid \lim_{N\to\infty} S_N(\mathbf{x}) = a\}$ have also that property.

Consider the Fourier transform[33] of the distribution of the variable X_N; we have:

$$\mu(\exp(-ikX_N)) = \prod_{j=1}^{N} \mu_j\left(\exp\left(\frac{-ikf_j(x_j)}{\sqrt{N}}\right)\right) = \left(\mu\left(\exp\left(\frac{-ikf(x)}{\sqrt{N}}\right)\right)\right)^N,$$

(2.B.7)

since μ is a product measure. Let us write the Taylor expansion of $\ln\left(\mu\left(\exp\left(\frac{-ikf(x)}{\sqrt{N}}\right)\right)\right)$:

$$\ln\left(\mu\left(\exp\left(\frac{-ikf(x)}{\sqrt{N}}\right)\right)\right) = -\frac{k^2}{2N}\mathbb{E}(f(x))^2 + \mathcal{O}\left(\frac{k^3}{N^{3/2}}\right)$$

since by assumption $\mathbb{E}(f) = 0$. Inserting this in (2.B.7), we get:

$$\mu(\exp(-ikX_N)) = \exp\left(-\frac{\sigma^2 k^2}{2} + \mathcal{O}\left(\frac{k^3}{N^{1/2}}\right)\right),$$

with $\sigma^2 = \mathbb{E}(f(x))^2$.

If we neglect the term $\mathcal{O}(\frac{k^3}{N^{1/2}})$ (which vanishes as $N \to \infty$, but we leave aside the rigorous proof of that), we get that the Fourier transform of the distribution of X_N tends to a Gaussian. Since the inverse Fourier transform of $\frac{\exp(-\frac{\sigma^2 k^2}{2})}{\sqrt{2\pi}}$ is $\frac{\exp(-\frac{x^2}{2\sigma^2})}{\sqrt{2\pi}\sigma}$, we get (at least formally) (2.3.17).

---

[33] We define the Fourier transform of an integrable function $f(x)$ as:

$$\hat{f}(k) = \frac{1}{\sqrt{2\pi}}\int_{\mathbb{R}} \exp(-ikx)f(x)dx$$

Then, the inverse Fourier transform (assuming $\hat{f}$ to be integrable) is given by

$$f(x) = \frac{1}{\sqrt{2\pi}}\int_{\mathbb{R}} \exp(ikx)\hat{f}(k)dk.$$

Chapter 3
Classical Mechanics

3.1 Introduction

This chapter is not meant to be a course, even a crash course, in classical mechanics. We will only discuss the results of classical mechanics that we will need later, namely the notion of Hamiltonian flow, the conservation of energy, Liouville's theorem and the time reversibility of the equations of motion. Since we are not interested in mathematical rigor or generality, when we assume below that a function is "smooth" we mean that it is sufficiently differentiable for the results that we mention to hold (we will sometimes add details in the footnotes).

In the next chapter, we will discuss more general dynamical systems of which classical mechanics is an important but special case.

3.2 Newton's Laws

In classical mechanics one starts by defining frames of reference and by distinguishing between inertial and non inertial frames.

A *frame of reference* is simply a coordinate system that ascribes a set of numbers specifying the positions of all the particles in the system under consideration at any given time. We will use here cartesian coordinates. If these positions change in time, those coordinates, as functions of time, describe the trajectories of the particles.

An *inertial frame of reference* is one in which a particle which is not subjected to any force moves along a straight line at constant velocity. Instead of saying "which is not subjected to any force" one could say "is infinitely distant from any other particle." If we describe the motion of such a particle in a frame of reference attached to a merry-go-round or to an accelerating rocket, that motion will no longer appear to be on a straight line or to move at constant speed. Such frames are called *non inertial*.

© Springer Nature Switzerland AG 2022
J. Bricmont, *Making Sense of Statistical Mechanics*, Undergraduate Lecture
Notes in Physics, https://doi.org/10.1007/978-3-030-91794-4_3

It is obvious from the definition given here that we are dealing with an idealization, which is nevertheless approximately realized in many situations: the fact that the Earth rotates around the Sun and around itself makes a frame of reference attached to the Earth, strictly speaking, non inertial; but it can nevertheless be considered inertial for most experiments performed in laboratories.

A basic principle of mechanics is the equivalence of all inertial frames of reference, also called Galilean invariance: the laws of motion take the same form in all inertial frames of reference and the transformations between such frames consist of (constant in time) rotations and translations on a straight line at constant velocity of the origin of coordinates. This invariance implies conservation laws for total momentum, total angular momentum and energy (checking the first conservation laws will be left as exercises).[1]

Here we will always work in a fixed inertial frame, so we will not be concerned with Galilean invariance. Moreover, we will not discuss conservations laws apart form the conservation of energy. Newton's first law, says, in modern terminology, that there exist inertial reference frames; since we decided to work in one such frame, we will not discuss it further.

Consider N particles in $\mathbb{R}^3$ of masses $m_1, m_2, \ldots, m_N$. The position of the ith particle is represented by a vector $\vec{q}_i \in \mathbb{R}^3$ and the positions of all the particles of the system by a vector $\mathbf{q} = (\vec{q}_1, \vec{q}_2, \ldots, \vec{q}_N) \in \mathbb{R}^{3N}$.

Newton's second law states that:

$$m_i \frac{d^2 \vec{q}_i}{dt^2} = \sum_{j=1, j \neq i}^{N} \vec{F}_{ij}(\vec{q}_i, \vec{q}_j) + \sum_{i=1}^{N} \vec{F}_i(\vec{q}_i) \tag{3.2.1}$$

where $\vec{F}_{ij}(\vec{q}_i, \vec{q}_j)$ is the force exerted on the particle of index i by the one of index j and $F_i(\vec{q}_i)$ represents the force exerted on the system by bodies located outside of it.[2] We will assume that the forces are "conservative" or "derive from a potential", namely that, for each pair i, j, there are smooth functions $V_{ij} : \mathbb{R}^6 \to \mathbb{R}$, $V_i : \mathbb{R}^3 \to \mathbb{R}$, such that:

$$\vec{F}_{ij}(\vec{q}_i, \vec{q}_j) = -\nabla_{\vec{q}_i} V_{ij}(\vec{q}_i, \vec{q}_j), \tag{3.2.2}$$

$$\vec{F}_i(\vec{q}_i) = -\nabla_{\vec{q}_i} V_i(\vec{q}_i), \tag{3.2.3}$$

where $\nabla_{\vec{q}_i}$ is the gradient with respect to $\vec{q}_i$.

The best known potential V_{ij} is the gravitational one:

---

[1] The laws of motion can also be formulated in non inertial frames of reference, but at the expense of introducing in the equations of motion "fictional" forces like the centrifugal or the Coriolis forces which simply take into account the accelerating motion of the reference frame (relative to inertial ones).

[2] We will consider here only two-body and one-body forces.

$$V_{ij}(\vec{q}_i, \vec{q}_j) = \frac{Gm_i m_j}{\|\vec{q}_i - \vec{q}_j\|}, \tag{3.2.4}$$

with G Newton's constant and $\| \cdot \|$ the Euclidean norm in $\mathbb{R}^3$.

Let us now define the potential function of the system $V : \mathbb{R}^{3N} \to \mathbb{R}$:

$$V(\mathbf{q}) = \sum_{i<j=1}^{N} V_{ij}(\vec{q}_i, \vec{q}_j) + \sum_{i=1}^{N} V_i(\vec{q}_i), \tag{3.2.5}$$

One could also add to (3.2.5) an arbitrary constant since the potential enters the equations of motion (3.2.1) only through their derivatives, see (3.2.2), (3.2.3).

Then, (3.2.1) can be written as:

$$m_i \frac{d^2 \vec{q}_i}{dt^2} = -\nabla_{\vec{q}_i} V(\mathbf{q}). \tag{3.2.6}$$

We will assume that $V_{ij}(\vec{q}_i, \vec{q}_j) = \bar{V}_{ij}(\vec{q}_i - \vec{q}_j)$, where $\bar{V}_{ij} : \mathbb{R}^3 \to \mathbb{R}$. So, V_{ij} is a function only of the differences between the positions of the particles (as is (3.2.4)).

This assumption implies that Newton's third law is verified:

$$\vec{F}(\vec{q}_i, \vec{q}_j) = \nabla_{\vec{q}_i} \bar{V}_{ij}(\vec{q}_i - \vec{q}_j) = -\nabla_{\vec{q}_j} \bar{V}_{ij}(\vec{q}_i - \vec{q}_j) = -\vec{F}(\vec{q}_j, \vec{q}_i) \tag{3.2.7}$$

Actually, as in (3.2.4), the potential often depends only on the distance between $\vec{q}_i$ and $\vec{q}_j$: $V_{ij}(\vec{q}_i, \vec{q}_j) = \tilde{V}_{ij}(\|\vec{q}_i - \vec{q}_j\|)$, where $\tilde{V}_{ij} : \mathbb{R}_+ \to \mathbb{R}$.

The first fundamental question about (3.2.6) is whether they have solutions. Since these equations are second order in time, to get a unique solution, we need to specify a pair $(\mathbf{q}(t_0), \frac{d\mathbf{q}}{dt}(t_0))$, where $t_0 \in \mathbb{R}$ is some initial time. Then, under very general conditions on the functions V_{ij}, V_i and for any pair $(\mathbf{q}_0, \dot{\mathbf{q}}_0) \in \mathbb{R}^{6N}$, there exists an interval $I \ni t_0$ and a unique smooth function $\mathbf{q} : I \to \mathbb{R}^{3N}$ such that, $\forall t \in I$ the function $\mathbf{q}(t)$ satisfies (3.2.6) and such that $\mathbf{q}(t_0) = \mathbf{q}_0$, $\frac{d\mathbf{q}}{dt}(t_0) = \dot{\mathbf{q}}_0$ (see e.g. Kolmogorov and Fomin [197, Sect. 2.4]).

But we will be interested below in situations where the unique solution $\mathbf{q}(t)$ satisfies (3.2.5) *for all times* $t \in \mathbb{R}$. Such existence results exist but they are more tricky than the previous existence theorem because the equations themselves may become ill-defined during collisions or the particles may "conspire" so that some of them accelerate so much that they run away to infinity in a finite time.[3]

---

[3] To understand why this may happen, consider the simple equation (which is not of the form (3.2.6)):

$$\frac{dx}{dt} = x^2(t), \tag{3.2.8}$$

with $x(t) \in \mathbb{R}$ and initial condition at $t = 0$ $x(0) > 0$. It is easy to see, by direct integration, that the solution is

In what follows we will simply assume that the potentials V_{ij} are "nice" enough so that there is a unique solution $\mathbf{q}(t)$ satisfying (3.2.6) *for all times* $t \in \mathbb{R}$.

3.3 Hamilton's Equations

It is often convenient to rewrite Newton's equations (3.2.6) in Lagrangian form or in Hamiltonian form. We will only use the latter one.[4] To do so, we will introduce the *phase space* $\mathbb{R}^{6N}$, and write a vector $\mathbf{x} \in \mathbb{R}^{6N}$ as a pair $\mathbf{x} = (\mathbf{q}, \mathbf{p})$, with $\mathbf{q} = (\vec{q}_1, \vec{q}_2, \ldots, \vec{q}_N) \in \mathbb{R}^{3N}$, $\mathbf{p} = (\vec{p}_1, \vec{p}_2, \ldots, \vec{p}_N) \in \mathbb{R}^{3N}$.

The Hamiltonian is a function $H : \mathbb{R}^{6N} \to \mathbb{R}$:

$$H(\mathbf{q}, \mathbf{p}) = K(\mathbf{p}) + V(\mathbf{q}) \tag{3.3.1}$$

with a *kinetic energy*

$$K(\mathbf{p}) = \sum_{i=1}^{N} \frac{\|\vec{p}_i\|^2}{2m_i} \tag{3.3.2}$$

and a *potential energy* $V(\mathbf{q})$ given by (3.2.5).

Then *Hamilton's equations* are given by the following pair:

$$\frac{d\vec{q}_i(t)}{dt} = \nabla_{\vec{p}_i} H(\mathbf{q}(t), \mathbf{p}(t)), \tag{3.3.3}$$

and

$$\frac{d\vec{p}_i(t)}{dt} = -\nabla_{\vec{q}_i} H(\mathbf{q}(t), \mathbf{p}(t)), \tag{3.3.4}$$

for $i = 1, \ldots, N$.

With H defined by (3.3.1), (3.3.2), these equations are:

$$\frac{d\vec{q}_i(t)}{dt} = \frac{\vec{p}_i(t)}{m_i} \tag{3.3.5}$$

and

$$\frac{d\vec{p}_i(t)}{dt} = -\nabla_{\vec{q}_i} V(\mathbf{q}(t)). \tag{3.3.6}$$

$$x(t) = \frac{x(0)}{1 - tx(0)}, \tag{3.2.9}$$

which goes to ∞ as $t \to \frac{1}{x(0)}$.

[4] The Lagrangian will be defined in Appendix 5.B.2.

Taking the time derivative of (3.3.5), and inserting the result in (3.3.6), one gets

$$\frac{d^2 \vec{q}_i(t)}{dt^2} = -\frac{\nabla_{\vec{q}_i} V(\mathbf{q}(t))}{m_i},$$

which is Newton's equation (3.2.6). Thus, any solution of the pair (3.3.5), (3.3.6) will be such that the function $\mathbf{q}(t)$ is a solution of (3.2.6). One can also easily show that, for any solution $\mathbf{q}(t)$ of (3.2.6), the pair $(\mathbf{q}(t), \mathbf{p}(t))$ with $\vec{p}_i(t) = \frac{1}{m_i}\frac{d\vec{q}_i(t)}{dt}$ will satisfy (3.3.5), (3.3.6). Thus, Newton's equations and Hamilton's equations are *equivalent*.

One consequence of Hamilton's equations is that, if we consider a function $F : \mathbb{R} \times \mathbb{R}^{6N} \to \mathbb{R}$, the following expression for the time derivative of $F = F(t, \mathbf{q}(t), \mathbf{p}(t))$ follows immediately from (3.3.3) and (3.3.4):

$$\frac{dF(t, \mathbf{q}(t), \mathbf{p}(t))}{dt} = \frac{\partial F}{\partial t} + \sum_{i=1}^{N} \nabla_{\vec{q}_i} F \cdot \nabla_{\vec{p}_i} H - \nabla_{\vec{p}_i} F \cdot \nabla_{\vec{q}_i} H, \tag{3.3.7}$$

where $\cdot$ denotes the scalar product in $\mathbb{R}^{3N}$. The sum in the right hand side of (3.3.7) is called the Poisson bracket $\{F, H\}$ of F and H, but we will not use that notion here. If F does not depend explicitly on t, $\frac{\partial F}{\partial t} = 0$, we can write (3.3.7) as:

$$\frac{dF(\mathbf{q}(t), \mathbf{p}(t))}{dt} = \mathcal{L}F(\mathbf{q}(t), \mathbf{p}(t)), \tag{3.3.8}$$

where $\mathcal{L}$ is the linear operator defined sum in the right hand side of (3.3.7). Its solution is formally given by:

$$F(\mathbf{q}(t), \mathbf{p}(t)) = \exp(\mathcal{L}t) F(\mathbf{q}(0), \mathbf{p}(0)) \tag{3.3.9}$$

and is called the *Liouville evolution*.

3.3.1 The Hamiltonian Flow

Since (3.2.6) and (3.3.5), (3.3.6) are equivalent, if we assume that the potentials are such that a unique solution $\mathbf{q}(t)$ of (3.2.6) exist for all times, we also have, for the pair of (3.3.5), (3.3.6), for any $t_0 \in \mathbb{R}$ and any initial conditions $(\mathbf{q}_0, \mathbf{p}_0)$, a unique solution $(\mathbf{q}(t), \mathbf{p}(t))$ satisfying (3.3.5), (3.3.6) for all times and such that $(\mathbf{q}(t_0), \mathbf{p}(t_0)) = (\mathbf{q}_0, \mathbf{p}_0)$.

It will be convenient to associate to such solutions a family of maps $T^t : \mathbb{R}^{6N} \to \mathbb{R}^{6N}$, for $t \in \mathbb{R}$, defined by:

$$T^t(\mathbf{q}, \mathbf{p}) = (\mathbf{q}(t), \mathbf{p}(t)), \tag{3.3.10}$$

where $(\mathbf{q}(t), \mathbf{p}(t))$ is the unique solution of (3.3.5), (3.3.6) satisfying $(\mathbf{q}(0), \mathbf{p}(0)) = (\mathbf{q}, \mathbf{p})$.

So, the map T^t associates to every pair $(\mathbf{q}, \mathbf{p}) \in \mathbb{R}^{6N}$ the value at time t of the unique solution $(\mathbf{q}(t), \mathbf{p}(t))$ of (3.3.5), (3.3.6) that "passes" through the point $(\mathbf{q}, \mathbf{p})$ at time 0. Since the solutions are assumed to exist for all $t \in \mathbb{R}$, T^t is invertible: $T^t T^{-t} = \mathrm{Id}$, where Id is the identity operator.

The family of maps $(T^t)_{t \in \mathbb{R}}$ is called the *Hamiltonian flow*.

3.3.2 Conservation of Energy

The *energy* of a mechanical system of the type considered here is a function $E : \mathbb{R}^{6N} \to \mathbb{R}$ defined by

$$E(\mathbf{q}, \mathbf{p}) = K(\mathbf{p}) + V(\mathbf{q}) \tag{3.3.11}$$

with the kinetic energy $K(\mathbf{p})$ defined in (3.3.2) and the potential energy $V(\mathbf{q})$ defined in (3.2.5). This is of course identical to the Hamiltonian function and the energy, like the potential, is defined up to an additive constant. We have:

Theorem 3.1 (Conservation of energy) *Let* $(\mathbf{q}(t), \mathbf{p}(t))$ *be a solution of (3.3.5), (3.3.6). Then,* $\forall t \in \mathbb{R}$:

$$\frac{dE(t)}{dt} = 0, \tag{3.3.12}$$

where $E(t) = E(\mathbf{q}(t), \mathbf{p}(t))$.

Proof It is enough to compute the time derivative of $E(t)$, using (3.3.7), (3.3.1), (3.3.2):

$$\frac{dE(\mathbf{q}(t), \mathbf{p}(t))}{dt} = \sum_{i=1}^{N} \frac{\vec{p}_i \cdot \frac{d\vec{p}_i(t)}{dt}}{m_i} + \sum_{i=1}^{N} \nabla_{\vec{q}_i} V(\mathbf{q}(t)) \cdot \frac{d\vec{q}_i(t)}{dt}.$$

By (3.3.5), (3.3.6), this equals 0.

Remark 3.2 One may rewrite the energy in a more familiar form, using $\vec{p}_i(t) = \frac{1}{m_i} \frac{d\vec{q}_i(t)}{dt}$:

$$E(t) = \sum_{i=1}^{N} \frac{m_i \|\vec{v}_i(t)\|^2}{2} + V(\mathbf{q}(t)),$$

with $\vec{v}_i(t) = \frac{d\vec{q}_i(t)}{dt}$.

Theorem 3.1 implies an important corollary. Let, for $E_0 \in \mathbb{R}_+$, S_{E_0} be a surface in $\mathbb{R}^{6N}$ of constant energy:

$$S_{E_0} = \{(\mathbf{q}, \mathbf{p}) | E(\mathbf{q}, \mathbf{p}) = E_0\}$$

Corollary 3.3 *For all* $t \in \mathbb{R}$, *the Hamiltonian flow* T^t *defined in Sect. 3.3.1 maps* S_{E_0} *into itself, for a certain* E_0 *determined by the initial conditions:* $E_0 = E(\mathbf{q}(0), \mathbf{p}(0))$.

One also expresses this result by saying that the surface S_{E_0} is invariant under the Hamiltonian flow.

3.4 Liouville's Theorem and Measure

In words, Liouville's theorem says that the Lebesgue measure (introduced in Appendix 2.A.2) of a set is invariant under the Hamiltonian flow. Let D be a subset of finite measure of $\mathbb{R}^{6N}$ and $T^t(D)$ be the image of D under the Hamiltonian flow: $T^t(D) = \{(\mathbf{q}(t), \mathbf{p}(t)) | (\mathbf{q}(0), \mathbf{p}(0)) \in D\}$. Finally, let $V(t)$ be the size of $T^t(D)$: $V(t) = |T^t(D)|$.

3.4.1 Liouville's Theorem

Theorem 3.4 (Liouville's theorem) *With the above notation, we have,* $\forall t \in \mathbb{R}$:

$$\frac{dV(t)}{dt} = 0 \tag{3.4.1}$$

In order to prove Liouville's theorem, let us consider a more general system:

$$\dot{\mathbf{x}} = \mathbf{f}(\mathbf{x}), \tag{3.4.2}$$

where $\mathbf{x} \in \mathbb{R}^M$, and $\mathbf{f} : \mathbb{R}^M \to \mathbb{R}^M$. We assume that $\mathbf{f}$ is such that solutions of (3.4.2) exist for all times. Let Φ^t be the flow, similar to the Hamiltonian flow, but associated to (3.4.2) and let, as before, D be a subset of finite measure of $\mathbb{R}^M$ and $V(t) = |\Phi^t(D)|$. Then, we have the

Theorem 3.5 (Generalized Liouville's theorem) *With the above notation, we have,* $\forall t \in \mathbb{R}$:

$$\frac{dV(t))}{dt} = \int_{\Phi^t(D)} \text{div } \mathbf{f} dx \tag{3.4.3}$$

where div $\mathbf{f} = \sum_{i=1}^{m} \dfrac{\partial f_i}{\partial x_i}$.

Proof First of all, it is enough to prove (3.4.3) for $t = 0$, since, to prove it for all t, one may simply replace D by $\Phi^t(D)$. By the change of variables formula, one has:

$$V(t) = \int_{\Phi^t(D)} d\mathbf{x} = \int_D \det(\frac{\partial \Phi^t(\mathbf{x})}{\partial \mathbf{x}}) d\mathbf{x} \tag{3.4.4}$$

Equation (3.4.2) implies $\frac{d}{dt} \Phi^t(\mathbf{x})_{t=0} = \mathbf{f}(\mathbf{x})$. Thus,

$$\frac{\partial \Phi^t(\mathbf{x})}{\partial \mathbf{x}} = 1 + t\frac{\partial \mathbf{f}}{\partial \mathbf{x}} + \mathcal{O}(t^2) \tag{3.4.5}$$

as $t \to 0$. By definition of the determinant, one has for a square matrix A:

$$\det(1 + tA + \mathcal{O}(t^2)) = 1 + t \text{ trace } A + \mathcal{O}(t^2) \tag{3.4.6}$$

as $t \to 0$, where $\text{trace} A = \sum_{i=1}^{m} A_{ii}$ because only the diagonal elements of A give a contribution of order t to the determinant (all other terms in the determinant are at least of order t^2; check that as an exercise). So,

$$\frac{d}{dt} \det(1 + tA + \mathcal{O}(t^2))|_{t=0} = \text{trace } A$$

But if we let $A = \frac{\partial \mathbf{f}}{\partial \mathbf{x}}$, $\text{trace} A = \text{div } \mathbf{f}$, and, combining this with (3.4.4) and (3.4.5) one obtains (3.4.3).

Now, to prove Liouville's Theorem 3.4, set in (3.4.2) $M = 2N$, $\mathbf{x} = (\mathbf{q}, \mathbf{p})$ and let $\mathbf{f}$ be the right hand side of (3.3.3, 3.3.4). One has:

$$\text{div } \mathbf{f} = \sum_{j=1}^{n} \frac{\partial^2 H}{\partial q_j \partial p_j} - \frac{\partial^2 H}{\partial p_j \partial q_j} = 0, \tag{3.4.7}$$

by interchanging derivatives.[5] So, by (3.4.3) we have (3.4.1).

Corollary 3.6 *If* $F \in L^1(\mathbb{R}^{6N})$, *then*

$$\int_{\mathbb{R}^{6N}} F(\mathbf{q}(t), \mathbf{p}(t)) d\mathbf{q} d\mathbf{p}$$

is constant $\forall t \in \mathbb{R}$.

---

[5] Assuming H to be of class $\mathcal{C}^2$.

Proof To prove the corollary, just observe that, for a function of the form $F(\mathbf{x}) = \mathbb{1}_A(\mathbf{x})$, for $A \subset R^{6N}$, $\mu_{\text{Leb}}(A) < \infty$, $\mathbf{x} = (\mathbf{q}, \mathbf{p})$, and $\mathbb{1}_A$ the indicator function of the set A,

$$\int_{\mathbb{R}^{6N}} F(\mathbf{q}(t), \mathbf{p}(t)) d\mathbf{q} d\mathbf{p} = \int_{T^{-t}A} d\mathbf{q} d\mathbf{p}$$

and Liouville's theorem applied to $-t$ implies the constancy of that integral.

For more general functions, one writes, see (2.A.9),

$$\int_{\mathbb{R}^{6N}} F(\mathbf{q}(t), \mathbf{p}(t)) d\mathbf{q} d\mathbf{p} = \lim_{n \to \infty} \int_{\mathbb{R}^{6N}} F_n(\mathbf{q}(t), \mathbf{p}(t)) d\mathbf{q} d\mathbf{p}$$

where $F_n = \sum_{m=1}^{\infty} c_m \mathbb{1}_{A_m}$ with $c_m \in \mathbb{R}$. Since, for each n, $\int_{\mathbb{R}^{6N}} F_n(\mathbf{q}(t), \mathbf{p}(t)) d\mathbf{q} d\mathbf{p}$ is constant in time by the previous observation, the corollary is proven.

Consider now the evolution of probability densities under the Hamiltonian flow. Let $\rho_0(\mathbf{q}, \mathbf{p})$ be an initial probability density in $\mathbb{R}^{6N}$ at time $t_0 = 0$, so that $P(\mathbf{q}, \mathbf{p} \in A) = \int_A \rho_0(\mathbf{q}, \mathbf{p}) d\mathbf{q} d\mathbf{p}$. Then, the time evolution of that density will be: $\rho_t(\mathbf{q}, \mathbf{p}) = \rho_0(T^{-t}(\mathbf{q}, \mathbf{p}))$, so that, with the abbreviation $\mathbf{x}(t) = (\mathbf{q}(t), \mathbf{p}(t))$,

$$P(\mathbf{x}(t) \in A) = \int_A \rho_t(\mathbf{x}) d\mathbf{x} = \int_{T^{-t}A} \rho_0(\mathbf{y}) d\mathbf{y}.$$

Changing variables in the first integral $\mathbf{x} = T^t \mathbf{y}$, we get:

$$\int_A \rho_t(\mathbf{x}) d\mathbf{x} = \int_{T^{-t}A} \rho_t(T^t(\mathbf{y})) \det\left(\frac{\partial T^t(\mathbf{y})}{\partial \mathbf{y}}\right) d\mathbf{y} = \int_{T^{-t}A} \rho_0(\mathbf{y}) d\mathbf{y}.$$

Since A is arbitrary, the integrands must be equal:

$$\rho_t(T^t(\mathbf{y})) \det\left(\frac{\partial T^t(\mathbf{y})}{\partial \mathbf{y}}\right) = \rho_0(\mathbf{y}).$$

For the Hamiltonian flow, $\frac{d}{dt} \det(\frac{\partial T^t(\mathbf{y})}{\partial \mathbf{y}}) |_{t=0} = 0$, see (3.4.5), (3.4.7), and, since the right hand side is independent of t, we get:

$$\frac{d\rho_t(T^t(\mathbf{q}, \mathbf{p}))}{dt} |_{t=0} = 0,$$

which means, by (3.3.7),

$$\left(\frac{\partial \rho_t(\mathbf{q}(t), \mathbf{p}(t))}{\partial t} + \sum_{i=1}^{N} \frac{\partial \rho(t, \mathbf{q}(t), \mathbf{p}(t))}{\partial \vec{q}_i} \cdot \nabla_{\vec{p}_i} H - \frac{\partial \rho(t, \mathbf{q}(t), \mathbf{p}(t))}{\partial \vec{p}_i} \cdot \nabla_{\vec{q}_i} H\right) |_{t=0} = 0,$$

Since we can replace $t = 0$ with $t = t_0$ for any t_0, we get, for all times:

$$\frac{\partial \rho_t(\mathbf{q}(t), \mathbf{p}(t))}{\partial t} = \sum_{i=1}^{N} \frac{\partial \rho(t, \mathbf{q}(t), \mathbf{p}(t))}{\partial \vec{p}_i} \cdot \nabla_{\vec{q}_i} H - \frac{\partial \rho(t, \mathbf{q}(t), \mathbf{p}(t))}{\partial \vec{q}_i} \cdot \nabla_{\vec{p}_i} H$$

$$(3.4.8)$$

which is *the Liouville equation for probability densities.*

3.4.2   The Liouville Measure

We noticed in Sect. 3.3.2 that the constant energy surfaces S_{E_0} are invariant under the Hamiltonian flow. We also know from Liouville's theorem that the Lebesgue measure of sets in $\mathbb{R}^{6N}$ is invariant under the Hamiltonian flow. But any constant energy surface is of zero Lebesgue measure in $\mathbb{R}^{6N}$ for non-trivial Hamiltonians, which makes Liouville's theorem rather vacuous if one applies it to the subsets of a constant energy surface S_{E_0}.

Fortunately one can define a measure, the Liouville measure, that is concentrated on any given constant energy surface S_{E_0} and such that the Liouville measure of sets in S_{E_0} is invariant under the Hamiltonian flow.

To define that measure, we need the notion of a surface measure dS on any smooth manifold $\mathcal{M}$ so that, for $A \subset \mathcal{M}$, $\int_A dS$ gives the area of A. For example, if $\mathcal{M}$ is the $n - 1$-sphere of radius 1 (the unit sphere in $\mathbb{R}^n$) which can be parametrized by $n - 1$ angles $\phi_1, \ldots, \phi_{n-1}$, the measure dS equals:

$$dS = \sin^{n-2}(\phi_{n-1}) \sin^{n-3}(\phi_{n-2}) \ldots \sin(\phi_2) d\phi_1 \ldots d\phi_{n-1}, \qquad (3.4.9)$$

which, for $n = 2$ gives the familiar formula on the circle $dS = d\phi$ and, for $n = 3$ gives, for the sphere, $dS = \sin(\phi_2) d\phi_1 d\phi_2$ (where ϕ_2 is often denoted θ). For a $n - 1$-sphere of radius R, dS is given by (3.4.9) multiplied by R^{n-1}.

Now, fix E_0. The Liouville measure on S_{E_0} is, at least formally:

$$\delta(H(\mathbf{q}, \mathbf{p}) - E_0) d\mathbf{q} d\mathbf{p},$$

with δ the delta measure defined in (2.A.4). But one can define it more explicitly: let $\epsilon > 0$, and $\mathcal{A}(E_0, \epsilon) = \{\mathbf{q}, \mathbf{p} \mid E_0 - \epsilon \leq H(\mathbf{q}, \mathbf{p}) \leq E_0 + \epsilon\}$ be a "slice" of configurations in $\mathbb{R}^{6N}$ whose energy is close to E_0. We will define the Liouville measure on S_{E_0} as the limit when $\epsilon \to 0$ of the Lebesgue measure $d\mathbf{q} d\mathbf{p}$ in $\mathcal{A}(E_0, \epsilon)$.

Rewrite that Lebesgue measure as:

$$\int \mathbb{1}_{\mathcal{A}(E_0,\epsilon)} d\mathbf{q} d\mathbf{p} = \int_{E_0-\epsilon}^{E_0+\epsilon} \delta(H(\mathbf{q},\mathbf{p}) - E) dS_E dE. \tag{3.4.10}$$

For $H(\mathbf{q},\mathbf{p}) = E$, with $E_0 - \epsilon \leq E \leq E_0 + \epsilon$, we can write

$$H(\mathbf{q},\mathbf{p}) = E_0 + \vec{\nabla} H(E_0) \cdot \vec{n}(E - E_0) + \mathcal{O}(\epsilon^2),$$

where $\vec{n}$ is the unit vector normal to S_E. Inserting this in (3.4.10), and changing variables $E \rightarrow \vec{\nabla} H(E_0) \cdot \vec{n}(E - E_0)$, whose Jacobian is $\|\vec{\nabla} H(E_0)\|$, we get:

$$\lim_{\epsilon \to 0} \int_{E_0-\epsilon}^{E_0+\epsilon} \delta(H(\mathbf{q},\mathbf{p}) - E) dS_E dE = \frac{dS_{E_0}}{\|\vec{\nabla} H(E_0)\|}, \tag{3.4.11}$$

which defines the Liouville measure $d\mu_{\text{Liouville}}$ on S_{E_0}. Since the Lebesgue measure in $\mathcal{A}(E_0, \epsilon)$ is invariant under the time evolution by Liouville's theorem, so is the Liouville measure.

Consider for example a Hamiltonian equal to its kinetic energy term (see (3.3.2)), plus an harmonic potential:

$$H = \sum_{i=1}^{N} \frac{\|\vec{p}_i\|^2}{2} + \sum_{i=1}^{N} \frac{\|\vec{q}_i\|^2}{2},$$

where we set for simplicity all $m_i = 1$. Then the surfaces S_E are spheres in $\mathbb{R}^{6N}$ of radius $\sqrt{E}$; since $\nabla H = (\vec{p}_1, \ldots, \vec{p}_N, \vec{q}_1, \ldots, \vec{q}_N)$ has a constant norm, $\|\nabla H(E)\| = (\sum_{i=1}^{N}(\|\vec{p}_i\|^2 + \|\vec{q}_i\|^2))^{\frac{1}{2}} = \sqrt{E}$, on that sphere, we get that the Liouville measure in that example is proportional to dS given by (3.4.9) with $n = 6N$, and with a factor of proportionality $(\sqrt{E})^{6N-2}$, one power $\sqrt{E}^{-1}$ coming from the factor $\|\vec{\nabla} H(E_0)\|$ in (3.4.11).

3.5 Time Reversibility

Another fundamental property of the equations of motion (3.2.1) or (3.3.5, 3.3.6) is that they are invariant under time reversal. Define the operator $I : \mathbb{R}^{6N} \rightarrow \mathbb{R}^{6N}$:

$$I(\mathbf{x}) = I(\mathbf{q},\mathbf{p}) = (\mathbf{q}, -\mathbf{p}),$$

which reverses the momenta and leaves the positions unchanged. If we apply that operator at any time t to a solution of the equations of motion $(\mathbf{q}(t), \mathbf{p}(t))$ and then, let the system evolve according to the same equations of motion during the same amount of time t, we obtain the initial conditions of the motion, with momenta reversed: $\forall (\mathbf{q}(0), \mathbf{p}(0)) \in \mathbb{R}^{6N}$,

$$T^t I T^t(\mathbf{q}(0), \mathbf{p}(0)) = (\mathbf{q}(0), -\mathbf{p}(0)), \tag{3.5.1}$$

or:

$$I T^t I T^t(\mathbf{q}(0), \mathbf{p}(0)) = (\mathbf{q}(0), \mathbf{p}(0)). \tag{3.5.2}$$

Identities (3.5.1), (3.5.2) are often written as: $T^t I T^t = I$ or $I T^t I T^t = \text{Id}$, which follows from $I^2 = \text{Id}$, Id being the identity operator;.

To prove (3.5.1), consider the function $(\tilde{\mathbf{q}}(\tau), \tilde{\mathbf{p}}(\tau)) = (\mathbf{q}(t - \tau), -\mathbf{p}(t - \tau))$. We have by definition $(\tilde{\mathbf{q}}(0), \tilde{\mathbf{p}}(0)) = (\mathbf{q}(t), -\mathbf{p}(t))$ and $(\tilde{\mathbf{q}}(t), \tilde{\mathbf{p}}(t)) = (\mathbf{q}(0), -\mathbf{p}(0))$. Moreover it is easy to see that the function $(\tilde{\mathbf{q}}(\tau), \tilde{\mathbf{p}}(\tau))$ satisfies equations (3.3.5), (3.3.6), with derivatives taken with respect to τ instead of t. This proves (3.5.1).

This result has the somewhat curious but very important consequence (specially for Chap. 8): consider a gas enclosed in a box Λ where, initially, all the particles are concentrated in a small subset of the box and let the system evolve for some time t. We expect that the gas will expand in the box until its density becomes more or less uniform, and we never see the gas undergoing the reverse motion: starting from a uniform distribution and going to a small subset of the box. Yet, what (3.5.1) shows is that, *for every initial condition* of the gas initially concentrated in a small subset of the box and expanding into the whole box, there is another initial condition undergoing the reverse motion. So, there is a one-to-one correspondence between the expected motions and the unexpected ones. This apparent paradox will be discussed in Sect. 8.1.3.

3.6   Summary

An isolated classical dynamical system is defined by a point $\mathbf{x} = (\mathbf{q}, \mathbf{p})$ in phase space $\mathbb{R}^{6N}$, with $\mathbf{q} = (\vec{q}_1, \vec{q}_2, \ldots, \vec{q}_N) \in \mathbb{R}^{3N}$, $\mathbf{p} = (\vec{p}_1, \vec{p}_2, \ldots, \vec{p}_N) \in \mathbb{R}^{3N}$ The time evolution of this point is given by Hamilton's equations (3.3.3), (3.3.4); this gives rise to a family of maps T^t (3.3.10) of the phase space into itself, called the Hamiltonian flow. It has the following properties:

1. The energy function is constant along the solutions of Hamilton's equations, see (3.3.12).
2. The Hamiltonian flow preserves the volume in phase space (3.4.1)
3. This property allows us to define a measure invariant under the Hamiltonian flow on any surface in phase space of constant energy, the Liouville measure (3.4.11).
4. The Hamiltonian flow is invariant under time reversal (3.5.2).

3.7 Exercises

3.1. Consider Newton's equation for a particle moving on a line:

$$\frac{d^2x(t)}{dt^2} = -\frac{dV(x(t))}{dx}$$

where $x(t) \in \mathbb{R}$ and $V(x)$ is smooth. Using the law of conservation of energy, derive the following formula for the time t taken by a trajectory to go from $x(0) = x_0$ with initial velocity $v_0 = \frac{dx(t)}{dt}|_{t=0}$ to $x(t) = x_1$, with $\frac{dx(t)}{dt} > 0 \ \forall t$:

$$t = \int_{x_0}^{x_1} \frac{dx}{\sqrt{2(E - V(x))}} \tag{3.7.1}$$

with $E = \frac{1}{2}v_0^2 + V(x_0)$.

3.2. * Let $\mathbb{T}^2$ be the two dimensional torus with coordinates (x, y) and consider the equations

$$\frac{dx(t)}{dt} = a$$

$$\frac{dy(t)}{dt} = b$$

Show that, if $\frac{a}{b} \in \mathbb{Q}$, the motion is periodic, while if $\frac{a}{b} \notin \mathbb{Q}$, the solution is dense in $\mathbb{T}^2$ (see footnote 29 in Chap. 2). For the second part, one (easy) way to prove it is to use the Corollary 4.9 of the ergodic theorem and the ergodicity of irrational rotations of the circle, proven in Sect. 4.3.1.

3.3. Consider the equations for $x(t)$, $y(t) \in \mathbb{R}$

$$\frac{d^2x(t)}{dt^2} = -\omega_1^2 x(t)$$

$$\frac{d^2y(t)}{dt^2} = -\omega_2^2 y(t) \tag{3.7.2}$$

with initial condition $x(0) = y(0) = 0$.

Show that, if $\frac{\omega_1}{\omega_2} \in \mathbb{Q}$, the motion is periodic, while if $\frac{\omega_1}{\omega_2} \notin \mathbb{Q}$, the solution is dense in $\mathbb{T}^2$ (use the result of exercise 3.2).

3.4. Show that for a system satisfying (3.2.6) with $V_i = 0 \ \forall i = 1, \dots, N$ in (3.2.5), and $V_{ij}(\vec{q}_i, \vec{q}_j) = \bar{V}_{ij}(\vec{q}_i - \vec{q}_j)$, with $\bar{V}_{ij} : \mathbb{R}^3 \to \mathbb{R}$, namely $V_{ij}(\vec{q}_i, \vec{q}_j)$ is only a function of the differences between the positions of the particles, the total momentum $\mathbf{P}(t) = \sum_{i=1}^{N} \vec{p}_i(t)$ is conserved, i.e. is constant in time.

3.5. Show that for a system satisfying (3.2.6), with V_i and $V_{ij}(\vec{q}_i, \vec{q}_j)$ as in exercise 3.4, the total angular momentum $\mathbf{M}(t) = \sum_{i=1}^{N} \vec{r}_i(t) \wedge \vec{p}_i(t)$ where $\wedge$ is the vector product, is conserved, i.e. is constant in time.

3.6. Check (3.4.6).

3.7. Two particles of mass m et m' follow the same trajectory under the action of the same potential and take respectively a time t and t' to cover those trajectories. Show that $\frac{t'}{t} = \sqrt{\frac{m'}{m}}$.

3.8. Give, if there is one, the potential corresponding to the following forces $F : \mathbb{R}^3 \to \mathbb{R}^3$:

(1) $F(x, y, z) = (x, y, z)$

(2) $F(x, y, z) = (cx, cy, 0)$

(3) $F(x, y, z) = (yz, xz, xy)$

(4) $F(x, y, z) = (x, x, 0)$

(5) $F(x, y, z) = (0, 0, -mg)$

Hint: we see, from (3.2.3) that, if there is a potential function which is twice continuously differentiable $V(x, y, z)$, then $\partial_i \partial_j V = \partial_j \partial_i V$, where $i, j = x, y, z$ and so, $\partial_i F_j = \partial_j F_i$.

3.9. Use (3.7.1) to find the solution of the equation of motion for $V(x) = \frac{x^2}{2}$, with $x_0 = 0$ for simplicity. Hint: use the change of variables $x = a \sin \phi$ for a suitable a.

Chapter 4
Dynamical Systems

4.1 Introduction

We introduced in the preceding chapter the Hamiltonian flow which is a family of maps from a constant energy surface into itself, under which the Liouville measure of a set is invariant. It is useful to consider a more general setting of transformations for which there is no energy function that is conserved, for example if the physical system is not isolated, but for which there is nevertheless an invariant measure.

The fundamental objects that we will study are maps $T : \Omega \to \Omega$, where (Ω, Σ, μ) is a measure space. We will always assume, for simplicity and because that is what we are interested in, that the measure of the space Ω is finite: $\mu(\Omega) < \infty$. We will then always normalize it so that $\mu(\Omega) = 1$, so that our measure space is a *probability space*.

Given such a map, one can study the *trajectory* of any point $x \in \Omega$: $x, Tx, T^2x, T^3x, \ldots$. The point x is called the *initial condition* of the trajectory. We could also consider general differential equations of the form (3.4.2), that give rise to continuous time dynamical systems as opposed to the discrete time ones, but we will limit ourselves to the latter systems. They are sufficient to illustrate the main concepts that we want to discuss and they have the advantage over differential equations, that we do not have to prove that solutions (hence, trajectories) exist.

These dynamical systems satisfy general properties, mainly the Poincaré recurrence theorem, discussed in Sect. 4.2 and Birkhoff's ergodic theorem (Sect. 4.3). The latter theorem invites us to classify dynamical systems according to their ergodic properties introduced in Sects. 4.3 and 4.4.

We will be interested particularly in "chaotic" dynamical systems, which means systems whose trajectories are, in practice, unpredictable (see Sect. 4.5). Although "chaotic" Hamiltonian systems exist, it is easier to study simple models that are not Hamiltonian, and we will do that here.

We will see in Sect. 4.6 that for these unpredictable systems one can nevertheless use statistical methods in order to study the long time behavior of the trajectories. We

© Springer Nature Switzerland AG 2022
J. Bricmont, *Making Sense of Statistical Mechanics*, Undergraduate Lecture Notes in Physics, https://doi.org/10.1007/978-3-030-91794-4_4

will end this chapter with a brief introduction, in Sect. 4.7, to the notion of entropy of dynamical systems.

Another reason to study "chaotic" dynamical systems is that there is a line of thought (which we do not share) that links the unpredictability of some dynamical systems to foundations of statistical mechanics (see Sect. 8.3.3). We will discuss the relation between the probabilistic approach to dynamical systems and to statistical mechanics in Sect. 8.10.

Before going into all that, we refer the reader to the definitions of Appendix 2.A.5 about maps preserving measures and list a few examples of measure spaces (Ω, Σ, μ) and maps $T : \Omega \to \Omega$ that preserve the measure μ. Here, Σ will always be the Borel sets in $\mathbb{R}^n$ or the cylinder sets defined in (2.A.5).

1. Rotations of the circle or of the torus. Let $\Omega = [0, 1[$ with addition   mod 1, which can be identified with the unit circle, $d\mu = dx$ the Lebesgue measure, and let $T = T_\alpha$, for $\alpha \in [0, 1)$:

$$T_\alpha x = x + \alpha \quad \text{mod } 1. \tag{4.1.1}$$

   If one considers $\Omega = [0, 1[^n$, which can be identified with the n-dimensional torus, with $d\mu$ the Lebesgue measure, one can generalize the map (4.1.1), for $\alpha \in [0, 1)^n$ to:

$$T_\alpha \mathbf{x} = \mathbf{x} + \alpha \quad \text{mod } 1, \tag{4.1.2}$$

   for $\mathbf{x} \in [0, 1[^n$ and the addition is modulo 1 for each component.
   In both cases, the invariance of the Lebesgue measure is obvious since the transformation is just a translation (modulo 1).

2. Let $\Omega = [0, 1[$, $d\mu = dx$ and let (Fig. 4.1)

$$Tx = 2x \quad \text{mod } 1. \tag{4.1.3}$$

Then μ is invariant under T. Indeed, from the definition of T, we get $T^{-1}A = B_1 \cup B_2$, where $B_1 \subset [0, \frac{1}{2}[$ and $B_2 \subset [\frac{1}{2}, 1[$, with $\mu(B_1) = \mu(B_2) = \frac{1}{2}\mu(A)$. Viewed as a map from $[0, 1[$ into itself, that map is not continuous. But the bijection $\Phi : [0, 1[\to S^1$, $\Phi(x) = z = \exp(2\pi i x)$ identifies $[0, 1[$ with the unit circle; then the map $T' = \Phi T \Phi^{-1}$ maps S^1 into itself and is continuous and even analytic:

$$T' = \Phi T \Phi^{-1}(z) = \exp(2\pi i 2x) = z^2. \tag{4.1.4}$$

T' is called the *lift* of the map T. Since relations like (4.1.4) occur frequently, we introduce a

Definition 4.1 Two maps $T : \Omega \to \Omega$ and $T' : \Omega' \to \Omega'$ are *conjugated* when they are related as in (4.1.4) by a bijection $\Phi : \Omega \to \Omega'$:

$$T' = \Phi T \Phi^{-1} \tag{4.1.5}$$

Fig. 4.1 Graph of the map (4.1.3)

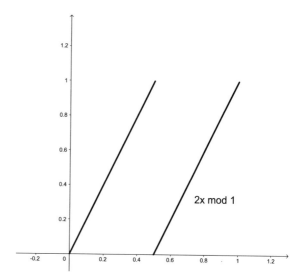

2x mod 1

If μ is a measure on Ω, Σ, using the conjugation Φ, one defines a measure $\mu^* = \Phi^*(\mu)$ on (Ω', Σ') with $\Sigma' = \{A \mid \exists B \in \Sigma, \Phi(B) = A\}: \forall A \in \Sigma'$

$$\mu^*(A) = \mu(\Phi^{-1}(A)) \tag{4.1.6}$$

This defines a conjugation between the measures μ and μ^*.

The map (4.1.3) can be generalized to

$$Tx = px \quad \mod 1. \tag{4.1.7}$$

with $p \in \mathbb{N}$, $p \geq 2$. We will study also two closely related maps:

a. The tent map:

$$Tx = 2x \quad \text{if } x \leq \frac{1}{2}$$

$$Tx = 2(1 - x) \quad \text{if } x \geq \frac{1}{2}$$

$$\tag{4.1.8}$$

  (to understand the name, just see the graph of the map, Fig. 4.2).
b. The logistic map:

$$Tx = \lambda x(1 - x) \tag{4.1.9}$$

which is a discrete version of the logistic equation: $\frac{dx}{dt} = \lambda x(1 - x)$ (which the reader may integrate as an exercise). We will however study (4.1.9) only for $\lambda = 4$ (see Fig. 4.3).

One can also extend the map (4.1.3) to automorphisms of $\Omega = [0, 1[^n$ (i.e. group isomorphism of the n-dimensional torus, with which $[0, 1[^n$ is identified, into itself), with $d\mu$ the Lebesgue measure. Such automorphisms are defined by $n \times n$ matrices $A = [a_{ij}]$, with integer entries, $a_{ij} \in \mathbb{Z}$, and with det $A = \pm 1$, acting on $\mathbf{x} \in [0, 1[^n$ as:

$$T\mathbf{x} = A\mathbf{x} \quad \text{mod } 1 \quad \text{in each coordinate.} \tag{4.1.10}$$

The most well-known example of such an automorphism is Arnold's cat map,[1] with $n = 2$ and

$$T(x_1, x_2) = (2x_1 + x_2, x_1 + x_2) \quad \text{mod } 1 \quad \text{in each coordinate.}$$

Again, these maps are not continuous, but they can be lifted to maps of the n-torus into itself that are smooth, as we did for $n = 1$ in (4.1.4).

3. Let $\Omega = [0, 1] \times [0, 1], d\mu = dxdy$ the Lebesgue measure on the square Ω and T_b the *baker's transformation*: $T_b(x, y) = (x', y')$, with

$$x' = 2x \quad \text{mod } 1$$
$$y' = \frac{y}{2} \quad \text{if } x < \frac{1}{2}$$
$$y' = \frac{y+1}{2} \quad \text{if } x \geq \frac{1}{2}. \tag{4.1.11}$$

or, $T_b(x, y) = (2x \mod 1, \frac{y}{2} + \frac{[2x]}{2})$. where $[x]$ is the integer part of x. Geometrically, the transformation dilates a set by a factor 2 in the x direction and contracts it by a factor $\frac{1}{2}$ in the y direction. Then, the part with $x \geq 1$ is cut and shifted back on top of the part with $x < 1$, see Fig. 4.4.

This ressembles what a baker does with the dough. He presses it (dilation by a factor 2 in one direction and contraction by a factor $\frac{1}{2}$ in the other) and then puts one part on top of the other. This mixes efficiently the dough, and we will prove later that this transformation does exactly that.

The transformation T_b is invertible: $T_b^{-1}(x', y') = (x, y)$, with

$$y = 2y' \quad \text{mod } 1$$
$$x = \frac{x'}{2} \quad \text{if } y' < \frac{1}{2}$$
$$x = \frac{x'+1}{2} \quad \text{if } y' \geq \frac{1}{2}.$$

The invariance of μ under T is easy to see, because the area of a rectangle is obviously invariant under that transformation and one can write any measurable

---

[1] For pictures of the action of that map on an image of a cat, see e.g. https://galileo-unbound.blog/2019/06/16/vladimir-arnolds-cat-map/. The original pictures were published by Arnold and Avez [14, p. 6].

Fig. 4.2 Graph of the map (4.1.8)

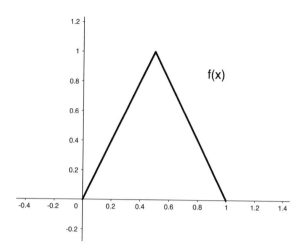

set as the union of a set of rectangles and a set of arbitrarily small measure (this follows from (2.A.11)).

4. This example is a bit more abstract than the previous ones: let $\Omega_k = \{0, 1, \ldots, k-1\}^{\mathbb{Z}}$, Σ be the σ-algebra generated by the cylinder sets of Ω_k and μ be any product measure on (Ω_k, Σ). Then the map, which is a special case of the map (2.A.16), $T_{\text{shift}} : \Omega_k \to \Omega_k$ is defined by

$$(T_{\text{shift}}\mathbf{x})_i = x_{i+1}, \quad \forall i \in \mathbb{Z}, \tag{4.1.12}$$

with $\mathbf{x} = (\ldots, x_{-2}, x_{-1}, x_0, x_1, x_2, \ldots) \in \Omega_k$ is μ invariant. This map is called the *Bernoulli shift on k symbols*. T_{shift} is invertible: $T_{\text{shift}}^{-1}(\mathbf{x})_i = x_{i-1}, \forall i \in \mathbb{Z}$. A similar map can be defined on $\Omega_k^+ = \{0, 1, \ldots, k-1\}^{\mathbb{N}}$, which is μ invariant but not invertible. T_{shift} is continuous for the metric (2.A.17), see (2.A.18). We will sometimes consider those maps to be defined on the space $\tilde{\Omega}_2$ which is obtained by identifying sequences that ends by $01111\ldots$ and by $10000\ldots$, which corresponds to the binary expansions of the same real number, see (2.A.7).

5. Finally, the most important and general example of measure preserving transformations was introduced in Chap. 3: $\Omega = S_E$ is a constant energy surface for a Hamiltonian dynamical system, $\mu_{\text{Liouville}}$ is the Liouville measure on S_E defined in (3.4.1), and $T = T^t$ the Hamiltonian flow for any given t, see (3.3.10).

We will see in Sect. 4.6.2 other examples of measure preserving transformations, but whose construction or whose invariance of μ under T is less explicit than in the previous examples.

Measure preserving transformations satisfy two important theorems: the Poincaré recurrence theorem and Birkhoff's ergodic theorem.

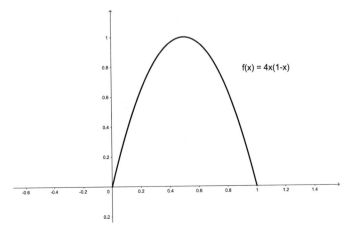

Fig. 4.3 Graph of the map (4.1.9) for $\lambda = 4$

Fig. 4.4 The baker's map.
The vertical rectangles are
mapped onto the horizontal
ones

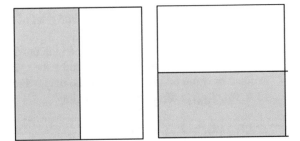

4.2 Poincaré's Recurrence Theorem, or The Eternal Return

In words, Poincaré's recurrence theorem says that, for any measure preserving transformation, almost every trajectory comes back arbitrarily close to its initial condition and does that infinitely often.

That property is called *recurrence*.

Since we know that the Hamiltonian flow on a constant energy surface preserves the Liouville measure on that surface, it follows that, for any mechanical system bounded (in phase space) almost all configurations will come back infinitely often to a configuration arbitrarily close to itself, or to any other configuration that it visits (since one could always define that configuration as an initial condition).

Thus, bounded mechanical systems do not necessarily have only periodic trajectories but almost all their trajectories have a property somewhat similar but weaker than periodicity, namely recurrence.

If one replaced the measure preserving transformation by a deterministic transformation on a *finite set*, then, obviously, all trajectories must eventually be periodic since some element of the finite set must be visited twice (over an infinite time) and from then on, the trajectory becomes periodic.

The genius of Poincaré was to extend this result to a weaker notion (recurrence) for measure preserving transformations on infinite sets but bounded in the sense that the measure of the space on which the transformation acts is finite:

Theorem 4.2 (Poincaré's recurrence theorem [261]) *Let (Ω, Σ, μ) be a measure space with $\mu(\Omega) < \infty$, where μ is T-invariant for $T : \Omega \to \Omega$. Then, $\forall A \in \Sigma$, with $\mu(A) > 0$, $\exists B \subset A$ with $\mu(A \backslash B) = 0$ and such that $\forall x \in B$, $\exists$ sequence $(n_i)_{i=1}^{\infty}$, $n_i \in \mathbb{N}$, with $n_1 < n_2 < n_3...$ and $T^{n_i} x \in A$.*

Remark 4.3 The Poincaré's recurrence theorem means what we said above: for any set $A \in \Sigma$ of positive measure, no matter how small, almost all points will return infinitely often to A. One could call it the eternal return theorem, the idea of eternal return to the present being a fashionable idea at the end of the 19th century, specially in the philosophy of Nietzsche (the theorem was published in 1890 [261]).

Remark 4.4 By Liouville's Theorem 3.4, the Poincaré's recurrence theorem applies to the Hamiltonian flow on a constant energy surface S_E, with the Liouville measure $\mu_{\text{Liouville}}$, provided that the surface is of finite Liouville measure. For that, we need the potential to be confining, for example that the system be enclosed in a finite box $\Lambda \subset \mathbb{R}^3$. This can be achieved by defining the one-body potential in (3.2.5) $V_i(\vec{q}_i) = 0$ if $\vec{q}_i \in \Lambda$ and $V_i(\vec{q}_i) = \infty$ if $\vec{q}_i \notin \Lambda$, $\forall i$. For the kinetic part of the Hamiltonian, (3.3.2), the set of momenta such that $K(\mathbf{p}) \leq E$ is a ball in $\mathbb{R}^{3N}$, hence has a finite area.

Remark 4.5 Applied to the Hamiltonian flow, the Poincaré's recurrence theorem has a curious consequence: take a gas in an isolated box Λ of total energy E and confine the gas in a small subset Λ_0 of that box. Then, let the gas expand in the box: intuitively the gas will spread itself in it and remain in a state of more or less uniform density in Λ. This will be explained in Sect. 8.1. But, since the set of configurations where all the particles are confined in Λ_0 has a positive Liouville measure in S_E (minuscule but positive),[2] Poincaré's recurrence theorem implies that almost every configuration in Λ_0 will return to Λ_0 infinitely often.

Let us stress that this means that *all* particles that were initially in Λ_0 return to Λ_0 at the same time (and infinitely often), not simply that each particle initially in Λ_0 returns to Λ_0 at some time. Indeed, we are applying the recurrence theorem to $S_E \subset \mathbb{R}^{6N}$ and each point of $\mathbb{R}^{6N}$ represents the set of positions and momenta of all the particles of the gas.

Of course, this sort of behavior is never observed in real physical systems, one reason being that nothing is said in the recurrence theorem about the time it would

---

[2] If there are N particles, one can estimate that measure as being of the order of $(\frac{|\Lambda_0|}{|\Lambda|})^N$, which is indeed very small for N of the order of Avogadro's number $N \approx 10^{23}$, even if $\frac{|\Lambda_0|}{|\Lambda|} \approx \frac{1}{2}$.

take for the gas to return to Λ_0. Rough estimates show that it would be much larger than the age of the Universe.[3]

The physicist Chandrasekhar has estimated the recurrence time for a fluctuation of one percent in the density of a ball of air of radius a which is part of a larger container of air under the usual conditions of temperature and pressure, as a function of a. He got (see [77, table VIII]; quoted by Dorfman [109, p. 55]):

$a \approx 10^{-5}$ cm, $t \approx 10^{-11}$ s.
$a \approx 2.5 \ 10^{-5}$ cm, $t \approx 1$ s.
$a \approx 3 \ 10^{-5}$ cm, $t \approx 10^6$ s.
$a \approx 5 \ 10^{-5}$ cm, $t \approx 10^{68}$ s.
$a \approx 1$ cm, $t \approx 10^{10^{14}}$ s.

While the recurrence theorem implies that, if all the particles are initially in Λ_0, they will all return to Λ_0, it does not imply that all the particles will be at some time in the complement of Λ_0, $\Lambda \setminus \Lambda_0$ or in any other subset of Λ_0. To prove that sort of property, we need an ergodic theorem, see Sect. 4.3.

Remark 4.6 It is easy to see why the hypothesis $\mu(\Omega) < \infty$ is necessary for the recurrence theorem to be true: let $\Omega = \mathbb{R}$, let μ be the Lebesgue measure (one has $\mu(\mathbb{R}) = \infty$), and let $T(x) = x + 1$, (μ is T-invariant for that T); finally, let $A = [0, \frac{1}{2}]$. No point of A comes back to A: $\forall n \geq 1$ $T^n(x) \notin A, \forall x \in A$.

And it is trivial to see why the assumption that μ is T-invariant is essential: if $T : [0, 1] \to [0, 1]$ equals $Tx = \frac{x}{2}$, then all trajectories tend to 0 and do not return close to their origin. But here, of course the Lebesgue measure is not invariant, and the only invariant measure is the delta measure at 0, see (2.A.2), to which Poincaré's theorem applies but is trivial.

Let us now give the

4.2.1 Proof of Poincaré's Recurrence Theorem

Let, for $N \geq 0$ $A_N = \bigcup_{n=N}^{\infty} T^{-n} A$. Elements of A_N are those that are sent into A

by the map T^n for some $n \geq N$. Thus, $B = A \bigcap (\bigcap_{N=0}^{\infty} A_N)$ are the elements of A

that come back to A infinitely often, namely that are such that $\forall x \in B$, there exist a

---

[3] If the system is ergodic (see (4.3.9) below), then the average time spent by the system in Λ_0 is of the order of $(\frac{|\Lambda_0|}{|\Lambda|})^N$; this means that, on average, the system will take a time which is of the order of the inverse of that time, namely of the order of $(\frac{|\Lambda_0|}{|\Lambda|})^{-N} \approx 2^N$, for $\frac{|\Lambda_0|}{|\Lambda|} \approx \frac{1}{2}$; if $N \approx 10^{23}$ this time is indeed very large (we do not specify here the unit of time here, but whether we count in seconds or hours will not make much difference as long as this unit is reasonable). In exercise 4.5, the reader can estimate another example of recurrence times. .

sequence (n_i) as in the theorem: since, for each $x \in B$, there exist arbitrarily large $n's$ with $T^n x \in A$, one may construct the sequence inductively by taking n_{i+1} to be the first integer n with $T^n x \in A$ strictly larger than n_i.

Let us show that $\mu(B) = \mu(A)$, which, since $\mu(\Omega) < \infty$, is equivalent to $\mu(A \backslash B) = 0$ (write $\mu(A) = \mu(B) + \mu(A \backslash B)$). One has $T^{-1} A_N = A_{N+1}$ and thus $\mu(A_N) = \mu(A_{N+1})$, since μ is T-invariant. So, $\mu(A_0) = \mu(A_N)$, $\forall N$. Since $A_0 \supset A_1 \supset \dots \supset A_N$, and since $\mu(\Omega) < \infty$, one has $\mu(A_N) \leq \mu(A_0) \leq \mu(\Omega) < \infty$ and thus $\mu(A_0 \backslash A_N) = \mu(A_0) - \mu(A_N) = 0$, $\forall N$. Since a countable union of sets of measure zero is of measure zero $\mu(A_0 \backslash (\bigcap_{N=0}^{\infty} A_N)) = \mu(\bigcup_{N=0}^{\infty} (A_0 \backslash A_N)) = 0$. Thus, since $\bigcap_{N=0}^{\infty} A_N \subset A_0$, $\mu(B) = \mu(A \bigcap (\bigcap_{N=0}^{\infty} A_N)) = \mu(A \bigcap A_0)$ and $\mu(A \bigcap A_0) = \mu(A)$, since $A \subset A_0$. So, $\mu(B) = \mu(A)$. $\qquad\qquad\square$

4.3 Ergodic Theorems

Although we have not yet defined what unpredictable dynamical systems are, we said that, when a dynamical system is unpredictable, one should study its trajectories by statistical methods. A first step in that direction is to study certain time averages of trajectories, for example the average time spent by the trajectory in a set $A \in \Sigma$. We will study a slightly more general object.

Given a measure space (Ω, Σ, μ), a map $T : \Omega \to \Omega$ such that μ is T-invariant, and a μ-integrable function $F : \Omega \to \mathbb{R}$, one may consider the temporal averages:

$$\frac{1}{N} \sum_{n=0}^{N-1} F(T^n x) \qquad\qquad (4.3.1)$$

and ask whether the limits $N \to \infty$ of those quantities exist.

If one takes $F = \mathbb{1}_A$, the indicator function of a set $A \in \Sigma$, formula (4.3.1) gives the average time, up to time N, spent by the system in A, and the limit $N \to \infty$, if it exists, gives the average time τ_A spent in A.

The following theorem, that we will not prove because the proof is a bit long and not particularly illuminating (see Walters [327, Sect. 1.6] or Cornfeld, Fomin and Sinai [87, Appendix 3]) asserts that all these limits actually exist[4]:

Theorem 4.7 (Birkhoff's ergodic theorem) *Let (Ω, Σ, μ) be a measure space and T be a map from $\Omega \to \Omega$ such that μ is T-invariant. If $F \in L^1(\Omega, d\mu)$, then the limit*

---

[4] Another version of this theorem, due to von Neumann, asserts that, for $F \in L^2(\Omega, d\mu)$, the limit (4.3.2), exists, but in the sense of convergence in $L^2(\Omega, d\mu)$. The proof of this version is rather simple, see e.g. Reed and Simon [274, Sect. II.5].

$$\lim_{N \to \infty} \frac{1}{N} \sum_{n=0}^{N-1} F(T^n x) \qquad (4.3.2)$$

exists almost everywhere, and that limit defines a function $F^*(x) \in L^1(\Omega, d\mu)$ satisfying:

$$F^*(Tx) = F^*(x) \quad almost\ everywhere, \qquad (4.3.3)$$

and

$$\int_\Omega F^*(x) d\mu(x) = \int_\Omega F(x) d\mu(x). \qquad (4.3.4)$$

The ergodic theorem claims the existence of a huge number of limits, for all the sums of the type (4.3.2), for all functions $F \in L^1(\Omega, d\mu)$, and for all most all initial conditions x with respect to μ. But it does not tell us anything about the value of those limits and, under hypotheses as general as those of the ergodic theorem, one cannot expect to be able to say anything concrete about those limits.

The form of convergence in (4.3.2) is called Cesaro convergence. A sequence $(a_n)_{n \in \mathbb{N}}, a_n \in \mathbb{R}$, converges to a in the Cesaro sense if $\frac{1}{N} \sum_{n=0}^N a_n \to a$, as $N \to \infty$. A sequence may converge in the Cesaro sense but not in the usual one, for example the sequence $a_n = (-1)^n$; obviously that sequence does not converge in the usual sense, but $|\frac{1}{N} \sum_{n=0}^N a_n| \le \frac{1}{N} \to 0$, as $N \to \infty$. If a sequence converges to zero in the usual sense, then it converges also in the Cesaro sense.[5]

But there exist a class of transformations T, the ergodic ones, for which the limiting function F^* is rather easy to compute.

Definition 4.8 Let (Ω, Σ, μ) be a probability space and T be a map from $\Omega \to \Omega$ such that μ is T-invariant. T is ergodic with respect to μ if any function $F \in L^1(\Omega, d\mu)$ or $F \in L^2(\Omega, d\mu)$ satisfying

$$F(Tx) = F(x) \quad \mu\ almost\ everywhere \qquad (4.3.7)$$

is a constant function almost everywhere.

---

[5] That is because, if $a_n \to 0$ as $n \to \infty$, then, by definition, $\forall \epsilon > 0 \, \exists M$ such that $|a_n| \le \frac{\epsilon}{2} \, \forall n \ge M$. Then, if $K = \max_{n < M} |a_n|$, we get $|\frac{1}{N} \sum_{n=0}^N a_n| \le \frac{KM}{N} + \frac{\epsilon}{2} \le \epsilon$ for $N \ge \frac{2KM}{\epsilon}$.

For an example of a bounded sequence $(a_n)_{n \in \mathbb{N}}$ that does not converge even in the Cesaro sense, let

$$a_n = 1 \quad for \quad 10^{2p} \le n < 10^{2p+1} \qquad (4.3.5)$$

$$a_n = -1 \quad for \quad 10^{2p+1} \le n < 10^{2p+2} \qquad (4.3.6)$$

Then, $\frac{1}{10^{2p}} \sum_{n=0}^{10^{2p}} a_n \to -\frac{9}{11}$, and $\frac{1}{10^{2p+1}} \sum_{n=0}^{10^{2p+1}} a_n \to \frac{9}{11}$, as $p \to \infty$. .

An alternative definition of ergodicity relies on invariant sets: T is ergodic with respect to μ if the only sets $A \in \Sigma$ such that $T^{-1}(A) = A$ almost everywhere satisfy $\mu(A) = 0$ or $\mu(A) = 1$.

Using this definition, one can also define what it means for a measure μ to be *ergodic with respect to* T for a given transformation T: that a function in $L^1(\Omega, d\mu)$ satisfying (4.3.7) is necessarily constant.

But this definition and the ergodic theorem implies, since F^* satisfies (4.3.7) by (4.3.3), and $\mu(\Omega) = 1$, that the function F^* must be almost everywhere constant if T is ergodic with respect to μ: $F^*(x) = F^*$ for almost all x with respect to μ. But, by (4.3.4), we have $\int_\Omega F^*(x)d\mu(x) = F^* = \int_\Omega F(x)d\mu(x)$. Thus, we have the

Corollary 4.9 *Let* (Ω, Σ, μ) *be a probability space and* T *be a map from* $\Omega \to \Omega$ *which is ergodic with respect to* μ. *Then,* $\forall F \in L^1(\Omega, d\mu)$,

$$\lim_{N \to \infty} \frac{1}{N} \sum_{n=0}^{N-1} F(T^n x) = \int_\Omega F(x)d\mu(x) \quad \mu \text{ almost everywhere} \tag{4.3.8}$$

Let us remark that, if $F = \mathbb{1}_A$, for $A \in \Sigma$, the average time spent in A, $\tau_A = \lim_{N \to \infty} \frac{1}{N} \sum_{n=0}^{N-1} \mathbb{1}_A(T^n x)$ is easy to compute for a transformation T that is ergodic with respect to μ:

$$\tau_A = \mu(A), \tag{4.3.9}$$

So, in order to obtain interesting applications of the ergodic theorem, it "suffices" to check that (4.3.7) implies that F is constant and then to compute $\int_\Omega F(x)d\mu(x)$ for suitable F's.

4.3.1 Examples and Applications

Let us consider some of the examples given in Sect. 4.1 and see whether they are ergodic or not.

1. For rotations of the circle, (4.1.1), one must distinguish between $\alpha \in \mathbb{Q}$ and $\alpha \notin \mathbb{Q}$ If $\alpha \in \mathbb{Q}$, i.e. $\alpha = \frac{p}{q}$ with $p, q \in \mathbb{N}$, functions of the form $F_m(x) = \exp(2\pi imqx)$ with $m \in \mathbb{Z}$ are invariant under T_α:

$$F_m(T_\alpha x)$$
$$= \exp(2\pi imq(x + \frac{p}{q}))$$
$$= \exp(2\pi imqx) = F_m(x),$$

since $mq\frac{p}{q} = mp \in \mathbb{Z}$ and $e^{2\pi i\ell} = 1$ $\forall \ell \in \mathbb{Z}$. Of course, these functions are not constant for $m \neq 0$, so T_α is not ergodic for $\alpha \in \mathbb{Q}$.

If $\alpha \notin \mathbb{Q}$, we can expand any $F \in L^2(\Omega, d\mu)$ into a Fourier series:

$$F(x) = \sum_{n \in \mathbb{Z}} c_n \exp(2\pi i n x) \tag{4.3.10}$$

where the series converges in $L^2(\Omega, d\mu)$ and where the coefficients c_n are given by

$$c_n = \int_0^1 F(x) \exp(-2\pi i n x) dx \tag{4.3.11}$$

If $F(x) = F(T_\alpha x)$, using (4.3.10), one gets:

$$\sum_{n \in \mathbb{Z}} c_n e^{2\pi i n x} = \sum_{n \in \mathbb{Z}} c_n e^{2\pi i n x} e^{2\pi i n \alpha},$$

and, since the coefficients of the Fourier series are given by (4.3.11), we have:

$$c_n(e^{2\pi i n \alpha} - 1) = 0 \quad \forall n \in \mathbb{Z}. \tag{4.3.12}$$

But, for $\alpha \notin \mathbb{Q}$, $\nexists n \neq 0$ such that $\exp(2\pi i n \alpha) = 1$, since the only solutions of that latter equation are of the form $\alpha = \frac{m}{n}$, with $m \in \mathbb{Z}$. Thus (4.3.12) implies $c_n = 0$, for $n \neq 0$, which, by (4.3.10), implies

$$F(x) = c_0$$

namely F is constant and therefore T_α is ergodic.

The reader can check, using Fourier series on the n-dimensional torus, that the map T_α defined by (4.1.2) is ergodic if and only if the equation $\sum_{i=1}^n l_i \alpha_i = m$ with $l_i, m \in \mathbb{Z}$ has no solution except $l_i = 0 \, \forall i$, and $m = 0$ (in other words, if the set of real numbers $(\alpha_i)_{i=1}^n$ are independent over the rationals).

2. Consider now T defined by (4.1.3). Using (4.3.10) and $F(x) = F(Tx)$, one obtains:

$$\sum_{n \in \mathbb{Z}} c_n \exp(2\pi i n x) = \sum_{n \in \mathbb{Z}} c_n \exp(2\pi i 2 n x) \tag{4.3.13}$$

which again, since the coefficients are determined by (4.3.11), implies

$$c_n = c_{2n} \quad \forall n \in \mathbb{Z}. \tag{4.3.14}$$

Iterating (4.3.14), one gets:

$$c_n = c_{2^k n} \quad \forall n \in \mathbb{Z}, \forall k \in \mathbb{N}. \tag{4.3.15}$$

But, since $F \in L^2(\Omega, d\mu)$, one knows that $\sum_{n \in \mathbb{Z}} |c_n|^2 < \infty$, which implies that $c_n \to 0$ as $|n| \to +\infty$. Since $|2^k n| \to \infty$ for $n \neq 0$ as $k \to \infty$, (4.3.15) implies $c_n = 0$, $\forall n \neq 0$, which again, by (4.3.10), means that $F = c_0$, a constant, and thus that T is ergodic.

A more subtle exercise is to check that automorphisms of the n-dimensional torus is ergodic if and only if the matrix A in (4.1.10) does not have roots of unity as eigenvalues (see Walters [327, Sect. 1.5]).

3. The baker's map.[6] Suppose first that F is continuously differentiable. Since $T_b(x, y) = (2x \bmod 1, \frac{y}{2} + \frac{[2x]}{2})$, if we differentiate with respect to y the identity $F(T_b(x, y)) = F(x, y)$, one gets:

$$\partial_y F(T_b(x, y)) = \frac{1}{2} D_2 F(T_b(x, y)) = D_2 F(x, y), \qquad (4.3.16)$$

where D_2 denotes the derivative with respect to the second component. Taking the sup over (x, y) of (4.3.16), one obtains:

$$\sup_{x,y} |D_2 F(x, y)| \leq \frac{1}{2} \sup_{x,y} |D_2 F(x, y)|$$

and thus that $D_2 F(x, y) = 0$. Approximating functions in $F \in L^2([0, 1] \times [0, 1], d\mu)$ by C^1 functions, one can show that if $F \in L^2(\Omega, d\mu)$ is invariant under T_b, then F is independent of y. But if F is only a function of x, one has $F(T_b(x)) = F(2x \bmod 1)$ and we are back to the previous case. Thus, the baker's map is also ergodic.

4. The Bernoulli shift on k symbols defined by (4.1.12) is ergodic. We will prove that below by proving a stronger property of the Bernoulli shift (mixing), and for the more general map (2.A.16), see (4.4.5).

5. Finally the issue of ergodicity of Hamiltonian systems is such a vast and complicated subject that we cannot do justice to it. Ergodicity and stronger chaotic properties hold for some Hamiltonian systems on billards (see e.g. Bunimovich [64] and references therein), but ergodicity is not expected to hold in general for smooth Hamiltonian systems. Indeed, a theorem by Kolmogorov, Arnol'd and Moser implies that, at least for small perturbations of integrable systems,[7] there exists a subset A of surfaces S_E of constant energy, invariant under the Hamiltonian flow, and such that $0 < \mu_{\text{Liouville}}(A) < \mu_{\text{Liouville}}(S_E)$, which contradicts the notion of ergodicity, by the Definition 4.8 (see e.g. Arnold [15, Appendix 8], Arnol'd, Kozlov and Neishtadt [16], Chierchia and Mather [83] and references

---

[6] We only sketch the proof of ergodicity of the baker's map, since we will prove stronger properties of that map like mixing later, see the end of Sect. 4.6.1.

[7] It would be too long to define precisely that notion, but, roughly speaking, integrable dynamical systems are systems that can be reduced to a set of non-interacting subsystems by appropriate transformations.

therein, or De la Llave [95]. Those results were anticipated numerically by Fermi, Pasta, Ulam and Tsingou, see [125]).

Let us consider some elementary applications of the ergodicity of example 1 and 2 above.

1. Distribution of the first digits of the powers of 2. Consider the sequence 2^n of the powers of two:

$$2, 4, 8, 16, 32, 64, 128, 256, 512, 1024, \ldots,$$

and the sequence composed of the first digits of those numbers:

$$2, 4, 8, 1, 3, 6, 1, 2, 5, 1, \ldots. \tag{4.3.17}$$

What is the frequency with which a given $p \in \{1, 2, \ldots, 9\}$ appears in that sequence? This question is apparently complicated and seems unrelated to ergodic theory; yet that theory provides a very simple answer. Indeed, saying that p is the first digit of 2^n means that for some $r \in \mathbb{N}$,

$$p \, 10^r \le 2^n < (p + 1)10^r.$$

Taking the logarithm in base 10 of those inequalities, we get

$$n \log_{10} 2 \in J_p \equiv [\log_{10} p, \log_{10}(p + 1)[\quad \mod 1. \tag{4.3.18}$$

So, the frequency of apparition of p in the sequence (4.3.17) is

$$\lim_{N \to \infty} \frac{1}{N} \sum_{n=0}^{N-1} \mathbb{1}(T_\alpha^n(0) \in J_p) \tag{4.3.19}$$

where $\mathbb{1}$ is the indicator function, $T_\alpha(x) = x + \alpha \mod 1$ and $\alpha = \log_{10} 2$. If $\alpha \notin \mathbb{Q}$, one can apply the ergodic theorem, and the limit (4.3.19) equals:

$$\mu(J_p) = \int_{J_p} dx = \log_{10} \left(1 + \frac{1}{p}\right). \tag{4.3.20}$$

A small caveat: the ergodic theorem states that, if one replaces 0 in (4.3.19) by x, the limit (4.3.19) equals (4.3.20) for almost all x, not a priori for $x = 0$. However, for the transformation $T_\alpha(x) = x + \alpha \mod 1$, it is easy to see that all points are equivalent (by rotation invariance of the map on the circle) and, thus, the limit (4.3.19) equals (4.3.20) for all x and in particular for $x = 0$. It remains to show that $\alpha \notin \mathbb{Q}$. If $\alpha = \frac{p}{q}$, with $p, q \in \mathbb{N}, q > p$, one has $2^q = 10^p = 2^p 5^p$, i.e. $2^{q-p} = 5^p$, which is impossible since 2^{q-p} is even and 5^p is odd.

2. Almost all real numbers are normal. Let $x \in [0, 1[$. We can write the binary expansion of x: $x = \sum_{n=1}^{\infty} a_n 2^{-n}$, and ask with which frequency do the digits 0 and 1 appear? It is easy to see that $a_n = 0$ if and only if $T^{n-1}x \in [0, \frac{1}{2}[$, where T is defined by (4.1.3). Thus, the frequency with which $a_n = 0$ appears is:

$$\lim_{N \to \infty} \frac{1}{N} \sum_{n=0}^{N-1} \mathbb{1}(T^n x \in [0, \frac{1}{2}[) = \int_0^{1/2} dx = \frac{1}{2}$$

almost everywhere, by the ergodic theorem and the ergodicity of T. One obtains obviously the same result for $a_n = 1$ and thus, for almost all real numbers, the symbols 0 et 1 appear with a frequency $\frac{1}{2}$ in their binary expansion. One may generalize this result to the expansion in base p, $x = \sum_{n=1}^{\infty} a_n p^{-n}$, where $a_n = 0, 1, \ldots, p-1$, using the map $T : x \to px \mod 1$. One gets that, for almost all real numbers, the symbols $0, 1, \ldots, p-1$ appear with a frequency $\frac{1}{p}$. One calls *normal in base* p such a number and *normal* a number that is normal in base p for all $p = 2, 3, \ldots$.

Let A_p be the set of numbers that are not normal in base p. We know that $\mu_{\text{Leb}}(A_p) = 0$. But, since a countable union of sets of measure zero is of measure zero, it follows that almost every number is normal. Yet, giving a concrete example of a normal number is not completely obvious.[8] On the other hand, for each p, A_p, is uncountable and dense (see footnote 29 in Chap. 2).[9] In the language of Sect. 2.3.1, normal numbers are typical.

4.3.2 Ergodicity and the Law of Large Numbers

There is a connection between the law of large numbers (2.3.9) and the ergodicity of the shift map T_{shift}, defined in (2.A.16), on product spaces.[10]

---

[8] One may take, for example, in base 2, $x = 0.0100100111000\ldots$, where we concatenate all finite sequences of 0's and 1's. The same construction can be done in any base.

[9] It is dense because it contains all the numbers for which the sequence $(a_n)_{n=1}^{\infty}$ in its binary expansion is periodic. To see that it is uncountable, consider the map $x = \sum_{n=1}^{\infty} a_n 2^{-n} \to y = \sum_{n=1}^{\infty} b_n 2^{-n}$, where $b_{2n} = a_n$ and $b_{2n+1} = 1$, $\forall n$ where x is normal in base 2. Obviously the frequency of appearance of 1 in y is $\frac{3}{4}$ ($\frac{1}{2}$ because of $b_{2n+1} = 1$ plus $\frac{1}{4}$ because half of the $b_{2n} = a_n$ equal 1 since x is normal), so $y \in A_2$. Since the set of normal numbers in base 2 is of measure one, it is uncountable and so is the set of y's constructed from those x's.

[10] The ergodicity of T_{shift} follows from the fact that this map is mixing and this will be proven in the next section.

Let $\Omega \subset \mathbb{R}^n$ be a bounded set (to simplify matters), μ a probability measure on Ω and let $F :\in L^1(\Omega, d\mu)$. Let $\boldsymbol{\Omega} = \prod_{n=0}^{\infty} \Omega_n$, where each Ω_n is a copy of Ω. But then, the sum (2.3.6) entering the law of large numbers (with F instead of f and $n = i - 1$), $\sum_{n=0}^{N-1} F(x_n)$, with $x_n \in \Omega_n$ can be rewritten as:

$$\sum_{n=0}^{N-1} F(x_n) = \sum_{n=0}^{N-1} F((T_{\text{shift}}^n \mathbf{x})_0) \tag{4.3.21}$$

where $T_{\text{shift}} : \boldsymbol{\Omega} \to \boldsymbol{\Omega}$ is defined in (2.A.16).

But since the shift map is ergodic with the respect to the product measure on $\boldsymbol{\Omega}$, $\boldsymbol{\mu} = \prod_{n=0}^{\infty} \mu_n$ each μ_n being a copy of μ, we have, by Corollary 4.9,

$$\lim_{N \to \infty} \frac{1}{N} \sum_{n=0}^{N-1} F(x_n) = \lim_{N \to \infty} \frac{1}{N} \sum_{n=0}^{N-1} F((T_{\text{shift}}^n \mathbf{x})_0)$$

$$= \int F(x_0) d\mu(x_0) = \int F(x) d\mu(x) = \mathbb{E}(F), \tag{4.3.22}$$

$\boldsymbol{\mu}$ almost everywhere, which is identical to the law of large numbers (2.3.9).

4.4 Mixing

Given a probability space (Ω, Σ, μ), one defines an important notion, the *correlation* between two random variables, $f, g : \Omega \to \mathbb{R}$:

$$C(f, g) = \int_{\Omega} f(x)g(x)d\mu(x) - \int_{\Omega} f(x)d\mu(x) \int_{\Omega} g(x)d\mu(x),$$

which measures the degree of dependence between those variables: if f and g are independent $C(f, g)$ equals zero by (2.A.22).

If we take two sets $A, B \in \Sigma$ and let $f = \mathbb{1}_A, g = \mathbb{1}_B$, the correlation $C(f, g) = \mu(A \cap B) - \mu(A)\mu(B)$ measures the degree of dependence between the events in A and those in B. If $\mu(B) \neq 0$, then we can write $\frac{\mu(A \cap B) - \mu(A)\mu(B)}{\mu(B)} = \frac{\mu(A \cap B)}{\mu(B)} - \mu(A) = \mu(A \mid B) - \mu(A)$, the difference between the conditional probability of A given B and the probability of A, that clearly measures the degree of dependence between A and B.

If one has a measure preserving map $T : \Omega \to \Omega$, one is interested in how much two random variables depend on each other when the map T acts n times on the argument of one of them:

$$C(f, g, n) = \int_\Omega f(x) g(T^n x) d\mu(x) - \int_\Omega f(x) d\mu(x) \int_\Omega g(T^n x) d\mu(x)$$

$$= \int_\Omega f(x) g(T^n x) d\mu(x) - \int_\Omega f(x) d\mu(x) \int_\Omega g(x) d\mu(x), \quad (4.4.1)$$

since $\int_\Omega g(T^n x) d\mu(x) = \int_\Omega g(x) d\mu(x)$, because μ is T-invariant. In particular, one wants to see how much correlation remains as $n \to \infty$, i.e. how $C(f, g, n)$ behaves in that limit.

First, consider the following Cesaro average of $C(f, g, n)$:

$$\frac{1}{N} \sum_{n=0}^{N} C(f, g, n) = \int_\Omega f(x) \left(\frac{1}{N} \sum_{n=0}^{N} g(T^n x)) d\mu(x) - \int_\Omega g(x) d\mu(x) \right) d\mu(x).$$

If T is ergodic with respect to μ, one knows that if, say, g is bounded, $\frac{1}{N} \sum_{n=0}^{N} g(T^n x))$ $d\mu(x) \to \int_\Omega g(x) d\mu(x)$ almost everywhere, as $N \to \infty$. So, if $f \in L^1(\Omega, d\mu)$, one gets, by the dominated convergence Theorem 2.6, that $\frac{1}{N} \sum_{n=0}^{N} C(f, g, n) \to 0$, as $N \to \infty$. This means that $C(f, g, n)$ converges to zero in the Cesaro sense.

One could ask for convergence of (4.4.1) in the usual sense, as $n \to \infty$, which leads to the following:

Definition 4.10 A measure preserving map $T : \Omega \to \Omega$ on a measure space (Ω, Σ, μ) is *mixing* if, $\forall f, g \in L^2(\Omega, \mu)$,

$$C(f, g, n) \to 0, \qquad (4.4.2)$$

as $n \to \infty$.

An alternative definition is, using in (4.4.2) $f = \mathbb{1}_A$, $g = \mathbb{1}_B$ and the fact that $\mathbb{1}_B(T^n x) = \mathbb{1}_{T^{-n}B}(x) \colon \forall A, B \in \Sigma$

$$\mu(A \cap T^{-n} B) - \mu(A) \mu(B) \to 0, \qquad (4.4.3)$$

as $n \to \infty$.[11]

Obviously, since ordinary convergence implies convergence in the Cesaro sense (see footnote 5 in Chap. 4), any mixing transformation is ergodic.

A consequence of that definition that will be useful later is that, if two measures μ and ν are related by $d\mu = \frac{f(x) d\nu(x)}{\int_\Omega f(x) d\nu}$, where f is a ν integrable function, and if the system $(\Omega, T, d\nu)$ is mixing, then for any bounded function g, we have:

$$\lim_{n \to \infty} < g(T^n x) >_\mu = < g(x) >_\nu . \qquad (4.4.4)$$

In the examples of Sect. 4.1, the ergodic rotations of the circle or of the torus are not mixing. Indeed, consider the two functions $f = \exp(2\pi i m x)$ and $g = \exp(-2\pi i m x)$

---

[11] If T is invertible, (4.4.3) implies $\mu(T^n A \cap B) - \mu(A) \mu(B) \to 0$, as $n \to \infty$.

for some $m \neq 0$; we have, since $\int_0^1 \exp(2\pi imx) = \int_0^1 \exp(-2\pi imx) = 0$,

$$C(f, g, n) = \int_0^1 f(x)g(T^n(\mathbf{x}))d\mu(x) - \int_0^1 f(x)d\mu(x)\int_0^1 g(x)d\mu(x)$$

$$= \int_0^1 \exp(2\pi imx)\exp(-2\pi imx)\exp(-2\pi imn\alpha)dx = \exp(-2\pi imn\alpha),$$

and $\exp(-2\pi imn\alpha) \nrightarrow 0$ as $n \to \infty$. The reader can check that $\exp(-2\pi imn\alpha) \to 0$ as $n \to \infty$ in the Cesaro sense, for $\alpha \notin \mathbb{Q}$, as it must do, because of the ergodicity of irrational rotations of the circle.

On the other hand, all the transformations defined by (4.1.3), (4.1.7), (4.1.11) are mixing. To see that the map (4.1.3) is mixing, expand the two functions f and g in (4.4.1) into Fourier's series: $f = \sum_{k=-\infty}^{\infty} c_k \exp(2\pi ikx), g = \sum_{k=-\infty}^{\infty} d_l \exp(2\pi ilx)$. We can assume, by changing f and g by a constant, that $\int_\Omega f(x)d\mu(x) = \int_\Omega g(x)d\mu(x) = 0$, which means $c_0 = d_0 = 0$. We get:

$$C(f, g, n) = \sum_{k,l=-\infty}^{\infty} c_k d_l \int_0^1 \exp(2\pi ikx)\exp(2\pi i2^n lx)dx$$

$$= \sum_{l=-\infty}^{\infty} c_{-2^n l} d_l.$$

Since $f, g \in L^2([0, 1[, dx), \sum_{l=-\infty}^{\infty} |c_l|^2 < \infty, \sum_{l=-\infty}^{\infty} |d_l|^2 < \infty$, and the series $\sum_{l=-\infty}^{\infty} |c_{-2^n l} d_l|$ converges by Schwartz' inequality. So, for every $\epsilon > 0$, $\exists N$ such that $\sum_{|l| \geq N} |c_{-2^n l} d_l| \leq \epsilon$. And, since each $c_{-2^n l} \to 0$ for $l \neq 0$ as $n \to \infty$ (by $\sum_{l=-\infty}^{\infty} |c_l|^2 < \infty$), we have $\lim_{n \to \infty} \sum_{|l| < N} |c_{-2^n l} d_l| = 0$ and we get that $C(f, g, n) \to 0$ as $n \to \infty$. A similar proof works for the map (4.1.7).

To prove mixing for the shift map (2.A.16) (and thus also for the map (4.1.12)), note that, by (2.A.12) with $|F(\mathbf{x}) - G(\mathbf{x})|$ replaced by $|f(\mathbf{x}) - g(\mathbf{x})|^2$, for $\forall f, g \in L^2(\Omega, d\mu)$ and any $\epsilon > 0$, one can find $N < \infty$ and functions f_0, g_0 that depend only on the variables $(x_{-N}, x_{-N+1}, \ldots, x_{N-1}, x_N)$ and such that the L^2 norms satisfy $\|f - f_0\|_2 \leq \epsilon$ and $\|g - g_0\|_2 \leq \epsilon$. Assume again, by changing f and g by a constant, that $\int_\Omega f(\mathbf{x})d\mu(\mathbf{x}) = \int_\Omega g(\mathbf{x})d\mu(\mathbf{x}) = 0$. We can write:

$$C(f, g, n) = \int_\Omega f(\mathbf{x})g(T_{\text{shift}}^n \mathbf{x})d\mu(\mathbf{x}) - \int_\Omega f(\mathbf{x})d\mu(\mathbf{x})\int_\Omega g(\mathbf{x})d\mu(\mathbf{x}),$$

$$= \int_\Omega f(\mathbf{x})g(T_{\text{shift}}^n \mathbf{x})d\mu(\mathbf{x})$$

$$= \int_\Omega (f(\mathbf{x}) - f_0(\mathbf{x}))g(T_{\text{shift}}^n \mathbf{x})d\mu(\mathbf{x}) + \int_\Omega f_0(\mathbf{x})(g(T_{\text{shift}}^n \mathbf{x}) - g_0(T_{\text{shift}}^n \mathbf{x}))d\mu(\mathbf{x})$$

$$+ \int_\Omega f_0(\mathbf{x})g_0(T_{\text{shift}}^n(\mathbf{x}))d\mu(\mathbf{x})$$

$$(4.4.5)$$

By Schwartz's inequality and the bounds on the L^2 norms $\|f - f_0\|_2 \leq \epsilon$ and $\|g - g_0\|_2 \leq \epsilon$ (note that, since μ is invariant under T_{shift}, $\|g \circ T_{\text{shift}}^n - g_0 \circ T_{\text{shift}}^n\|_2 = $

$\|g - g_0\|_2$), the first two terms are less that $C\epsilon$, with $C = \max(\|f_0\|_2, \|g\|_2)$ and the last one is small for n large: $g_0(T_{\text{shift}}^n \mathbf{x})$ will depend on the variables $(x_{-N+n}, x_{-N+1+n}, \ldots, x_{N-1+n}, x_{N+n})$, which are independent (with respect to the product measure μ) from the variables $(x_{-N}, x_{-N+1}, \ldots, x_{N-1}, x_N)$ if $-N + n > N$ on which f_0 depends.

Thus, for n large:

$$\int_\Omega f_0(\mathbf{x}) g_0(T_{\text{shift}}^n \mathbf{x}) d\mu(\mathbf{x}) = \int_\Omega f_0(\mathbf{x}) d\mu(\mathbf{x}) \int_\Omega g_0(T_{\text{shift}}^n \mathbf{x}) d\mu(\mathbf{x}) = \int_\Omega f_0(\mathbf{x}) d\mu(\mathbf{x}) \int_\Omega g_0(\mathbf{x}) d\mu(\mathbf{x}),$$

since μ is T_{shift} invariant.

By assumption, $\int_\Omega f(\mathbf{x}) d\mu(\mathbf{x}) = \int_\Omega g(T_{\text{shift}}^n(\mathbf{x})) d\mu(\mathbf{x}) = \int_\Omega g(\mathbf{x}) d\mu(\mathbf{x}) = 0$, and $\|f - f_0\|_2 \le \epsilon$, $\|g - g_0\|_2 \le \epsilon$; so, we have $|\int_\Omega f_0(\mathbf{x}) d\mu(\mathbf{x}) \int_\Omega g_0(T(\mathbf{x})) d\mu(\mathbf{x})| \le \epsilon^2$.

So, $C(f, g, n) \le C\epsilon + \epsilon^2$, for n large which proves mixing for the shift map.

4.5 Sensitive Dependence on Initial Conditions

Suppose that the space Ω is a bounded and closed subset of $\mathbb{R}^m$ for some m, and that the transformation $T : \Omega \to \Omega$ is Lipschitz continuous[12]; then, each trajectory depends continuously on the initial conditions: let x_0, y_0, be two initial conditions; continuous dependence on the initial conditions means that, $\forall n$, there exists a constant depending on n (but not on x_0, y_0),[13] $C(n)$, such that:

$$\text{dist}(T^n x_0, T^n y_0) \le C(n) \text{dist}(x_0, y_0). \tag{4.5.1}$$

where dist is the Euclidean distance in $\mathbb{R}^m$, dist $(T^n x_0, T^n y_0) = |T^n x_0 - T^n y_0|$.

But one might ask: how does $C(n)$ depend on n? Of course, since $C(n)$ in (4.5.1) enters into an upper bound, that bound remains true if one lets $C(n)$ grow very fast with n. But that is rather silly. Suppose that, instead, we try to find the function $C(n)$ growing as slowly as possible with n but fast enough so that (4.5.1) still holds.

If one compares the first two examples in Sect. 4.1, one observes very different behaviors for the "best" $C(n)$ (meaning the one growing with n as slowly as possible). For the map T_α given by (4.1.1), it is obvious that $T_\alpha(x_0) - T_\alpha(y_0) = x_0 - y_0$ and therefore one can take $C(n) = 1$, $\forall n$. The same holds for T_α in (4.1.2). This is of course the slowest conceivable growth rate.

---

[12] Lipschitz continuity is defined by (4.5.1) below for $n = 1$; any differentiable function is Lipschitz continuous. Among the examples of Sect. 4.1, the lifts on the circle or the n-torus (similar to T' in (4.1.4)) of the maps (4.1.3), (4.1.7), (4.1.10) are Lipschitz continuous. The baker's map is not continuous and thus not Lipschitz continuous.

[13] That is because, since Ω is a bounded and closed subset of $\mathbb{R}^m$, it is compact and thus the Lipschitz continuous maps T and T^n (for any n) are uniformly Lipschitz continuous on Ω and (4.5.1) simply expresses this uniform continuity.

But for T given by (4.1.3), the behavior is completely different. If $x_0 - y_0 = \epsilon$, then $T^n x_0 - T^n y_0 = 2^n \epsilon$, at least as long as $2^n \epsilon < 1$ (so that the mod 1 operation in the definition of T does not play a role). But then, the slowest growing $C(n)$ cannot grow slower than 2^n, and it is easy to see that the bound (4.5.1) holds with $C(n) = 2^n$.

This is called (obviously) an exponential growth for $C(n)$. It is easy to see that one cannot have a faster type of growth, like $C(n) \approx C^{n^2}$, for some C. Indeed, by Lipschitz continuity of T, one has, for some constant C,

$$\text{dist}\,(T^n x_0, T^n y_0) \leq C\text{dist}\,(T^{n-1} x_0, T^{n-1} y_0),$$

and, iterating that bound, one gets (4.5.1) with $C(n) = C^n$.

But, if the best $C(n)$ in (4.5.1) grows exponentially with n, then the system becomes quickly unpredictable in practice because of the way the exponential function grows. Indeed, suppose that we have a physical system; any measurement of the initial condition will unavoidably have an error attached to it, which is what we call dist (x_0, y_0), and denote that error by ϵ.

If we want our system to be predictable, we need dist $(T^n x_0, T^n y_0)$ to remain small. Suppose that we have a system where the error grows linearly with n: $C(n) \approx Cn$. For an example of such growth, consider the map from the unit disk D in $\mathbb{R}^2$ into itself given in polar coordinates (r, ϕ) by;

$$T(r, \phi) = (r, \phi + r). \tag{4.5.2}$$

For initial conditions (r, ϕ), (r', ϕ), $r \neq r'$, we have $|T^n(r, \phi) - T^n(r', \phi)| \approx n|r - r'|$, for $n|r - r'|$ not close to a multiple of 2π.

Suppose that we are willing to tolerate an error η in the precision of our predictions. This means that we want to have

$$Cn\epsilon \leq \eta, \tag{4.5.3}$$

or $n \leq \frac{\eta}{C\epsilon}$. Suppose now that we get better measuring devices that allow us to reduce the error in our initial conditions by a factor 10^k. How much longer will our predictions be reliable? The condition (4.5.3) becomes

$$\frac{Cn\epsilon}{10^k} \leq \eta. \tag{4.5.4}$$

which means that the time over which our predictions are reliable changes from $n \leq \frac{\eta}{C\epsilon}$, to (see (4.5.4)) $n \leq \frac{10^k \eta}{C\epsilon}$, so that it increases proportionally to the reduction in the error in our data. As an exercise, the reader can check that if $C(n)$ grows quadratically with n, $C(n) \approx Cn^2$, then the time over which our predictions are reliable will grow by a factor $10^{\frac{k}{2}}$.

Now, consider an exponential $C(n) \approx C^n$ with $C > 1$; condition (4.5.3) is replaced by

$$C^n \epsilon \le \eta,$$

or $n \le \frac{\ln \eta - \ln \epsilon}{\ln C}$. And, if we get better measuring devices that allow us to reduce the error in our initial conditions by a factor 10^k, the condition (4.5.4) is replaced by

$$\frac{C^n \epsilon}{10^k} \le \eta,$$

which means that the time over which our predictions are reliable changes from $n \le \frac{\ln \eta - \ln \epsilon}{\ln C}$, to $n \le \frac{\ln \eta - \ln \epsilon + k \ln 10}{\ln C}$. So, that time is not multiplied by a factor 10^k or $10^{\frac{k}{2}}$ but is increased only by an amount of order k.

In order to characterize systems where $C(n)$ grows exponentially, we define:

Definition 4.11 (*Sensitive dependence on initial conditions*) A dynamical system defined by a map $T : \Omega \to \Omega$ exhibits sensitive dependence on initial conditions if the smallest growing constant $C(n)$ in (4.5.1) grows exponentially with n.

Formally, the sensitive dependence on initial conditions is defined by:
$\exists d > 0, \exists C < \infty$ such that $\forall x \in \Omega, \forall \epsilon > 0, \exists y \in \Omega, \exists n \le C |\ln \epsilon|$ with dist $(y, x) \le \epsilon$ and:

$$\text{dist}(T^n(x), T^n(y)) \ge d. \tag{4.5.5}$$

It is useful to compare this definition with the one of uniform continuity of T^n:
$\forall n, \forall d > 0, \exists \epsilon > 0$, such that $\forall x, y, \in \Omega, |x - y| \le \epsilon$ implies:

$$\text{dist}(T^n(x), T^n(y)) \le d. \tag{4.5.6}$$

Note that the two properties (4.5.5) and (4.5.6) are perfectly compatible: for (4.5.5), one chooses n large but not too large as a function of ϵ (we want $n \le C|\ln \epsilon|$), while in (4.5.6) one fixes n and one chooses ϵ small as a function of n.

The sensitive dependence on initial conditions is metaphorically described by the *butterfly effect*: the meteorologist Edward Lorenz who showed numerically that the weather system exhibits sensitive dependence on initial conditions, introduced that metaphor in a 1972 lecture entitled: "Predictability: Does the Flap of a Butterfly's Wings in Brazil set off a Tornado in Texas?"[14] The idea is that the flap of a butterfly's wings corresponds to a very small change in the initial conditions of the weather system that may lead to a big difference in the future: a tornado or no tornado in Texas.

However, the idea that this "butterfly effect" may occur in the weather system is not so new. In fact Henri Poincaré wrote already in 1909:

> Why have the meteorologists such difficulty in predicting the weather with any certainty? Why do the rains, the tempests themselves seem to us to come by chance, so that many

---

[14] Address at the American Association for this Advancement of Science, December 29, 1972. See http://news.mit.edu/2008/obit-lorenz-0416.

persons find it quite natural to pray for rain or shine, when they would think it ridiculous to pray for an eclipse? We see that great perturbations generally happen in regions where the atmosphere is in unstable equilibrium. The meteorologists are aware that this equilibrium is unstable, that a cyclone is arising somewhere; but where they can not tell; one-tenth of a degree more or less at any point, and the cyclone bursts here and not there, and spreads its ravages over countries it would have spared. This we could have foreseen if we had known that tenth of a degree, but the observations were neither sufficiently close nor sufficiently precise, and for this reason all seems due to the agency of chance. Here again we find the same contrast between a very slight cause, unappreciable to the observer, and important effects, which are sometimes tremendous disasters.

Henri Poincaré, [264] p. 398 (original: [263]).

Let us now see which of the maps introduced in Sect. 4.1 exhibit sensitive dependence on initial conditions.

Obviously the rotations of the circle or of the torus, (4.1.1), (4.1.2) do not possess this sensitive dependence since they preserve the distance between two points.

But the map (4.1.3) is sensitive to initial conditions. Let $x \in [0, 1[$ and let $\epsilon > 0$. Choose n so that $2^{-(n+1)} < \epsilon$ (for example by choosing $n = [|\log_2 \epsilon|]$), so that $n \leq C|\ln \epsilon|$ and write the binary expansion of $x = \sum_{k=1}^{\infty} \frac{a_k}{2^k}$, $a_k \in \{0, 1\}$. Define $y = \sum_{k=1}^{n} \frac{a_k}{2^k} + \frac{(1-a_{n+1})}{2^{n+1}} + \sum_{k=n+2}^{\infty} \frac{a_k}{2^k}$ whose binary expansion differs from the one of x only for $k = n + 1$. Obviously, $|x - y| = 2^{-n-1} < \epsilon$ but $|T^n(x) - T^n(y)| = \frac{1}{2}$. One can easily check that this fits the definition of sensitive dependence with $d = \frac{1}{2}$ in (4.5.5).

A similar proof works for the map (4.1.7).

The proof of the sensitive dependence for the shift map (4.1.12) is very similar to the one for the map (4.1.3). Let $\mathbf{x} \in \Omega_k$ and let $\epsilon > 0$. Choose n as above so that $2^{-(n+1)} < \epsilon$ and let $\mathbf{y} \in \Omega_k$ be defined by $y_k = x_k \ \forall k \neq n + 1$ and $|y_{n+1} - x_{n+1}| = 1$. Then, with the metric (2.A.17), we have dist $(\mathbf{x}, \mathbf{y}) = 2^{-(n+1)}$ and dist $(T_{\text{shift}}^n(\mathbf{x}), T_{\text{shift}}^n(\mathbf{y})) = 1$, which again fits the definition of sensitive dependence with $d = 1$ in (4.5.5). Extending this property of sensitive dependence to the general map (2.A.16) is left as an exercise.

A way to precisely quantify the exponential growth characteristic of the sensitive dependence on initial conditions is via its Lyapunov exponent: we want to determine the constant C in the growth rate $C(n) = C^n$. If $T : \Omega \to \Omega$ is differentiable we have $|T^n(x_0 + \epsilon) - T^n(x_0)| \approx \epsilon (T^n)'(x_0)$ and $(T^n)'(x_0) = \prod_{k=0}^{n-1} T'(T^k(x)) = \prod_{k=0}^{n-1} T'(x_k)$, with $x_k = T^k(x_0)$, which follows from iterating the formula $f(g(x))' = f'(g(x))g'(x)$.

It is convenient to pass to logarithms: $\ln |(T^n)'(x_0)| = \sum_{k=0}^{n-1} \ln |T'(x_k)|$ and to introduce:

Definition 4.12 (*The Lyapunov exponent*) The (largest) Lyapunov exponent[15] λ of a dynamical system defined by a C^1 map $T : \Omega \to \Omega$ is:

---

[15] One can define a spectrum of such exponents, but we will introduce only the largest one.

$$\lambda = \lim_{n \to \infty} \frac{1}{n} \ln |\frac{d}{dx} T^n(x)| = \lim_{n \to \infty} \frac{1}{n} \sum_{k=0}^{n-1} \ln |T'(x_k)|. \tag{4.5.7}$$

One interest of that definition is that the ergodic theorem guarantees the existence of the limit (4.5.7) almost everywhere. We are only interested in the situations where the Lyapunov exponent is positive, which is a signature of the exponential growth of $C(n)$ in (4.5.1). If that exponent is negative, then trajectories will tend to get closer to each other rather than diverge.

Formally one can write

$$|T^n(x_0 + \epsilon) - T^n(x_0)| \approx \epsilon \exp(\lambda n), \tag{4.5.8}$$

so that $\lambda > 0$ characterizes the growth rate. Of course, to be rigorous, (4.5.8) has to be understood in the sense of (4.5.7), namely dividing both sides by ϵ and letting $\epsilon \to 0$. Otherwise, if one fixes ϵ and let $n \to \infty$, $\epsilon \exp(\lambda n) \to \infty$, and then (4.5.8) makes no sense if Ω is a bounded set. But physicists often write (4.5.8) as a heuristic formula, which is valid as long as the right hand side is not too large.

Since we cannot hope to increase the precisions of our measurements indefinitely, the inverse of λ sets a *temporal limitation or horizon* to our ability to predict the future. Indeed, for $n \approx \lambda^{-1}$ the factor in (4.5.8) multiplying ϵ is of order one, and for multiples of that time, $n \approx k\lambda^{-1}$, we get an exponential growth: $|T^n(x_0 + \epsilon) - T(x_0)| \approx \epsilon \exp(k)$. Estimates on the Lyapunov exponent for the atmosphere suggests that we will never be able to predict the weather accurately over more than a couple of weeks (which would still be much longer than what we can do now). However, for other systems, such as planets or other parts of the solar system, the temporal horizon can be of the order of millions of years, see e.g. Gaspard [142, p. 7].

Finally, we have to mention an important caveat: several authors (e.g. Devaney [98]) define sensitive dependence on initial conditions by saying that a small error in the initial conditions in the state of a system may lead to a large difference in that state later on. But if nothing is said about the length of time necessary for this large change to occur, this definition is incorrect: indeed it would apply to almost all systems, including those where the errors grows only linearly as for the map (4.5.2), for which an initial error of order ϵ becomes an error of order one, only after a time $n \approx \epsilon^{-1}$. The only meaningful definition of sensitive dependence on initial conditions is when the errors grows exponentially, or that the time necessary to go from an error of order ϵ to an error of order one grows only logarithmically ($n \approx |\ln \epsilon|$) when ϵ is small.

We saw that the map T given by (4.1.3) exhibits sensitive dependence on initial conditions. In the form given here, it is not differentiable, hence we cannot define its Lyapunov exponent, but one can do it for the equivalent map $z \to z^2$, with $z = \exp(2\pi i x) \in S^1$. We have $T^n(z) = \exp(2\pi i 2^n x)$ and its Lyapunov exponent is equal to $\ln 2$. It is equal to $\ln p$ for the map T defined by (4.1.7). For the baker's map (4.1.11), it is also not differentiable and there is no differentiable equivalent, but since the map reduces itself to $x \to 2x$ for the x variable, the sensitive dependence on initial conditions of the latter map implies the one of the baker map.

4.6   Statistical Theory of Dynamical Systems

Given a system with sensitive dependence on initial conditions, what can one do? We have seen that one cannot predict trajectories beyond a certain "temporal horizon". The next best thing one can try to do is to predict statistical properties of the trajectories of that system, which is similar to what one does in statistical mechanics. For example, one can try to compute the average time τ_A spent in a region $A \in \Sigma$, given by (4.3.8) with $F = \mathbb{1}_A$. But we already know how to do that: use the ergodic theorem, at least for ergodic transformations.

For example, for the maps T defined by (4.1.3), (4.1.7), or (4.1.11), that are sensitive with respect to initial conditions, but also ergodic with respect to the Lebesgue measure, the average time τ_A spent in a region $A \in \Sigma$ is given, see (4.3.9), by $\tau_A = \mu_{\text{Leb}}(A)$.

Or consider the Bernoulli shift on k symbols defined by (4.1.12), and ask with which frequency does a given finite string of symbols $(\alpha_0, \alpha_1, \ldots, \alpha_{n-1})$, with $\alpha_i \in \{0, 1, \ldots, k-1\}, \forall i = 0, 1, \ldots, n-1$ occur in an element $\mathbf{x} \in \Omega_k$? Obviously any frequency will occur for *some element* $\mathbf{x} \in \Omega_k$, because we can simply construct such an element by inserting the finite sequence $(\alpha_0, \alpha_1 \ldots, \alpha_{n-1})$ in the infinite sequence $\mathbf{x}$ with the desired frequency.

But if one asks the same question for *almost all* $\mathbf{x}$ with respect to to the product measure μ on Ω_k, then there is a unique answer. Let $A_{\alpha_0, \alpha_1, \ldots, \alpha_{n-1}} = \{\mathbf{x} = (x_n)_{n \in \mathbb{Z}}, x_0 = \alpha_0, x_1 = \alpha_1, \ldots, x_{n-1} = \alpha_n\}$. Then the frequency of appearance of the sequence $(\alpha_0, \alpha_1, \ldots, \alpha_{n-1})$ in $\mathbf{x} \in \Omega_k$ is given by:

$$\lim_{N \to \infty} \frac{1}{N} \sum_{m=0}^{N-1} \mathbb{1}_{A_{\alpha_0, \alpha_1, \ldots, \alpha_{n-1}}} (T_{\text{shift}}^m \mathbf{x}), \tag{4.6.1}$$

with T_{shift} defined in (4.1.12); by the ergodic theorem the limit in (4.6.1) exists for almost all $\mathbf{x}$ with respect to to the product measure μ on Ω_k, and, by ergodicity of T_{shift}, is equal to $\mu(A_{\alpha_0, \alpha_1, \ldots, \alpha_{n-1}}) = \prod_{i=0}^{n-1} \mu(\alpha_i)$.

4.6.1   Itineraries and Coding

An important tool in the study of dynamical systems is the notion of coding, which means a correspondance between a trajectory of a dynamical system and a sequence of symbols. Let T be a μ invariant map on Ω and let $(\Omega_0, \Omega_1, \ldots, \Omega_{k-1})$ be a partition of Ω: $\Omega = \cup_{i=0}^{k-1} \Omega_i$, $\Omega_i \cap \Omega_j = \emptyset$, $i \neq j$. Let us set $k = 2$ for simplicity.

Given $x \in \Omega$, one defines the following map from $\sigma : \Omega \to \{0, 1\}^{\mathbb{N}} = \Omega_2^+$:

$$\sigma(x)_n = 0 \quad \text{if} \quad T^n(x) \in \Omega_0$$
$$\sigma(x)_n = 1 \quad \text{if} \quad T^n(x) \in \Omega_1.$$

$\sigma(x)$ is called the *itinerary* of x. The word is natural, since $\sigma(x)_n$ corresponds to the index of the subsets Ω_0 or Ω_1 visited by the trajectory $(T^n x)_{n \in \mathbb{N}}$ at time n. We get, by definition,

$$\sigma(Tx) = T_{\text{shift}}\sigma(x) \tag{4.6.2}$$

where T_{shift} is the shift on Ω_2^+ defined in (4.1.12). Since σ is bijective, one can write (4.6.2) as $T_{\text{shift}} = \sigma T \sigma^{-1}$; in other words, T and T_{shift} are conjugated, see (4.1.5).

Since the map T_{shift} is simple to study, the properties of T are linked to the properties of σ: is σ injective, which means that two different point $x, y \in \Omega$ are mapped onto different itineraries? Is σ surjective, which means that *any* sequence of symbols in $\{0, 1\}^{\mathbb{Z}}$ is the itinerary of some $x \in \Omega$? What are the properties of the measure $\mu^* = \sigma^*(\mu)$ on $\{0, 1\}^{\mathbb{Z}}$ defined by (see Definition 4.1):

$$\forall A \in \Sigma(\{0, 1\}^{\mathbb{Z}}),$$
$$\mu^*(A) = \sigma^*(\mu) = \mu(\sigma^{-1}(A)),$$

where $\Sigma(\{0, 1\}^{\mathbb{Z}})$ is the σ-algebra generated by the cylinder sets on $\{0, 1\}^{\mathbb{Z}}$? Is the measure μ^* a product measure, as in the Bernoulli shift defined in (4.1.12)?

Obviously, if $T = \text{Id}$, then every trajectory is constant and the map σ has only two points in its image: the sequence with all 0's or all 1's. This is of course neither injective nor surjective. On the other hand, if σ is injective, it indicates that any two distinct points will diverge since they will not have the same itinerary. And if σ is surjective, it means that, for any sequence of choices of visits to Ω_0 or Ω_1 at given times one can find an x that will visit those sets at the prescribed times.[16] Both properties are indicative of a certain "chaoticity" of the map T.

Let us start with the example of the map T defined by (4.1.3). Since $\Omega = [0, 1[$, let $\Omega_0 = [0, \frac{1}{2}[$ and $\Omega_1 = [\frac{1}{2}, 1[$. Write the binary expansion of $x \in [0, 1[$, $x = \sum_{n=0}^{\infty} \frac{a_n}{2^{n+1}}$, with $a_n \in \{0, 1\}$, $n \in \mathbb{N}$, with the same convention as the one used after (2.A.6).

It is easy to check that $T^i x = \sum_{n=0}^{\infty} \frac{a_{n+i}}{2^{n+1}}$ and, since, with our convention for the binary expansion, $\sum_{n=0}^{\infty} \frac{a_n}{2^{n+1}} < \frac{1}{2}$ iff $a_0 = 0$, we get:

$$\sigma(x)_n = 0 \quad \text{if} \quad a_n = 0$$
$$\sigma(x)_n = 1 \quad \text{if} \quad a_n = 1.$$

This implies immediately that the map σ is both injective (if $\sigma(x) = \sigma(y)$, then x and y have the same binary expansion and are therefore equal) and surjective on the set Ω defined in (2.A.7).

---

[16] It means, for example, that there exists an x so that $T^n x$ is in Ω_0 if n is a prime and in Ω_1 if n is not a prime.

As for the map $\mu^* = \sigma^*(\mu)$ it is easy to see that it is equal to the product measure on $\tilde{\Omega}$, with equal measure $\frac{1}{2}$ for each symbol 0 and 1: let $(\alpha_0, \alpha_1, \ldots, \alpha_{n-1})$, with $\alpha_i \in \{0, 1\}$ $\forall i = 0, 2, \ldots, n-1$ be a sequence of binary symbols; a cylinder in $\tilde{\Omega}$ is of the form $A_{\alpha_0, \alpha_1, \ldots, \alpha_{n-1}} = \{\mathbf{a} = (a_n)_{n \in \mathbb{N}}, a_0 = \alpha_0, a_1 = \alpha_1, \ldots a_{n-1} = \alpha_{n-1}\}$ (or a translate of such sets). But the set of $x \in [0, 1[$ whose binary expansion belongs to $A_{\alpha_0, \alpha_1, \ldots, \alpha_{n-1}}$ is an interval[17] of length $\frac{1}{2^n}$ which is the same as the measure of $A_{\alpha_0, \alpha_1, \ldots, \alpha_{n-1}} \subset \tilde{\Omega}$, with respect to the product measure giving a weight $\frac{1}{2}$ for each symbol 0 and 1. Since a measure is determined by its values on cylinder sets, the measure $\mu^* = \sigma^*(\mu)$ induced from the Lebesgue measure on $[0, 1[$ by σ is that product measure.

We can do a similar coding for the baker's map defined in (4.1.11). Let us write $x = \sum_{n=0}^{\infty} \frac{a_n}{2^{n+1}}$ and $y = \sum_{n=1}^{\infty} \frac{a_{-n}}{2^n}$. Then, one can associate to each pair $(x, y) \in \Omega = [0, 1] \times [0, 1]$ a sequence $\sigma(x, y) \in \tilde{\Omega} = \{\mathbf{a} = (a_n)_{n \in \mathbb{Z}}\}$, where $\sigma(x, y)_n = a_n$ for $n \geq 0$ coming from $x = \sum_{n=0}^{\infty} \frac{a_n}{2^{n+1}}$, and, for $n < 0$, we write $\sigma(x, y)_{-n} = a_{-n}$, coming from $y = \sum_{n=1}^{\infty} \frac{a_{-n}}{2^n}$, with the same conventions as in (2.A.7) for numbers with two binary expansions.

One can check that the sequence associated to $T_b(x, y)$ is the shifted sequence $T_{\text{shift}}\mathbf{a}$, with T_{shift} defined in (4.1.12). In formulas:

$$\sigma(T(x, y)) = T_{\text{shift}}(\sigma(x, y)), \tag{4.6.3}$$

which means that σ conjugates T and T_{shift}. Again the measure $\sigma^*(\mu)$ on $\tilde{\Omega}$, where μ is the Lebesgue measure on the unit square, is the product measure on $\tilde{\Omega}$ with equal probability $\frac{1}{2}$ for each symbol 0 and 1.

As an exercise (exercise 4.4), the reader may prove both mixing and sensitive dependence on initial conditions for the baker's map, using the conjugation (4.6.3) and the corresponding properties for the shift map.

This idea of coding, which is illustrated here on elementary examples can be extended to many dynamical systems and be used to prove their "chaotic" dynamical properties, see Bowen [45], Bowen and Ruelle [46], Eckmann and Ruelle [120], Lanford [207] and Sinai [295].

4.6.2 Strange Attractors

When one studies the long term behavior of dynamical systems, the notion of attractor is often useful. A set $A \subset \Omega$ which is invariant under the dynamics is an *attractor* if there exist a open set $O \supset A$ such that, $\forall x \in O$, dist $(T^n x, A) \to 0$ as $n \to \infty$. The largest such open set is called the *basin of attraction* of A.

The simplest examples of attractors are isolated points (for example 0 for the map $T : \mathbb{R} \to \mathbb{R}: T : x \to \frac{x}{2}$) or periodic orbits: let $T : \mathbb{R}^2 \to \mathbb{R}^2$ be defined in polar

---

[17] To be precise, it is the interval $[\sum_{i=0}^{n-1} \frac{\alpha_i}{2^{i+1}}, \sum_{i=0}^{n-1} \frac{\alpha_i}{2^{i+1}} + \frac{1}{2^n}[$.

coordinates by: $T(r, \phi) = (\sqrt{r}, \phi + \alpha \mod 2\pi)$, with α rational. Then it is easy to see that the unit circle S^1 is invariant, is an attractor and that the dynamical system on S^1 is periodic.

A new kind of attractor was discovered through the study of chaotic dynamical systems: attractors that are neither points nor smooth surfaces but rather Cantor sets and on which the dynamics depends sensitively to initial conditions. Because of these two properties, they have been labelled "strange".[18] We will give here some elementary examples of such attractors.

4.6.2.1 The Modified Baker's Map

Let us first consider a modified baker's map. Let $D = [0, 1[\times [0, 1[$ and $f : D \to D$ defined by $T(x, y) = (x', y')$, where

$$x' = 2x \mod 1$$
$$y' = \frac{y}{3} \text{ if } 0 \leq x < \frac{1}{2}$$
$$y' = \frac{y}{3} + \frac{2}{3} \text{ if } \frac{1}{2} \leq x < 1. \tag{4.6.4}$$

That map is not measure preserving, since the contraction in the y direction (by a factor $\frac{1}{3}$) is bigger than the dilatation (by a factor 2) in the x direction. The map is invertible on its image $f(D)$, which means that it is injective, but, unlike the situation of the usual baker map, the image is not the unit square: $f(D) \subsetneq D$. Let $\Lambda = \bigcap_{n \in \mathbb{N}} f^n(D)$. That set is obviously invariant under f and the dynamics on Λ is interesting.

It can again be studied through a coding: write, as before $x = \sum_{n=0}^{\infty} \frac{a_n}{2^{n+1}}$, but write y in base 3: $y = \sum_{n=1}^{\infty} \frac{a_{-n}}{3^n}$, and let $\sigma(x, y) = (a_n)_{n \in \mathbb{Z}}$, with that convention for the symbols a_n.

Then, we have from (4.6.4) $x' = \sum_{n=0}^{\infty} \frac{a_{n+1}}{2^{n+1}}$ (the sequence $(a_n)_{n=0}^{\infty}$ is shifted to the left and the first term a_0 disappears from the expansion of x'). and $y' = \frac{2a_0}{3} + \sum_{n=1}^{\infty} \frac{a_{-(n+1)}}{3^n}$ (the sequence $(a_{-n})_{n=1}^{\infty}$ is shifted to the left and the first term a_0 in the expansion of x appears, multiplied by 2, as the first term in the expansion of y').

This means that the first term in the expansion of the second coordinate (y') of $T(x, y)$ is equal to 0 or 2. But since, if we apply T to $T(x, y)$, we will simply shift that first term to the left, and the second coordinate of all points of the form $T^2(x, y)$ will have their first two terms (a_{-1}, a_{-2}) equal to 0 or 2.

By induction, one sees that all the points in $\Lambda = \bigcap_{n \in \mathbb{N}} f^n(D)$ have the ternary expansion of their y coordinate made up only of 0's and 2's. This is the Cantor set C studied in Appendix 2.A.7.

---

[18] The two properties, being a Cantor set and having sensitive dependence on initial conditions, do not always go together: one can have attractors with sensitive dependence on initial conditions that are smooth surfaces, but in many examples they are Cantor sets.

To summarize we have shown that $\Lambda = \bigcap_{n \in \mathbb{N}} T^n(D) = [0, 1] \times C$ and can be identified with the set of sequences $\tilde{\Omega} = \{(a_n)_{n=-\infty}^{\infty} | a_n \in \{0, 1\}, n \geq 0, a_n \in \{0, 2\}, n < 0\}$. Since C is of measure 0 in $[0, 1]$, Λ is of measure 0 in $[0, 1[\times [0, 1[$.

The map σ sends Λ onto $\tilde{\Omega}$ and, if T_{shift} denotes the shift to the left on $\tilde{\Omega}$, we have again the conjugation:

$$\sigma(T(x, y)) = T_{\text{shift}}(\sigma(x, y)),$$

It is easy to construct a measure on $\tilde{\Omega}$ invariant under T_{shift}: take the product measure that give equal weight $\frac{1}{2}$ to each symbol a_n (equal to 0 or 1 if $n \geq 0$ and to 0 or 2 if $n < 0$) and denote it by ν.

What is the relationship between that measure ν and the Lebesgue measure on $D = [0, 1[\times [0, 1[$? Obviously, this cannot be as simple as before, since Λ is a set of measure zero in D. We do not have $\sigma^* \mu = \nu$ (see Definition 4.1), but rather:

$$\lim_{n \to \infty} T_{\text{shift}}^{*n} \sigma^* \mu_{\text{Leb}} = \nu, \tag{4.6.5}$$

where T_{shift}^* is defined by formula (2.A.7) with T instead of Φ.

To see this, write the elements of D using the binary expansion for their x component and the ternary expansion for their y component, then $\sigma^* \mu_{\text{Leb}}$ gives equal weight to each symbol $\frac{1}{2}$ for each value of $a_n \in \{0, 1\}, n \geq 0$ and $\frac{1}{3}$ for each value of $a_n \in \{0, 1, 2\}, n < 0$. But then $T_{\text{shift}}^* \sigma^* \mu_{\text{Leb}}$ gives weight $\frac{1}{2}$ for each value of $a_n \in \{0, 1\}, n \geq 0$, $\frac{1}{2}$ for each value of $a_{-1} \in \{0, 2\}$ and $\frac{1}{3}$ for each value of $a_n \in \{0, 1, 2\}, n < -1$. For $T_{\text{shift}}^{*2} \sigma^* \mu_{\text{Leb}}$, we get weight $\frac{1}{2}$ for each value of $a_{-1} \in \{0, 2\}$ and of $a_{-2} \in \{0, 2\}$, and the rest unchanged. By induction, we get (4.6.5).

This example is important because it provides an elementary example of maps for which one can prove a *Sinai–Ruelle–Bowen (SRB) ergodic theorem*, [45, 46, 207, 295]: $\forall F \in L^1(D, d\mu_{\text{Leb}})$,

$$\lim_{N \to \infty} \frac{1}{N} \sum_{n=0}^{N-1} F(T^n x) = \int_{\tilde{\Omega}} F(x) d\nu(x), \quad \mu_{\text{Leb}} \text{ almost everywhere} \tag{4.6.6}$$

What is important here is that the measure with respect to which the limit holds almost everywhere (the Lebesgue measure μ_{Leb}) is *not* the one with respect to which the integral is taken (ν) and is *not invariant* under the dynamics. The measure ν is an elementary example of a Sinai–Ruelle–Bowen (SRB) measure. The difference between μ_{Leb} and ν will be relevant in Sect. 6.8.

Fig. 4.5 The map (4.6.7).
Source https://upload.
wikimedia.org/wikipedia/
commons/5/57/Smale-
Williams_Solenoid.png, Ilya
Voyager, CCO via
Wikimedia Commons

4.6.2.2 The Solenoid

The reader might worry that, in the previous example, the map T is not continuous. That is why we will also consider the example of the *solenoid*, which is closely related to the modified baker's map, but where this problem is avoided. We will follow Lanford [207] in the presentation of this example.

Let $\mathbf{T}$ be a solid torus in three dimensions: $\mathbf{T} = S^1 \times D^2$, with S^1 the unit circle, $S^1 = \{z \in \mathbb{C}, |z| = 1\}$ and D^2 the unit disk $D^2 = \{w \in \mathbb{C}, |w| \le 1\}$. We define the solenoid map:

$$T(z, w) = \left(z^2, \frac{1}{2}z + \frac{1}{4}w \right) \tag{4.6.7}$$

where, if we write $z = \exp(2\pi i x)$, we see that $z \to z^2$ is just the map $x \to 2x$ mod 1 defined in (4.1.3). The image of the torus under T is a tube inside the torus of transverse radius $\frac{1}{4}$ winding twice around the torus, see Fig. 4.5. The intersection of this tube with the transverse plane $\{z = 1\}$ is made of two disjoint disk, each of radius $\frac{1}{4}$. Iterating, we see that the image of the torus under T^n is a tube inside the torus of transverse radius $\frac{1}{4^n}$ winding 2^n times around the torus. The intersection of this tube with the transverse plane $z = 1$ is made of 2^n disjoint disks, each of radius $\frac{1}{4^n}$.

One defines the *solenoid* as $\Lambda = \bigcap_{n \in \mathbb{N}} T^n(\mathbf{T})$. It is composed of a set of lines (with no thickness) wrapping around the torus. Its intersection with the transverse plane $\{z = 1\}$ is a Cantor set. Since that set is uncountable and each line in the solenoid has only countably many intersections with $\{z = 1\}$, the solenoid is not a single line but an uncountable union of lines. Cutting the solenoid along the $\{z = 1\}$ plane splits Λ into uncountably many loops going once around the torus, that can be labelled by points in a Cantor set.

We shall now introduce a coding on Λ through which this dynamics will be converted into a shift on a space of sequence of symbols and a measure on that space for which an SRB ergodic theorem of the form (4.6.6) holds (with μ being the Lebesgue measure on $\mathbf{T}$).

We write $z = \exp(2\pi i x)$ and write the binary expansion of $x = \sum_{n=0}^{\infty} \frac{a_n}{2^{n+1}}$. By introducing a coding $(a_n)_{n=-1}^{\infty}$ of the Cantor set which is the intersection of Λ and the plane $\{z = 1\}$, we can again introduce a conjugation σ satisfying $\sigma \circ T = T_{\text{shift}} \circ \sigma$ between the map T on Λ and the shift map T_{shift} on $\tilde{\mathbf{\Omega}}_2$, defined after (4.1.12).

The SRB measure ν is simply the product measure on $\tilde{\mathbf{\Omega}}$ with equal weight $\frac{1}{2}$ for each symbol. In terms of loops, it means that one gives equal probabilities to all loops and one puts the Lebesgue measure on each loop, see Lanford [207, Sect. 5] for the details of that construction.

4.6.2.3 The Logistic Map

Let us admit that the tent map (4.1.8) is ergodic with respect to the Lebesgue measure (this is true but not very easy to prove). One can check, by explicit computation that, if we denote the logistic map $4x(1 - x) = g(x)$, and write $x = C(y)$, with:

$$C(y) = \frac{1 - \cos \pi y}{2}. \tag{4.6.8}$$

Then, $g(x) = C \circ f \circ C^{-1}(x)$ with f the tent map: indeed, $g(C(y)) = 4(\frac{1-\cos \pi y}{2})$ $(\frac{1+\cos \pi y}{2}) = \sin^2 \pi y$, and, for $0 \le y \le \frac{1}{2}$, $C(f(y)) = \frac{1-\cos 2\pi y}{2} = \sin^2 \pi y$, while for $\frac{1}{2} \le y \le 1$, $C(f(y)) = \frac{1-\cos 2\pi(1-y)}{2} = \sin^2 \pi y$.

So, $g^n(x) = C \circ f^n \circ C^{-1}(x)$ and, if one wants to compute the average time spent by the orbit of x in an interval J for the map g:

$$\lim_{N \to \infty} \frac{1}{N} \sum_{n=0}^{N-1} I(g^n(x) \in J),$$

it is enough to compute

$$\lim_{N \to \infty} \frac{1}{N} \sum_{n=0}^{N-1} I(f^n(y) \in C^{-1}(J)) \tag{4.6.9}$$

with $y = C^{-1}(x)$. But, since f is ergodic with respect to the Lebesgue measure, (4.6.9) equals $\int_{C^{-1}(J)} dy = \int_J (C^{-1})'(x)dx$, for almost all $y = C^{-1}(x)$ with respect to the Lebesgue measure. From (4.6.8) one gets $C'(y) = \pi \frac{\sin \pi y}{2} = \frac{\pi}{2}\sqrt{1 - \cos^2 \pi y} = \frac{\pi}{2}\sqrt{1 - \cos \pi y}\sqrt{1 + \cos \pi y} = \pi\sqrt{x(1 - x)}$ with $x = \frac{1-\cos \pi y}{2} = C(y)$. Thus, the limit (4.6.9) equals, almost all $y = C^{-1}(x)$,

Fig. 4.6 The density of the
invariant measure in (4.6.10)
(times π) for the logistic map

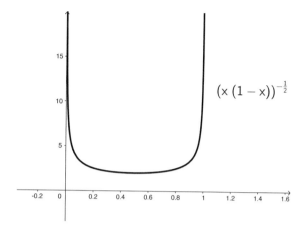

$$\int_J \frac{dx}{\pi\sqrt{x(1-x)}},\tag{4.6.10}$$

which gives us the invariant measure for the logistic map (see Fig. 4.6). This measure
is not the Lebesgue measure, but is absolutely continuous with respect to it.

For computer simulations showing that (4.6.10) corresponds to the asymptotic
distribution of trajectories for the logistic map, see Collet and Eckmann [85]. Finally,
one can show that this result is also true for almost all $x = C(y)$ with respect to the
Lebesgue measure, using the fact (that we will not prove) that the image by a function
whose derivative is bounded (such as C), of a set of zero Lebesgue measure is also
of zero Lebesgue measure.

When λ in (4.1.9) is not equal to 4 the logistic map has a very complicated and
interesting behavior, see e.g. Collet and Eckmann [85] or Strogatz [306].

4.7 Dynamical Entropies

Since the notion of entropy is central in this book, and since Kolmogorov and Sinai
have introduced a notion of entropy for dynamical systems, named after them, we
shall briefly discuss this notion; for more information about the Kolmogorov-Sinai
entropy, see Billingsley [28], Sinai [296], Cornfeld, Fomin and Sinai [87], or Walters
[327].

Let (Ω, Σ, μ, T) be a dynamical system with $\mu(\Omega) = 1$, and let $\tilde{\Omega} = (\Omega_1, \ldots, \Omega_k)$
be a finite partition of Ω. The entropy of that partition is, by definition,

$$S(\tilde{\Omega}, \mu) = -\sum_{i=1}^{k} \mu(\Omega_i) \ln \mu(\Omega_i)\tag{4.7.1}$$

If one interprets entropy as measuring an amount of information (see Chap. 7), then (4.7.1) is the amount of information that one obtains by knowing to which element of the partition $x \in \Omega$ belongs.

Given two partitions $\tilde{\Omega} = (\Omega_1, \ldots, \Omega_k)$ and $\tilde{\Omega}' = (\Omega'_1, \ldots, \Omega'_l)$, one defines their common refinement $\tilde{\Omega} \vee \tilde{\Omega}' = (\Omega_i \cap \Omega'_j)_{i=1,\ldots,k, j=1,\ldots,l}$.

The entropy of the partition $\tilde{\Omega} \vee T^{-1}\tilde{\Omega} \vee \cdots \vee T^{-n+1}\tilde{\Omega}$ is:

$$S(\tilde{\Omega}, \mu, T, n) \equiv S(\tilde{\Omega} \vee T^{-1}\tilde{\Omega} \vee \cdots \vee T^{-n+1}\tilde{\Omega}, \mu)$$

$$= - \sum_{i_0,\ldots,i_{n-1}\in\{1,\ldots,k\}} \mu(\Omega_{i_0} \cap T^{-1}\Omega_{i_1} \cap \cdots \cap T^{-n+1}\Omega_{i_{n-1}})$$

$$\ln \mu(\Omega_{i_0} \cap T^{-1}\Omega_{i_1} \cap \cdots \cap T^{-n+1}\Omega_{i_{n-1}})$$

This represents the amount of information obtained by knowing to which element of the partition all the points $x, Tx \ldots T^{n-1}x$ belong.

To understand these notions, it is good to have in example in mind: let $\Omega = \Omega_k = \{0, 1, \ldots, k-1\}^{\mathbb{Z}}$, Σ be the sigma algebra generated by the cylinder sets of Ω_k, μ any product measure on (Ω_k, Σ) and T_{shift} the shift defined by (4.1.12). Consider the partition $\tilde{\Omega}$ where $\Omega_i = \{\mathbf{x} \in \Omega_k \mid x_0 = i\}, i = 1, \ldots, k$ and denote by $p_i = \mu(\Omega_i)$. Then we have:

$$S(\tilde{\Omega}, \mu) = - \sum_{i=1}^{k} p_i \ln p_i$$

$$S(\tilde{\Omega}, \mu, T_{\text{shift}}, n) = - \sum_{i_0,\ldots,i_{n-1}\in\{1,\ldots,k\}} \prod_{\alpha=0}^{n-1} p_{i_\alpha} \ln \prod_{\alpha=0}^{n-1} p_{i_\alpha} \tag{4.7.2}$$

An important property of $S(\tilde{\Omega}, T, n)$ is that the following limit exists:

$$S(\tilde{\Omega}, \mu, T) = \lim_{n\to\infty} \frac{1}{n} S(\tilde{\Omega}, \mu, T, n) \tag{4.7.3}$$

Since $S(\tilde{\Omega}, \mu, T, n)$ is the amount of information obtained by knowing in which element of the partition the points $x, Tx \ldots T^{n-1}x$ belong, $S(\tilde{\Omega}, \mu, T)$ is the average information obtained by knowing in which element of the partition the infinite sequence of points $x, Tx, T^2x, \ldots$belong.

In the example given by (4.7.2), that limit is easy to compute, since

$$S(\tilde{\Omega}, \mu, T, n) = - \sum_{i_0,\ldots,i_{n-1}\in\{1,\ldots,k\}} \prod_{\alpha=0}^{n-1} p_{i_\alpha} \ln \prod_{\alpha=0}^{n-1} p_{i_\alpha}$$

$$= - \sum_{\alpha'=0}^{n-1} \sum_{i_0,\ldots,i_{n-1}\in\{1,\ldots,k\}} \prod_{\alpha=0}^{n-1} p_{i_\alpha} \ln p_{i_{\alpha'}}$$

$$= -n \sum_{i_{\alpha'}}^{k} p_{i_{\alpha'}} \ln p_{i_{\alpha'}} = -n \sum_{i}^{k} p_i \ln p_i, \tag{4.7.4}$$

since, for $\alpha \neq \alpha'$, we can use $\sum_{i_\alpha=1}^{k} p_{i_\alpha} = 1$. For (4.7.4), the limit (4.7.3) is trivial and equals $-\sum_{i=1}^{k} p_i \ln p_i$.

One can then define the *entropy* of the transformation T:

$$S(\mu, T) = \sup_{\tilde{\Omega}} S(\tilde{\Omega}, \mu, T),$$

where the supremum is taken over all finite partitions of Ω.

A fundamental result makes the computation of that supremum easier: if $\tilde{\Omega}$ is a *generating partition*, namely if $\vee_{n=-\infty}^{\infty} T^n \tilde{\Omega} = \Sigma$ almost everywhere, then (see Cornfeld, Fomin and Sinai [87, Sect. 10.6]):

$$S(\mu, T) = S(\tilde{\Omega}, \mu, T).$$

For the shift map, since the cylinder sets are the elements of the partition $\vee_{n=-N}^{N} T_{\text{shift}}^n$ $\tilde{\Omega}$ with $N \in \mathbb{N}$, the partition $\tilde{\Omega} = (\Omega_i = \{\mathbf{x} \in \Omega \mid x_0 = i\})_{i=1,\dots,k}$ is generating; so, in that example:

$$S(\mu, T_{\text{shift}}) = -\sum_{i=1}^{k} p_i \ln p_i$$

One may contrast this result with what happens with a non-generating partition[19]: consider for example the k-shift with $k = 3$ and $\tilde{\Omega} = (\Omega_1 = \{\mathbf{x} \in \Omega \mid x_0 = 1\}, \Omega_2 = \{\mathbf{x} \in \Omega \mid x_0 = 2, 3\})$. We get:

$$S(\tilde{\Omega}, \mu, T_{\text{shift}}) = -(p_1 \ln p_1 + (p_2 + p_3) \ln(p_2 + p_3)) \tag{4.7.5}$$

which is strictly less than $S(\mu, T_{\text{shift}}) = -\sum_{i=1}^{3} p_i \ln p_i$, since $\ln(p_2 + p_3) > \ln p_i$, $i = 2, 3$.

It is easy to see that, if two transformations T on Ω and T' on Ω' are conjugated (see Definition 4.1), then $S(\mu, T) = S(\mu', T')$, since $\mu(\Omega_{i_0} \cap T^{-1}\Omega_{i_1} \cap \cdots \cap T^{-n+1}\Omega_{i_{n-1}}) = \mu'(\Omega'_{i_0} \cap T'^{-1}\Omega'_{i_1} \cap \cdots \cap T'^{-n+1}\Omega'_{i_{n-1}})$, where $\Omega'_{i_\alpha} = \Phi(\Omega_{i_\alpha})$.

Then, using the conjugations between the maps (4.1.3), (4.1.11) and the shift on Ω_2, with $p_1 = p_2 = \frac{1}{2}$, one shows that, for both maps $S(\mu, T) = \ln 2$. It also implies that the shifts on Ω_k with $p_i = \frac{1}{k}$, $\forall i = 1, \dots, k$, whose entropies equal $\ln k$ are not conjugated for different values of k.

4.8 Determinism and Predictability

We want to finish this chapter by a short discussion of a frequent confusion that occurs in the popular, but also in the scientific literature, between determinism and

---

[19] That partition is not generating because $\vee_{n=-\infty}^{\infty} T^n \tilde{\Omega}$ will not contain, for example sets of the form $\times_{i<0}\{1, 2, 3\}_i \times \{2\} \times_{i>0} \{1, 2, 3\}_i$, where $\{1, 2, 3\}_i$ is the copy of index i of $\{1, 2, 3\}$ in $\{1, 2, 3\}^{\mathbb{Z}}$.

predictability, confusion which is often caused by a lack of precise definitions. In Sect. 2.1 we mentioned Laplace's very clear expression of the idea of universal determinism. We also remarked that Laplace clearly distinguished between what nature does and the knowledge we have of it or between determinism and our ability to predict the future.

However, determinism is often confused with predictability. So, according to that view, a process is deterministic if we, humans, can predict it, or, maybe, if we, humans, will be able to predict it in the future. For example, in an often quoted lecture[20] to the Royal Society, on the three hundredth anniversary of Newton's Principia, the distinguished British mathematician Sir James Lighthill gave a perfect example of how to confuse predictability and determinism:

> We are all deeply conscious today that the enthusiasm of our forebears for the marvelous achievements of Newtonian mechanics led them to make generalizations in this area of *predictability* which, indeed, we may have generally tended to believe before 1960, but which we now recognize were false. We collectively wish to apologize for having misled the general educated public by spreading ideas about *determinism* of systems satisfying Newton's laws of motion that, after 1960, were to be proved incorrect [...].

> James Lighthill, [231], (Italics added by J.B.)

Of course, nobody who has ever defended universal determinism (in particular Laplace) meant it to be equated with predictability. Everybody agrees that not everything in the world is predictable, and it is somewhat surprising to see how many people present that truism as if it was a recent discovery.

To illustrate the problem posed by the conflation of the two terms, consider, for example, a perfectly regular, deterministic *and* in principle predictable mechanism, like a clock, but put on the top of a mountain, or in a locked drawer, so that its state (its initial conditions) become inaccessible to us. This renders the system trivially unpredictable, yet it seems difficult to claim that it becomes non-deterministic.

So, one has to admit that *some* physical phenomena obey deterministic laws and, yet, are not predictable, possibly for reasons that have nothing to do with indeterminism, like the example of the clock here. But, once this is admitted, how does one show that *any* unpredictable system is *truly* non-deterministic, and that the lack of predictability is not merely due to some limitation of our knowledge or of our abilities? We cannot infer indeterminism from ignorance alone. One needs other arguments.

The major scientific argument in favor of indeterminism is of course quantum mechanics whose formalism includes an element of "pure chance": when a measurement is made, the state vector is randomly reduced to one of the eigenvectors of the observable being measured, and that process is, according to the standard interpretation of quantum mechanics, intrinsically random. However, we have decided not to discuss quantum mechanics in this book, so we will not consider it here, except to mention that there are still many discussions about the real status of that theory, see footnote 23 in Chap. 2.

---

[20] Quoted e.g. by Reichl [276], p. 3, and by Prigogine and Stengers, ([272], pp. 93–94, and [273], pp. 41–42).

Confusing determinism and predictability is an instance of what the physicist E. T. Jaynes calls the "Mind Projection Fallacy"[21]:

> We are all under an ego-driven temptation to project our private thoughts out onto the real world, by supposing that the creations of one's own imagination are real properties of Nature, or that one's own ignorance signifies some kind of indecision on the part of Nature.

Edwin T. Jaynes, [182], p. 7.

This brings us to the second definition of determinism, that tries to be independent of human abilities; consider a physical system whose state is characterized by some numbers that change over time; let us say that it is deterministic if there exists a function F that maps the values taken by set of variables characterizing a system at a given time t_1, to those at a later time, say t_2; and, then, the latter to those at a later time, t_3, etc.[22] This corresponds pretty much to Laplace's conception, namely to the idea of predictability, but "in principle", i.e. putting aside limitations imposed by human abilities. The word 'exist' should be taken here in a literal sense: it does not refer to our *knowledge*; the function in question may be unknown, or unknowable, or so complicated that, even if it were known, it would be, in practice, impossible to use it in order to make predictions.

In this chapter, we have encountered many examples of such functions F and also in Chap. 3 (except that there one dealt with continuous rather than discrete time).

Of course, the notion of determinism introduced here has very little to do with the goals of science, which are not simply to find a function like F. In a sense, scientists do look for such functions, but with extra properties: simplicity, explanatory power, and, of course, the possibility, using F, to make at least some predictions. So, in a sense, the question of the existence of F is "metaphysical" and of no scientific interest. But, so is the question of "determinism": either it is identified with predictability by us, humans, and determinism is trivially false, or it is defined as above, through the existence of the function F, and it is most likely true (putting aside the issue of quantum mechanics, see footnote 23 in Chap. 2), but uninteresting. It is difficult to see how to formulate the issue of "determinism" in a sense that makes it both interesting and decidable.

It is likely that the hostility to determinism comes from a desire to "save free will". Namely, to find a description of the physical universe that can be reconciled with our deep feeling that, at least on some occasions, "we" choose to do X and not Y. That is, that Y was possible, but did not happen because of our free choice. But, if everything is caused by anterior events, ultimately going back to the Big Bang, Y was not really possible (it only appeared to be so because of our ignorance) and free will is an illusion. Since most of our moral, legal and political philosophies assume some kind of free will, a lot appears to be at stake.

---

[21] Which is related to idealism, since the latter tends to identify what exists and what is in our mind.

[22] This idea is essentially the one proposed by Bertrand Russell in [283]; see Earman [117] for a discussion.

But the problem is: what is the alternative to determinism *within physics*? Nothing has ever been proposed except *pure randomness*! Or, in other words, events with no cause. But that will not give us a picture of the world in which free will exists either. Our feeling of free will is not that there is some intrinsically random process at work in our minds, but that *conscious choices* are made. And that is simply something that no known physical theory accounts for. Our feeling of free will implies that there is a causal agent in the world, the 'I', that is simply 'above' all physical laws. It suggests a dualistic view of the world, which itself meets great difficulties. One solution is, as mentioned above, to declare that free will is an illusion. But if that is the case, it is a 'necessary illusion' in the sense that we cannot live without, in some sense, believing in it, unlike, say, believing in the dogma of the Immaculate Conception. It is not clear what would be a solution to that problem, but one should avoid using this problem to create within physics a prejudice in favor of indeterminism, since neither determinism nor indeterminism in physics can "save" free will.

As Bertrand Russell once observed, scientists should look for deterministic laws like mushroom seekers should look for mushrooms. Deterministic laws are preferable to non-deterministic ones because they give both a way to control things more efficiently (at least in principle) and because they give more satisfactory explanations of why things are the way they are. Looking for deterministic laws behind the apparent disorder of things is at the heart of the scientific enterprise. Whether we succeed or not depends in a complicated way both on the structure of the world and on the structure of our minds. But the opposition to determinism tends to make people feel that the project itself is doomed to fail; and that state of mind does run counter to the scientific spirit.

However, the discovery of "chaos" discussed in this chapter has repeatedly led people to consider it as an argument against determinism (in fact, this is the idea behind the quote of Lighthill mentioned above); but it can only be considered as such if one confuses determinism with predictability. Actually, the existence of chaotic dynamical systems supports universal determinism rather than contradicts it. Indeed, suppose for a moment that no classical mechanical system behaves chaotically. That is, suppose that there existed a theorem proving that any such system must eventually behave in a non-chaotic fashion. It is not completely obvious what the conclusion would be, but certainly *that* would be an embarrassment for the classical deterministic world-view. Indeed, so many physical systems (like the weather) seem to exhibit sensitivity to initial conditions that one would be tempted to conclude that classical mechanics cannot adequately describe those systems. One might suggest that there must be an inherent indeterminism in the basic laws of nature. Deterministic chaos increases the explanatory power of deterministic assumptions, and therefore, according to normal scientific reasoning, *strengthens* those assumptions (again, putting aside quantum mechanics).

Moreover, one has designed empirical tests in order to check whether a series of numbers is "random". But some deterministic chaotic systems (such as the function $f : x \to 2x \bmod 1$) produce sequences of numbers that can pass all these tests. This is a very strong argument against the idea that one can ever prove that some phenomenon is "intrinsically random", in the sense that there does not exist a deterministic mechanism underlying and explaining its behavior. So the recent discoveries about chaos do not force us to change a single word of what Laplace wrote and, in fact, support his position.

4.9 Summary

A natural way to generalize Hamiltonian systems is to consider dynamical systems defined by a measure preserving transformation, an example of which is the Hamiltonian flow together with the Liouville measure, see Sect. 3.4.

There are two basic theorems that are valid for such dynamical systems: the Poincaré recurrence, that asserts that almost every trajectory will come infinitely often arbitrarily close to any point that it has already visited (Sect. 4.2) and the ergodic theorem that asserts that time averages of arbitrary functions over almost any trajectory converges as time goes to infinity (Sect. 4.3). However, the theorem is not terribly useful as such: because of its generality it does not give us any idea about the value of these limits, it simply says that they exist.

But, for ergodic transformations, the limit can be identified as the expectation value, with respect to the measure invariant under the transformation, of the function whose time average over one of its trajectories is taken. In particular, that average is the same for almost all trajectories. In that (restricted) context, the ergodic theorem is very useful. Unfortunately, although it is easy to give examples of ergodic transformations and of applications of the ergodic theorem for those transformations (Sect. 4.3.1), it is difficult to give examples of ergodic Hamiltonian systems, which limits the physical applications of the notion of ergodicity.

After discussing ergodicity, we consider stronger properties: mixing (Sect. 4.4) and sensitive dependence on initial conditions (Sect. 4.5), one reason being that these notions are sometimes thought to be relevant in order to understand the approach to equilibrium (we will see why this is not the case in Sect. 8.3).

A dynamical systems has the property of sensitive dependence on initial conditions if different trajectories diverge exponentially in time. This makes those systems unpredictable at least beyond a given time scale determined by the coefficient of that exponential growth (the inverse of the coefficient λ in (4.5.8)).

But even if the trajectories of dynamical systems having sensitive dependence on initial conditions cannot be predicted, one can predict the statistical properties of trajectories, namely the average time spent in a given region of the space on which the dynamical system is defined. This uses ergodicity, but with respect to measures that are associated to the dynamical system under consideration and whose support may be a set of zero Lebesgue measure towards which the orbits of the dynamical system are attracted. These attractors are sometimes Cantor sets and are then called "strange" (Sect. 4.6.2). This aspect of statistical predictability for systems that are individually unpredictable bears some ressemblance to what is done in statistical mechanics (see Chaps. 6–9).

Finally, following Kolmogorov and Sinai, we introduced a notion of entropy for dynamical systems (Sect. 4.7) and discussed the distinction between determinism and predictability, two notions that are often confused in the literature (Sect. 4.8).

4.10 Exercises

4.1. Check, using Fourier series on the n-dimensional torus, that the map T_α defined by (4.1.2) is ergodic if and only if the equation $\sum_{i=1}^{n} l_i \alpha_i = m$ with $l_i, m \in \mathbb{Z}$ has no solution except $l_i = 0 \ \forall i$, and $m = 0$.

4.2. Show that the map T defined by (4.1.7) is ergodic by adapting the proof given for the map (4.1.3).

4.3. * Show that automorphisms of the n-dimensional torus are ergodic if and only if the matrix A in (4.1.10) does not have roots of unity as eigenvalues.

4.4. Prove both mixing and sensitive dependence on initial conditions for the baker's map defined by (4.1.11), using the conjugation (4.6.3) and the corresponding properties for the shift map.

   Hint: check first that, if two maps are conjugated (see (4.1.5)) and if Φ and its inverse are uniformly Lipschitz continuous (see (4.5.1) for the definition of Lipschitz continuity), meaning that $\exists L < \infty$, so that, $\forall w, w' \in \Omega$

$$\frac{1}{L}\text{dist}\,(w, w') \le \text{dist}\,(\Phi(w), \Phi(w')) \le L\text{dist}\,(w, w'), \qquad (4.10.1)$$

   then, if one of the two maps has the property of sensitive dependence on initial conditions, the other map has that property also with ϵ replaced by $L\epsilon$ and d replaced by $\frac{d}{L}$. This follows easily from the definition (4.5.5).

4.5. * Consider the map (4.1.1) and estimate the recurrence time to an interval of length ϵ, if $\alpha \notin \mathbb{Q}$, namely evaluate $n(\epsilon) = \min\{n \in \mathbb{N} \mid \exists\, p \in \mathbb{N}, |n\alpha - p| \le \epsilon\}$ as a function of ϵ. You may use the fact that, for any $\alpha \notin \mathbb{Q}$, there exists infinitely many pairs (p, q) so that $|\alpha - \frac{p}{q}| \le \frac{1}{q^2}$.

4.6. Consider the map $x \to 1 - 2x^2$ on $[-1, 1]$ and use the change of variable $x = C(y) = \sin \pi y$ and the ergodicity of the map (4.1.3) to show that the measure $\frac{dx}{\pi\sqrt{1-x^2}}$ is invariant and ergodic under the map $x \to 1 - 2x^2$.

4.7. Consider the first digits in the sequence $(3^n)_{n=1}^{\infty}$ of the powers of 3: 3, 9, 8, 2, 7.... What is the frequency of a digit p in that sequence? Is the same result true for all sequences $(a^n)_{n=1}^{\infty}, a \in N$?

4.8. Consider the following modified baker's map:

$$x' = 4x \quad \mathrm{mod}\ 1$$

$$y' = \frac{y}{4} \quad \text{if } x < \frac{1}{4}$$

$$y' = \frac{y+1}{4} \quad \text{if } \frac{1}{4} \le x < \frac{1}{2},$$

$$y' = \frac{y+1}{2} \quad \text{if } \frac{1}{2} \le x < \frac{3}{4},$$

$$y' = \frac{y+3}{4} \quad \text{if } \frac{3}{4} \le x \le 1.$$

Find a coding for that map and show that it is mixing and sensitive to initial conditions.

4.9. How many points have a periodic orbit of length n under the baker map (including points that may also have a periodic orbit of length less than n)?

4.10. * Show that the set of points $x \in [0, 1[$ whose orbit under the map (4.1.3) is not dense in $[0, 1[$ is of measure zero (see footnote 29 in Chap. 2).

4.11. Let $f : [0, 1[\to [0, 1[$ given by $f(x) = 3x \mod 1$. Show that there exists an $x \in [0, 1[$ such that, $\forall n \in \mathbb{N}$ $f^{(n)}(x) < \frac{1}{3}$ if n is a prime number or $n = 0, 1$, and $f^{(n)}(x) \ge \frac{2}{3}$ if n is not a prime number

Is this x unique ?

4.12. Show that, for the baker's map, there is a dense set of points in $D = [0, 1] \times [0, 1]$ whose orbit is periodic in D.

4.13. Let $x_{n+1} = f(x_n)$ where $f : [0, 1] \to [0, 1]$ is given by $f(x) = 4x(1 - x)$. Compute, for almost all $x_0 \in [0, 1]$ (with respect to the Lebesgue measure), the limit:

$$\lim_{N \to \infty} \frac{1}{N} \sum_{n=0}^{N-1} (x_n(1 - x_n))^{3/2}. \tag{4.10.2}$$

Hint: use Birkhoff's ergodic theorem and the result of Sect. 4.6.2.3.

4.14. Check (4.7.5).

4.15. Consider the space $\Omega = \{\mathbf{x} = (x_i)_{i \in \mathbb{Z}}, x_i = 1, \ldots, l\}$, a $l \times l$ matrix $\mathcal{P} = (P_{xy})_{x, y = 1, \ldots, l}$, and a vector $\mathbf{P} = (p_x)_{x=1, \ldots, l}$ satisfying:

$$p_x \ge 0 \ \forall x = 1, \ldots, l$$

$$\sum_{x=1}^{l} p_x = 1$$

$$P_{xy} \ge 0 \ \forall x, y = 1, \ldots, l$$

$$\sum_{y=1}^{l} P_{xy} = 1$$

$$\sum_{x=1}^{l} p_x P_{xy} = p_y$$

Define a measure on Ω through the following probability of the cylinder sets[23]:

$$P(x_0 = a_0, x_1 = a_1, \ldots x_N = a_N) = p_{a_0} \prod_{i=1}^{N-1} P_{a_i a_{i+1}} \tag{4.10.3}$$

Check that one can use (4.10.3) to define a measure invariant under the shift on Ω and compute the Kolmogorov-Sinai entropy of that dynamical system. You may assume that the partition $\tilde{\Omega} = (\Omega_i, i = 1, \ldots, l)$, with $\Omega_i = \{x \in \Omega \mid x_0 = i\}$ is generating.

---

[23] This is called a Markov chain, but we will not use this notion, except in Sect. 8.7.1.

Chapter 5
Thermodynamics

5.1 Introduction

Thermodynamics is a phenomenological and macroscopic science. By "phenomeno-logical" we mean that it gives a mathematical formulation of empirical regularities, but does not try to "explain" them. That will be the task of statistical mechanics, which was developed after thermodynamics and whose goal was to "derive" or "explain" the thermodynamical laws. By "macroscopic" we mean that thermodynamics applies to bodies made of a large number of atoms or molecules. But it would be more correct to say that thermodynamics as such does not care about the microscopic constituents of matter; indeed, it was developed long before the existence of atoms was established. So, thermodynamics dealt, at least originally, with bodies roughly the size of what we find in our environment, like liquids, gases, crystals. It was, after all, developed in connection with vapor engines at the beginning of the industrial revolution.

This chapter is included for the sake of completeness, but is elementary and standard; so, readers familiar with thermodynamics can easily skip it.[1] We are only interested in setting the stage, so to speak, for what will come later, namely the statistical derivation of the thermodynamical laws, specially the second one.

It is well-known that thermodynamics consists of basically two laws, the first one dealing with the conservation of energy and the second one with the increase of entropy. In actual fact, there are two other "laws", the zeroth one and the third one. The zeroth law was added as an afterthought, as its name indicates, and the third law is not really universal, unlike the two classical laws, and we will not discuss it.

Since the first law is, as we will see later, a straightforward consequence of the laws of mechanics (but was formulated independently of them) it is the second law, with its introduction of the somewhat mysterious notion of entropy that has caused most controversies and confusions.

---

[1] We will mostly follow in this chapter the book of Callen [66], to which we refer for a more complete introduction to thermodynamics.

© Springer Nature Switzerland AG 2022
J. Bricmont, *Making Sense of Statistical Mechanics*, Undergraduate Lecture
Notes in Physics, https://doi.org/10.1007/978-3-030-91794-4_5

After discussing the zeroth law (Sect. 5.2) we will define the two basic laws (Sects. 5.3 and 5.4), define the thermodynamic entropy (Sect. 5.5) and other thermodynamic functions such as the free energy and the grand potential (Sect. 5.6); we will finally illustrate all these notions by computing them for ideal gases (Sect. 5.7).

5.2   The Zeroth Law

The usual way to state the zeroth laws is to say that there exist equilibrium states. But that statement is quite empty if we do not know what "states" are or what "equilibrium" means. A state is simply the assignment of parameters to a macroscopic body that are characteristic of that body in the sense that one always obtains the same values of those parameters when some external conditions are given. This definition is vague because we do not specify what those external conditions are, but that is because, at this point, we want to be sufficiently general and include the possibility of non-equilibrium systems where the external conditions could include an exchange of energy or of particles with the surroundings of the body.

But the simplest external condition one can think of is the one of an isolated system, namely a system that does not exchange energy or particles with its surroundings. Then, an "equilibrium state" means the state that an isolated systems tends to after a "sufficiently long time", which is another vague notion. In practice and for all the systems that we will consider, that time is not very long on a human scale, but there are materials such as glass that are not in an equilibrium state and that can remain in that state for very long times, several centuries, as one can see for example in stained glass windows in churches.

The simplest example of an equilibrium state is given by a chemically homogenous fluid, for example a gas consisting of one type of molecule. We will focus on that simple example, because it is simple but sufficient for us to introduce the basic concepts of thermodynamics.

The equilibrium state of such a fluid can be characterized by the volume V occupied by the fluid, its pressure P and its temperature T. We will introduce later other parameters such as the total energy, the entropy, the number of particles, and the chemical potential.

One important property of equilibrium states is that the parameters characterizing it are not independent; for example, there exists a relation, depending of course on the properties of the body under consideration, linking P, V and T, which one writes abstractly as:

$$f(P, V, T) = 0, \qquad\qquad\qquad (5.2.1)$$

for a function f to be specified in each situation. To give a simple example of a relation of the form (5.2.1), we have the ideal gas law or the Boyle-Mariotte law:

$$PV = \alpha T, \tag{5.2.2}$$

where α is a constant (to be discussed later). The expression "ideal gas" refers in statistical mechanics to gases whose molecules do not interact with each other, but, in thermodynamics, it means gases that satisfy relations like (5.2.2) or (5.7.1), (5.7.2) below.

Here the pressure (which is a force per unit surface) and the volume are "mechanical" quantities, but one may ask how the temperature is defined. For the time being, we will take it as defined by thermometers, but when one looks at, say, a mercury thermometer, what one sees directly is the height of the mercury column, namely a volume and, if the pressure is constant, it means that one uses a relation of the form (5.2.2) to define the temperature. This seems to be circular and to make (5.2.2) a definition rather than a law.

We will see later, after introducing the notion of entropy, that there is another definition of the temperature and that this definition implies laws such as (5.2.2) and also that two bodies in thermal contact (i.e. that can exchange energy) tend to equilibrate their temperatures. So that, if we define the temperature by a relation like (5.2.2) for *one* material (say, and ideal gas, discussed in Sect. 5.7) we can use the equilibration of temperatures through thermal contact to verify (5.2.2) for other materials, for which this relation would then be a law.

Given (5.2.1) one can (if the function f is "nice")[2] solve it and express either P as a function of V and T or V as a function of T and P or T as a function of P and V.

It is often convenient to represent equilibrium states in a two dimensional space $\mathbb{R}^2$ with V on the horizontal axis and P on the vertical one; since both V and P are positive, one considers only the positive quadrant of $\mathbb{R}^2$. Each point on that space corresponds to an equilibrium state and its temperature is fixed by (5.2.1).

It is common to study *reversible* or *quasi-static* transformations in that space, meaning transformations that only go through equilibrium states. This is an idealization of course, but one can think of a piston compressing a gas so slowly that, at each moment, the gas remains (almost) in equilibrium. If the piston was pushed or removed quickly, the gas would not be in equilibrium, and would go back to equilibrium only some time after the end of the move.

A reversible transformation can, as the name indicates, be reversed simply by following the transformation backwards. This notion of thermodynamic reversibility is very different from other notions of reversibility, for example the mechanical one

---

[2] In general, one can rely on the *implicit function theorem*: Let $f : \mathbb{R}^{n+m} \to \mathbb{R}^m$ be a continuously differentiable function, and let $\mathbb{R}^{n+m}$ have coordinates $(\mathbf{x}, \mathbf{y})$. Fix a point $(\mathbf{a}, \mathbf{b})$ with $f(\mathbf{a}, \mathbf{b}) = \mathbf{0} \in \mathbb{R}^m$. If the Jacobian matrix $J_{f,\mathbf{y}}(\mathbf{a}, \mathbf{b}) = \left[\dfrac{\partial f_i}{\partial y_j}(\mathbf{a}, \mathbf{b}) \right], i, j = 1, \ldots, m$, is invertible, then there exists an open set $U \subset \mathbb{R}^n$ containing $\mathbf{a}$ such that there exists a unique continuously differentiable function $g : U \to \mathbb{R}^m$ satisfying $g(\mathbf{a}) = \mathbf{b}$ and $f(\mathbf{x}, g(\mathbf{x})) = \mathbf{0}$ for all $\mathbf{x} \in U$. The function g is the (explicit) function defined implicitly by f. If the function f satisfies these hypotheses in a given open set, we can extend the function g accordingly. But, in many instances, one can solve the implicit function explicitly, for example in the case of ideal gases, see Sect. 5.7.

introduced in Sect. 3.5. Thermodynamic reversibility deals with macroscopic bodies while the mechanical reversibility is a property of particle trajectories.

It is important to note that we will almost always deal with *idealized* situations, which means that the transformations considered here are in reality mathematical transformations in the space of equilibrium states. The notion of isolated system or the ideal gas law are also, by definition, idealizations. To what extend they can be realized in practice is not the most crucial issue. What matters is that the empirical consequences of theoretical reasonings based on those idealizations can be adequately tested.

5.3 The First Law

The first law of thermodynamics states that the total energy of an isolated system is constant. That may sound like a theorem of mechanics (and it is one, see Sect. 3.3.2), but that is not the way it was introduced in thermodynamics (which was developed before the existence of atoms was established and actually is, as a scientific discipline, independent of what matter consists of).

To state the first law in the language of thermodynamics, we must distinguish between *work* and *heat*. This distinction has an anthropomorphic aspect, because work means "work controlled by us, humans".

5.3.1 Work

Let us consider a simple example of work: a piston moves by an amount $\Delta h > 0$. The force exerted by the piston on its surrounding during that move is $F = PS$ where P is the pressure and S the surface of the piston. The work done by that force (which is an energy) is, if we assume the pressure and therefore the force to be constant during the move, $\Delta W = F\Delta h = PS\Delta h$. Since $S\Delta h = \Delta V$, where ΔV is the variation of volume, we can write:

$$\Delta W = P\Delta V. \tag{5.3.1}$$

But, if Δh and therefore ΔV and ΔW are negative, we have a work exerted *on the system* by its surrounding. This means that the energy of the system increases and, whenever we will be interested in the variations of the system's energy, it will be convenient to define ΔW as

$$\Delta W = -P\Delta V. \tag{5.3.2}$$

so that a positive ΔV means a loss of energy (the systems exerts work on its surroundings) and a negative ΔV means a gain of energy coming from the surroundings of the system. Below, we will carefully distinguish between the use of (5.3.1) and of (5.3.2).

It is customary in physics to consider "infinitesimal" or "infinitely small" displacements of the piston, where the notion of "infinitesimals" really goes back to 18th century mathematics, before the formalization of analysis by Cauchy, Weierstrass and others in the 19th century. In Appendix 5.A , we explain, without too many details, the rigorous meaning that one can give to those infinitesimals, using differential forms, but the readers who are confortable with this notion of infinitely small displacements do not have to read it. So, in terms of those infinitely small displacements, (5.3.1) is written as:

$$\delta W = PdV, \tag{5.3.3}$$

where δW is the infinitesimal work done by the system on its surrounding.

The point of writing it in that form is that it can be integrated and, if the volume expands from a volume V_1 to a volume V_2, we have

$$\Delta W = \int_{V_1}^{V_2} PdV \tag{5.3.4}$$

The advantage of (5.3.4) over (5.3.1) is that here we do not need to assume that the pressure is contant during the move and, in general, it is not. Indeed, using formula (5.2.2) and assuming the temperature to be constant during the move, we have:

$$\Delta W = \int_{V_1}^{V_2} PdV = \alpha T \int_{V_1}^{V_2} \frac{dV}{V} = \alpha T \ln \frac{V_2}{V_1}. \tag{5.3.5}$$

This is the simplest example of work, called mechanical. There are of course many other forms of work than this one, due to electric or magnetic forces for example, but the other form of mechanical work that we will consider now is due to steering the fluid or to friction.

5.3.2 Heat

There are other transformations that one can perform on a body, for example heating it up with a flame. Intuitively, this transfers "heat" to the body. But in order to define what it means, we will first consider *adiabatic transformations*, meaning transformations where no heat is exchanged with the surroundings. This is realized (approximately of course) by enclosing the body in adiabatic walls, namely walls that do not allow the transfer of heat, like a Dewar wall (made of two silvered glass sheets separated by an evacuated interspace).

We will accept the empirical fact that one can always pass from a thermodynamic state A of a body to a state B or from B to A by an exchange of mechanical energy between the body and its surroundings, either by compression or dilatation or by steering or friction, but without exchanging heat with the surroundings. Indeed, one method will change the volume and the other the temperature; since, by (5.2.1), those two variables are sufficient to specify the equilibrium state, the combination of those transformations will allow us to connect any two states, at least in one direction (from A to B or from B to A). Since the mechanical energy is a well-defined and measurable quantity, one can define the difference of the total energy between the equilibrium states A and B (if the possible transformation goes from A to B) by:

$$E_A - E_B = \Delta W_{A \to B}. \tag{5.3.6}$$

This means that the energy is a function of the equilibrium state. Note that, in (5.3.6) one has only energy differences. But one may choose arbitrarily a reference state, denoted 0, and define the energy of a state A by:

$$E_A = \Delta W_{0 \to A}.$$

Of course, that definition will depend on the choice of the state 0, but, whether in mechanics (see (3.3.11)) or here, the energy is always defined up to a constant term.

Since other transformations than mechanical ones are possible, like heating, we must see what happens to the energy in those transformations. One *defines* the amount of heat given to the system ΔQ in a general (non necessarily adiabatic) transformation that may involve both work and heat transfers, by:

$$\Delta Q = E_A - E_B - \Delta W_{A \to B}. \tag{5.3.7}$$

We can rewrite (5.3.7) as

$$\Delta E_{A \to B} = E_A - E_B = \Delta Q + \Delta W_{A \to B}, \tag{5.3.8}$$

which can be written in differential form[3]:

$$dE = \delta Q + \delta W = \delta Q - P dV, \tag{5.3.9}$$

where in the last equality we use (5.3.2), since we are interested in the variation of the energy of the system.

Formula (5.3.8), or (5.3.9) is often called the first law; but stated like that, it looks like a definition (of ΔQ) rather than a law. But the real meaning of the first law is that *a total energy E is associated* to each equilibrium state.[4]

---

[3] See Appendix 5.A.

[4] Depending on the independent variables that one chooses to parametrize the equilibrium states, one may have different functions:

The proof of the existence of that function is what we just said: one can always pass from a thermodynamic state A to a state B or from B to A by an exchange of mechanical energy, so that the difference of the energy of state A and state B can be expressed entirely in mechanical terms by (5.3.6).

The reason why one speaks of "the constancy of energy" is that the most general change of energy of a body, given by (5.3.8) is through exchanges of energy with the surroundings of the body. So, if we imagine an isolated box in which our body is enclosed together with a piston and a heater that can act on it, the energy of the body may change, but the total energy *in the box* will remain constant.

It is important to underline the fact that, while the total energy is a function of the equilibrium state of a system, there is no state function called "total work" or "total heat". Work and heat are different ways *to transfer energy* to or out of the system. It makes no sense to ask how much of that total energy is work and how much is heat, unless one specifies the process by which one arrives at a given equilibrium state.

Callen illustrates the thermodynamic definition of the total energy of a system by the following analogy [66, p. 20]: imagine a pond with a certain amount of water. The total amount of water in the pond is analogous to the total energy of a system. The pond can receive or lose water in two ways: it can come in the pond by one stream and go out by another. This is like the mechanical changes of energy due to work. But it can also receive water from rain and lose it from evaporation. This is like the changes of energy due to heat transfers.

It is easy to measure the amount of water coming or leaving the pond through the streams; therefore, one can also measure the difference in the total amount of water between two "states" of the pond due to the streams (the changes in energy that are purely mechanical), if we prevent both rain and evaporation in the pond, for example by putting a tarpaulin over the pond, which is analogous to enclosing the system in adiabatic walls.

If one starts from an empty pond (analogous to our reference state 0 for the energy), one fills it through the incoming stream and one measures the amount of water flowing through it, one then obtains the total amount of water in the pond.

Next, by removing the tarpaulin and measuring the variation in the amount of water in the pond over a certain period, and deducting from that the amount of water flowing in and out of the pond through the streams over that period, one gets the amount of water in the pond due to rain and to evaporation. This is the analogue of (5.3.7).

But it makes no sense, if one knows only the total amount of water in the pond, to ask how much of that water comes from the streams or from rain and evaporation. To

---

$E(P, V)$,

$E(P, T)$,

$E(T, V)$.

One passes from one function to another by using the relation (5.2.1). Suppose we start with $E(P, V)$. Using (5.2.1), we can express say, V as a function of P and T: $V(P, T)$. then one obtains $E(P, T) = E(P, V(P, T))$. One can similarly obtain $E(T, V) = E(P(T, V), V)$, where $P(T, V)$ is again obtained by solving (5.2.1).

answer that question, one has to know *the process* through which the pond arrived at its current water content. Since many different processes can lead to the same content, knowing only that content does not allow us to tell how much water came from the streams or from rain and evaporation. In other words, although the total amount of water is a property of the pond there is no property of the pond alone corresponding to the amount of water "coming from the streams" or "from rain and evaporation".

5.4   The Second Law

A customary way to present the second law is to say that "entropy increases." Actually, in this section, we will not mention the entropy. In the next section, we will use the second law to define an entropy function, a function defined on equilibrium states, which, in that sense, is similar to the energy. But we will also see that, unlike the latter, entropy increases or stays constant for isolated systems.

The first law was based on the empirical fact that one can always pass from one equilibrium state to another or vice-versa by mechanical means. The second law is based on another empirical fact: certain processes passing from one equilibrium state to another cannot be reversed. For example, if you stir a liquid with a spoon, you will heat it up, but how can you use the heat in the liquid to move the spoon?

The first law implies that processes that create energy out of nothing are impossible The second law states that certain processes that may conserve energy and thus do not violate the first law, may nevertheless be impossible (like using the heat of a liquid to move a spoon).

There are, historically, two different statements of the second law, which are actually equivalent, as we will see below (Sect. 5.4.2).

1. **Lord** Kelvin's[5] **version of the second law**: A transformation *whose sole result* would be to change heat into work by using a source at constant temperature is impossible.
2. **Clausius' version of the second law**: A transformation *whose sole result* would be to transfer heat from a low temperature source to a higher temperature one is impossible.

The important words here are those in italics, *whose sole result*. It is only with that condition that the law holds.

One could summarize Lord Kelvin's version of the second law by saying that there does not exist a perfect heat engine (that would convert entirely heat into work using only a source of heat at a constant temperature) and Clausius' version of the second law by saying that there does not exist a perfect refrigerator (that would take heat at a low temperature, inside the refrigerator, and release it at a higher temperature, outside the refrigerator).

---

[5] Before he became a Lord, Kelvin's name was William Thomson.

Fig. 5.1 A Carnot cycle.
Source https://upload.
wikimedia.org/wikipedia/
commons/4/4e/Carnot-
cycle-p-V-diagram.svg
Derkleinebauer, Public
domain, via Wikimedia
Commons

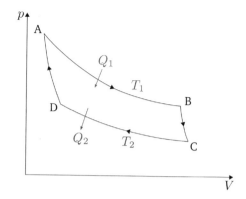

In order to prove the equivalence between those two versions, we need an important concept, the Carnot cycle:

5.4.1 The Carnot Cycle

We will present the Carnot cycle "abstractly" as a (piecewise smooth) curve in the set
of equilibrium states, say in the $P - V$ plane, and not directly as a concrete process,
although there exist of course approximations of the Carnot cycle by real physical
processes.

A Carnot cycle is the combination of four transformations, two transformations at
constant temperatures T_1 $(A \rightarrow B)$ and T_2 $(C \rightarrow D)$, with $T_1 > T_2$, and two adiabatic
transformations $(B \rightarrow C$ and $D \rightarrow A)$, see Fig. 5.1; here ΔW denotes the work done
by the system on its surroundings (which means that, if $\Delta W < 0$ that work is done
on the system).

1. During the transformation $A \rightarrow B$, the system absorbs a certain amount of heat
   $Q_1 > 0$ at a constant temperature T_1. Its volume increases and its pressure
   decreases. The system performs work on its surroundings: $\Delta W_{A \rightarrow B} = \int_A^B P dV$,
   where the integral is positive, since the volume increases.
2. During the transformation $B \rightarrow C$, the system continues to increase its volume
   and decrease its pressure, but is thermally isolated so does not receive or lose heat.
   The system performs work on its surroundings: $\Delta W_{B \rightarrow C} = \int_B^C P dV$, where the
   integral is positive, since the volume increases.
3. During the transformation $C \rightarrow D$, the system gives a certain amount of heat
   $Q_2 > 0$ at a constant temperature T_2. Its volume decreases and its pressure
   increases. Work is performed on the system: $\Delta W_{C \rightarrow D} = \int_C^D P dV$, where the
   integral is negative, since the volume decreases.
4. During the transformation $D \rightarrow A$, the system continues to decrease its volume
   and to increase its pressure, but is thermally isolated so does not receive or lose

heat. Work is performed on the system: $\Delta W_{D \to A} = \int_D^A P \, dV$, where the integral is negative, since the volume decreases.

The important property of the Carnot cycle is that it is a *cycle*, namely one ends up in the same equilibrium state that one started with (this property will allow us to show the impossibility of the two transformations having the *sole result* mentioned in Kelvin's and Clausius statements of the second law), and that it is *reversible*, namely one could run the steps 1 to 4 in the opposite order. From (5.3.8), we get:

$$\Delta E_{A \to A} = 0 = \Delta Q - \Delta W,$$

where we have $-\Delta W$, as opposed to (5.3.8), since ΔW denotes here the work done by the system (whose energy thus decreases). Since $\Delta Q = Q_1 - Q_2$, we get:

$$\Delta W = Q_1 - Q_2 > 0, \tag{5.4.1}$$

since the total amount of work performed $\Delta W > 0$ (ΔW equals the area enclosed in the region $ABCD$ in Fig. 5.1).

If we ran the circle in reverse, we would have:

$$\Delta \tilde{W} = Q_2 - Q_1 = -\Delta W,$$

which is negative, which means that we need work in order to transfer heath from a cold source to a hotter one.

We may define the *efficiency* η of a cycle by the ratio between the amount of work produced and the amount of heat initially given:

$$\eta = \frac{\Delta W}{Q_1} = 1 - \frac{Q_2}{Q_1} \tag{5.4.2}$$

A Carnot cycle is an example of a heat engine: it takes an amount of heat Q_1, at a high temperature T_1 and transforms part of it into work; but it needs a reservoir at a lower temperature T_2 in which it releases an amount of heat Q_2, which is "lost". So, it is compatible with Lord Kelvin's version of the second law (and also with Clausius' statement since it does not deal solely with transfers of heat).

A reversed Carnot cycle is an example of a refrigerator: it take amount of heat Q_2 at a low temperature T_2, inside the refrigerator, and releases it at a higher temperature T_1, outside the refrigerator. But in order to do that, it needs a certain amount of work equal to $Q_1 - Q_2$ (usually electric work used by a cooling engine inside the refrigerator) in order to function and that work is "lost" in the higher amount of heat $Q_1 > Q_2$ released at temperature T_1.

An important property of the Carnot cycle is that, since it is reversible, it is also optimal: one cannot build a better heat engine, namely one that would work between the two heat reservoirs at temperatures T_1 and T_2 and be more efficient than a Carnot engine operating between those same reservoirs (assuming of course the validity of the second law). Indeed, suppose that another engine absorbs an amount of heat Q_1, at a high temperature T_1 and releases an amount of heat Q_2' in a reservoir at a lower temperature T_2, but so that the work performed $\Delta W' = Q_1 - Q_2' > \Delta W$, with ΔW defined in (5.4.1), which means $Q_2' < Q_2$ (if the machine is more efficient it must release less heat at the lower temperature for a given amount of heat received at the higher temperature than the Carnot machine).

Now, let us run a reversed Carnot cycle (a refrigerator), that takes an amount of heat Q_2 in the reservoir at the lower temperature T_2 and uses an amount of work ΔW to release an amount of heat Q_1 in the reservoir at the higher temperature T_1. The net result of the combined cycles, the supposedly more efficient cycle + the reversed Carnot cycle is that the amount of heat taken from the reservoir at the higher temperature T_1 and given back to it are equal. So the net effect of the combined cycle would be to convert an amount of heat $Q_2 - Q_2'$ taken from the reservoir at temperature T_2 and convert it into an amount $\Delta W' - \Delta W$ of work. But this contradicts Kelvin's version of the second law, hence it is impossible.

Note that it is only the *reversible character* of the Carnot cycle that makes it more efficient than any other machine; hence, all reversible cycles have the same efficiency η defined in (5.4.2).

We will now use an ideal gas in our Carnot cycle and use properties of such a gas in order to derive the important relation (5.4.7) below, which holds for all Carnot cycles, since they all have the same efficiency.

We will first accept that, in a adiabatic transformation for an ideal gas, the following relation holds (this will be proven in Sect. 5.7.4):

$$TV^{\gamma-1} = \text{Constant}, \tag{5.4.3}$$

where $\gamma > 1$ is a constant depending on the gas.

Combining (5.4.3) with (5.2.2), i.e. $T \approx PV$, we get:

$$PV^{\gamma} = \text{Constant},$$

Since $\gamma > 1$, we can understand why the adiabatic curves are more steep that the isothermal ones where, by (5.2.2) $PV = \text{Constant}$. We will also accept the fact that, for an ideal gas, the energy is a function of the temperature and not of the volume (see (5.7.2) below).

Equation (5.4.3) allows us to compute an important relation between Q_1, Q_2 and T_1, T_2 in a Carnot cycle. Indeed, for the part $A \to B$ of the cycle, we have $\Delta E_{A \to B} = 0$, since the energy depends only on the temperature and the temperature is constant along that part of the cycle. So, we have $Q_1 = \Delta W_{A \to B}$, where $\Delta W_{A \to B}$ is the work done by the system and is positive, using (5.3.5) (since the temperature is constant), we get

$$Q_1 = \Delta W_{A \to B} = \alpha T_1 \ln \frac{V_B}{V_A} > 0 \tag{5.4.4}$$

Similarly, for the part $C \to D$ of the cycle, which is also at constant temperature, we get:

$$Q_2 = -\Delta W_{C \to D} = \alpha T_2 \ln \frac{V_C}{V_D} > 0, \tag{5.4.5}$$

because $\Delta W_{C \to D} = \int_C^D P dV$ is work done on the system and is negative. For the parts $B \to C$ and $D \to A$ of the cycle, since they are adiabatic, and since $T_A = T_B = T_1, T_C = T_D = T_2$, we get from (5.4.3):

$$T_1 V_B^{\gamma-1} = T_2 V_C^{\gamma-1}$$

and

$$T_1 V_A^{\gamma-1} = T_2 V_D^{\gamma-1}.$$

This implies:

$$\frac{V_B^{\gamma-1}}{V_A^{\gamma-1}} = \frac{V_C^{\gamma-1}}{V_D^{\gamma-1}},$$

or

$$\frac{V_B}{V_A} = \frac{V_C}{V_D}, \tag{5.4.6}$$

Combining (5.4.6) with (5.4.4), (5.4.5), we obtain the important relation:

$$\frac{Q_2}{Q_1} = \frac{T_2}{T_1}, \tag{5.4.7}$$

which implies that the efficiency defined in (5.4.2) $\eta = 1 - \frac{Q_2}{Q_1} = 1 - \frac{T_2}{T_1}$. Note that, since Carnot's engine has maximum efficiency, for any other transformation, with quantities of heat exchanged Q_1', Q_2', we must have $\eta' == 1 - \frac{Q_2'}{Q_1'} \leq \eta = 1 - \frac{T_2}{T_1}$, which implies:

$$\frac{Q_1'}{T_1} - \frac{Q_2'}{T_2} \leq 0. \tag{5.4.8}$$

5.4.2 Proof of the Equivalence Between Lord Kelvin's Version and Clausius' Version of the Second Law

To prove that the two statements are equivalent, we will use the elementary logical theorem: for two propositions P and Q, "P implies Q" is equivalent to "non Q implies non P".

1. **If Lord Kelvin's version is not true, then Clausius' version is not true.** If one can change heat into work at a fixed temperature T_2, then one can use this work to heat up (by friction or stirring) a body at any given temperature T_1. Now, choose $T_1 > T_2$; So, Clausius' version implies Lord Kelvin's version.
2. **If Clausius' version is not true, then Lord Kelvin's version is not true.** Suppose that one could transfer a quantity of heat Q from a source at a low temperature T_2 to a body at a higher temperature T_1. Then, via a Carnot cycle, one can use this amount of heat to produce a certain amount of work; the source at temperature T_2 gives and then receives the same amount of heat Q; so, the sole result of the operation would be to change heat into work by using a source at a constant temperature T_2. So, Lord Kelvin's version implies Clausius' version.

5.5 The Thermodynamic Entropy

5.5.1 Definition of the Entropy Function

We saw in (5.4.7) that, in a Carnot cycle, $\frac{Q_2}{Q_1} = \frac{T_2}{T_1}$, which can be written as $\frac{Q_1}{T_1} - \frac{Q_2}{T_2} = 0$ or[6]

$$\int_{\text{Carnot cycle}} \frac{\delta Q}{T} = 0, \qquad\qquad (5.5.1)$$

since:

1. On the part $A \to B$ of the cycle $\int_{A \to B} \frac{\delta Q}{T} = \frac{Q_1}{T_1}$, since $T = T_1$ is constant on that part of the cycle and the system absorbs a quantity of heat Q_1.
2. On the part $C \to D$ of the cycle $\int_{C \to D} \frac{\delta Q}{T} = -\frac{Q_2}{T_2}$, since $T = T_2$ is constant and the system gives to the source a quantity of heat Q_2.
3. On the parts $B \to C$ and $D \to A$ of the cycle we have $\int_{B \to C} \frac{\delta Q}{T} = \int_{D \to A} \frac{\delta Q}{T} = 0$, since on those parts the transformation is adiabatic, i.e. there is no exchange of heat.

This proves (5.5.1).

---

[6] See (5.A.3) below for a precise definition of such integrals.

Now consider any cycle C in the $P - V$ plane. For any $\epsilon > 0$, we can fill the interior of that cycle with a finite number of small Carnot cycles $C_1, C_2, \ldots, C_n$, adjacent to each other, so that:

$$\left| \int_C \frac{\delta Q}{T} - \sum_{i=1}^{n} \int_{C_i} \frac{\delta Q}{T} \right| \leq \epsilon. \tag{5.5.2}$$

Indeed, the integral over a part of a cycle in one direction equals minus the integral going over the same part of that cycle, but in another direction; thus, the integral over $\int_{C_i} \frac{\delta Q}{T}$ for one small cycle C_i is cancelled, in the sum (5.5.2), by the contributions of the integrals over the cycles adjacent to C_i. Thus, in the sum $\sum_{i=1}^{n} \int_{C_i} \frac{\delta Q}{T}$ we are left with the integral over the boundary of the union of the interiors of those cycles, which can be made arbitrarily close to the integral over C (We assume here that the differential form $\frac{\delta Q}{T}$ is continuous).

Since, $\forall i = 1, 2, \ldots, n$, $\int_{C_i} \frac{\delta Q}{T} = 0$ by (5.5.1) and since ϵ is arbitrary, we get, for any cycle C:

$$\int_C \frac{\delta Q}{T} = 0 \tag{5.5.3}$$

But that means (see 5.A) that the differential form $\frac{\delta Q}{T}$ is *exact*, i.e. that there exists a function of the equilibrium state, that satisfies:

$$dS = \frac{\delta Q}{T}. \tag{5.5.4}$$

That function is the *thermodynamic entropy*, and usually denoted by the letter S.

It will be useful to note that if we have any cycle $\tilde{C}$, not necessarily reversible, i.e. possibly going through non-equilibrium states, we can deduce from (5.4.8)

$$\int_{\tilde{C}} \frac{\delta Q}{T} \leq 0, \tag{5.5.5}$$

in the same way that we deduced (5.5.3) from (5.4.7).

To summarize, the first law says that there exists a total energy function of the equilibrium state and the second law implies that there exists another function of the equilibrium state, the entropy.

5.5.2 The Second Law or the Increase of the Thermodynamic Entropy

But we also obtain the following version of the second law:

The law of increase of the thermodynamic entropy.

Consider two equilibrium state A and B From (5.5.4), we get:

$$S(B) - S(A) = \int_\gamma \frac{\delta Q}{T} \qquad (5.5.6)$$

where the integral is taken over a path γ in the space of parameters of equilibrium states, namely a reversible path going from A to B.

Now consider any transformation or path γ' from A and B, which is not necessarily reversible, i.e. that may pass through non-equilibrium states, we want to show that:

$$\int_{\gamma'} \frac{\delta Q}{T} \leq S(B) - S(A) \qquad (5.5.7)$$

Indeed, we know, from (5.5.5) that $\int_{\tilde{C}} \frac{\delta Q}{T} \leq 0$ for any cycle, reversible or not. Now consider a cycle $\tilde{C}$ composed of an arbitrary path γ' going from A and B and a reversible path γ'' going from B and A, which is the reversed of the path γ in (5.5.6). We have $\int_{\tilde{C}} \frac{\delta Q}{T} = \int_{\gamma'} \frac{\delta Q}{T} + \int_{\gamma''} \frac{\delta Q}{T} \leq 0$. But since γ'' is the reversed of the path γ, we have, from (5.5.6), that $\int_{\gamma''} \frac{\delta Q}{T} = S(A) - S(B)$, which proves (5.5.7).

This establishes the following:

Proposition 5.1 (Second law of thermodynamics formulated in terms of entropy) *For any transformation in an isolated system going from an equilibrium state A to another equilibrium state B, the thermodynamic entropy defined by (5.5.4) cannot decrease.*

There are many examples of processes that will increase the entropy: the free expansion of a gas in a box, gas which is initially constrained by a wall to be in a subset of the box, the mixing of two fluids or putting into contact a hot and a cold body. In the latter case, if T_1 is the hot temperature and T_2 the colder one, a small amount of heat δQ will flow from the hot to the cold body; the hot body decreases its entropy by an amount $\frac{\delta Q}{T_1}$ while the cold body increases its entropy by an amount $\frac{\delta Q}{T_2}$ (we assume that the temperatures are constant during this exchange, because δQ is small). Since $T_1 > T_2$, $\frac{\delta Q}{T_2} > \frac{\delta Q}{T_1}$ and the total entropy of the system increases.

On the part $A \to B$ of a Carnot cycle, the entropy of the system increases by an amount $\int_{A \to B} \frac{\delta Q}{T} = \frac{Q_1}{T_1}$, while on the part $C \to D$ of a Carnot cycle, the entropy of the system decreases by an amount $\int_{C \to D} \frac{\delta Q}{T} = \frac{Q_2}{T_2}$, while on the parts $B \to C$ and $D \to A$ of a Carnot cycle, the entropy does not change and of course, it does not change in the entire Carnot cycle either by (5.5.1).

5.6 Other Thermodynamic Functions

5.6.1 The Particle Number

So far, we have done as if the total amount of material in a given body was fixed. But one may relax that contraint and consider that amount as a new variable. If we "accept" the existence of atoms (which of course we do), we may simply introduce the variable N, the total number of atoms or molecules in the body.

5.6.2 Fundamental Relations

So far, we have introduced six quantities: the temperature, the volume and the pressure that are directly measurable and the total energy, the entropy and the number of atoms or molecules that are measured indirectly.

We will distinguish now between extensive and intensive variables. A variable is *extensive* if it is proportional to the number of particles N; intuitively, both the energy and the volume are extensive: two liters of waters will contain twice as much energy as one liter of water under the same conditions. Entropy is also an extensive function.

Formally, if $\Psi(E, V, N)$ is an extensive function, we have:

$$\Psi(\lambda E, \lambda V, \lambda N) = \lambda \Psi(E, V, N), \tag{5.6.1}$$

for $\lambda > 0$.

A variable is *intensive* if it is independent of the number of particles N: the temperature and the pressure are intensive: two liters of waters at the same temperature will still have the same temperature when they are mixed together. And the same holds for the pressure.

We will now see that all the thermodynamic information about a system (including relations like (5.2.2) and (5.4.3)) is contained in its *fundamental relation*, namely in the concrete form of the function $S(E, V, N)$ or $E(S, V, N)$. The existence of the function $S(E, V, N)$ follows from (5.5.4) and the function $E(S, V, N)$ follows by inverting $S(E, V, N)$.

Using (5.3.9) and (5.5.4), we get (with N fixed here; we will consider N as a variable in Sect. 5.6.4):

$$dE = TdS - PdV, \tag{5.6.2}$$

which means:

$$\frac{\partial E(S, V)}{\partial S} = T(S, V),$$

$$\frac{\partial E(S, V)}{\partial V} = -P(S, V), \tag{5.6.3}$$

where $T = T(S, V)$, $P = P(S, V)$ are functions of the extensive variables. We can invert (5.6.2), to get $S = S(E, V)$ and, in differential form:

$$dS = \frac{1}{T}dE + \frac{P}{T}dV, \tag{5.6.4}$$

which means:

$$\frac{\partial S(E, V)}{\partial E} = \frac{1}{T(E, V)},$$

$$\frac{\partial S(E, V)}{\partial V} = \frac{P(E, V)}{T(E, V)}. \tag{5.6.5}$$

It is sometimes convenient to replace some extensive variables by intensive ones.

5.6.3   The Helmholtz Free Energy

We can define a new fundamental relation, by taking the Legendre transform of $E = E(S, V)$ (see Appendix 5.B for the definition and the mathematical properties of Legendre transforms) with respect to the variable S:

$$F(T, V) = E(T, V) - TS(T, V), \tag{5.6.6}$$

where we use

$$\frac{\partial E(S, V)}{\partial S} = T(S, V) \tag{5.6.7}$$

and invert that relation in order to obtain S as a function of T, V, $S = S(T, V)$, and then obtain $E = E(T, V) = E(S(T, V), V)$. One inserts $S = S(T, V)$, and $E = E(T, V)$ in (5.6.6) to get $F(T, V)$.

One could also start from the function $S = S(E, V)$ obtained from the inversion of (5.6.2), and use

$$\frac{\partial S(E, V)}{\partial E} = \frac{1}{T(S, V)} \tag{5.6.8}$$

in order to obtain E as a function of T, V, $E = E(T, V)$, and $S(T, V) = S(E(T, V), V)$ and then insert those functions in (5.6.6) to get $F(T, V)$.

In differential form, we get from (5.6.6):

$$dF(T, V) = -SdT - PdV, \tag{5.6.9}$$

by using (5.6.2) to write: $dF = dE - SdT - TdS = TdS - PdV - SdT - TdS$, i.e. $dF = -SdT - PdV$.

Relation (5.6.6) can be inverted:

$$E(S, V) = F + TS, \tag{5.6.10}$$

using

$$\frac{\partial F(T, V)}{\partial T} = -S(T, V), \tag{5.6.11}$$

in order to express T as a function of S, V, i.e. $T = T(S, V)$, and then $F = F(S, V) = F(T(S, V), V)$.

Since we can obtain the fundamental relation $E(S, V)$ from $F(T, V)$ thanks to (5.6.10), giving an explicit form for the function $F = F(T, V)$ is just another form of a fundamental relation.

For the sake of completeness, we mention some other Legendre transformations that we will not use later: The *enthalpy* is the Legendre transform $E = E(S, V)$ with respect to the pressure:

$$H(S, P) = E + PV, \tag{5.6.12}$$

where we use the second equality in (5.6.3)

$$\frac{\partial E(S, V)}{\partial V} = -P(S, V)$$

in order to express $V = V(S, P)$ and $E = E(S, P) = E(S, V(S, P))$.

One can also do a Legendre transform with respect to both S and V, which yields the *Gibbs potential* or *Gibbs free energy*:

$$G(T, P) = E - TS + PV \tag{5.6.13}$$

where we use (5.6.3)

$$\frac{\partial E(S, V)}{\partial S} = T(S, V)$$

and

$$\frac{\partial E(S, V)}{\partial V} = -P(S, V)$$

in order to express both S and V as functions of T and P, and then write $E(T, P) = E(S(T, P), V(T, P))$.

An explicit form for either $H(S, P)$ or $G(T, P)$ constitutes a fundamental relation, since one can recover $E(S, V)$ by inverting (5.6.12) or (5.6.13).

5.6.4 The Grand Potential

If we introduce the numbers of particles in a given body, N, or, if there are l types of particles, we introduce their respective numbers $N_1, \ldots, N_l$, we can express the fundamental relation as:

$$E = E(S, V, N_1, \ldots, N_l) \tag{5.6.14}$$

We must add to (5.6.3), the relations

$$\frac{\partial E}{\partial N_i} = \mu_i(S, V, N_1, \ldots, N_l), \tag{5.6.15}$$

where μ_i is, by definition, *the chemical potential* of the particles of index i. With these additional variables, (5.6.3) expresses T and V as functions of those variables: $T = T(S, V, N_1, \ldots, N_l)$ and $P = P(S, V, N_1, \ldots, N_l)$.

The intuitive meaning of (5.6.15) is that μ_i is the change of energy when one adds one particle of type i to the system (formally N_i is treated here as a continuous variable, which is justified if one considers a large number of particles).

This leads to the generalization of (5.6.2):

$$dE = TdS - PdV + \sum_{i=1}^{l} \mu_i dN_i, \tag{5.6.16}$$

where the variables are $S, V, N_1, \ldots, N_l$. And it leads also to the generalization of (5.6.4):

$$dS = \frac{1}{T}dE + \frac{P}{T}dV - \sum_{i=1}^{l} \frac{\mu_i}{T}dN_i, \tag{5.6.17}$$

where the variables are $E, V, N_1, \ldots, N_l$.

We also have the generalization of (5.6.9):

$$dF = dE - TdS - SdT = -SdT - PdV + \sum_{i=1}^{l} \mu_i dN_i, \tag{5.6.18}$$

where the variables are $T, V, N_1, \ldots, N_l$, which implies:

$$\frac{\partial F}{\partial N_i} = \mu_i(T, V, N_1, \ldots, N_l). \tag{5.6.19}$$

Note that, if $l = 1$,

$$\frac{\partial F}{\partial N} = \mu(T, V, N). \tag{5.6.20}$$

One can perform a new Legendre transform, this time with respect to the N_i variables:

$$\Phi(T, V, \mu_1, \ldots, \mu_m) = F - \sum_{i=1}^{l} \mu_i N_i = E - TS - \sum_{i=1}^{l} \mu_i N_i \tag{5.6.21}$$

where the variables $\{N_i\}_{i=1}^{l}$ are expressed in terms of $T, V, \mu_1, \ldots, \mu_m$ through (5.6.19) and $F(T, V, N_1, \ldots, N_l)$ is then expressed in terms of $T, V, \mu_1, \ldots, \mu_m$. Φ is called *the grand potential*.

In differential form, we have (exercise: check that!):

$$\begin{aligned} d\Phi(T, V, \mu_1, \ldots, \mu_m) = \\ - S(T, V, \mu_1, \ldots, \mu_m)dT - P(T, V, \mu_1, \ldots, \mu_m)dV \\ - \sum_{i=1}^{l} N_i(T, V, \mu_1, \ldots, \mu_m)d\mu_i. \end{aligned} \tag{5.6.22}$$

As before, (5.6.21) can be inverted:

$$E(S, V, N_1, \ldots, N_l) = \Phi + TS + \sum_{i=1}^{l} \mu_i N_i, \tag{5.6.23}$$

using (5.6.22), which implies:

$$\frac{\partial \Phi(T, V, \mu_1, \ldots, \mu_m)}{\partial T} = -S(T, V, \mu_1, \ldots, \mu_m) \tag{5.6.24}$$

and

$$\frac{\partial \Phi(T, V, \mu_1, \ldots, \mu_m)}{\partial \mu_i} = -N_i(T, V, \mu_1, \ldots, \mu_m)$$

which allows us to express $T = T(S, V, N_1, \ldots, N_l)$ and $\mu_i = \mu_i(S, V, N_1, \ldots, N_l)$ in terms of the variables $S, V, N_1, \ldots, N_l$. One then writes:

$$\Phi(T, V, \mu_1, \ldots, \mu_m) = \Phi(T(S, V, N_1, \ldots, N_l), V, \mu_1(S, V, N_1, \ldots, N_l), \ldots, \mu_m(S, V, N_1, \ldots, N_l)),$$

and insert this and $T = T(S, V, N_1, \ldots, N_l)$, $\mu_i = \mu_i(S, V, N_1, \ldots, N_l)$ in the right hand side of (5.6.23). We can obtain similarly $F(T, V, N_1, \ldots, N_l)$. Since we can recover $E(S, V, N_1, \ldots, N_l)$ and $F(T, V, N_1, \ldots, N_l)$ from an explicit knowledge of the function $\Phi(T, V, \mu_1, \ldots, \mu_m)$, that function is also a fundamental relation.

5.6.5 The Principle of Maximum Entropy and Equilibrium Conditions

Proposition 5.1 suggests (but does not prove) the following:

Principle of Maximum Entropy. Thermodynamic equilibrium states are characterized by values of the variables that maximizes the entropy function of the system, given the external constraints imposed on the latter.

If one accepts this principle, one can derive several easy consequences. First of all, this, combined with our experience that equilibrium states are homogeneous, implies that the entropy function is concave. Indeed, suppose there was a range of parameters on which it is strictly convex, and consider for simplicity only the energy variable; then, there would exist energies E_1 and E_2, and a λ, with $0 < \lambda < 1$, so that

$$S(\lambda E_1 + (1 - \lambda) E_2) < \lambda S(E_1) + (1 - \lambda) S(E_2) = S(\lambda E_1) + S((1 - \lambda) E_2),$$

$$(5.6.25)$$

where in the last equality we use the fact that the entropy is an extensive function, see (5.6.1). This is illustrated in Fig. 5.2.

The inequality (5.6.25) implies that a system with energy $\lambda E_1 + (1 - \lambda) E_2$ would have a higher entropy by separating itself into two subsystems of energies λE_1 and $(1 - \lambda) E_2$ than by staying homogeneous. But this is contrary to what one observes in equilibrium, so that the inequality opposite to the one in (5.6.25) must hold, $\forall E_1, E_2, \lambda$, which means that S is a concave function.

This means also that the function S has a unique maximum value (which may correspond to an interval of its parameters if it is not strictly concave).

From this and the maximum entropy principle, we can derive several conditions on the intensive parameters of systems in equilibrium.

Consider a system composed of two subsystems 1 and 2, with energies E_1 and E_2, and separated by a adiabatic wall. The system is isolated so that the total energy $E = E_1 + E_2$ is conserved. What will be the energies of both subsystems if we allow the wall to exchange energy? To answer that question, we must maximize $S(E_1, E_2)$ under the constraint $E = E_1 + E_2$, keeping the volumes V_1, V_2, and the number of particles N_1, N_2 of each subsystem fixed. Writing $E_2 = E - E_1$ and setting the derivative with respect to E_1 in $S(E_1, E - E_1)$ to zero, we get:

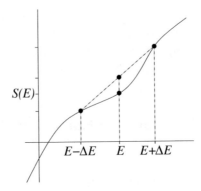

Fig. 5.2 If the graph of the entropy $S(E)$ has a part where it is not concave, as in this figure, then the maximum of the entropy would occur on the straight line in the figure, and not on the graph of $S(E)$. This would correspond to a separation of the system in two different parts, one with energy $E_1 = E - \Delta E$, and the other with energy $E_2 = E + \Delta E$. The entropy of system would be $\frac{1}{2}S(E - \Delta E) + \frac{1}{2}S(E + \Delta E) > S(E)$. This is contrary to what we observe, namely that systems in equilibrium are spatially homogeneous. Figure taken from Tumulka's lectures [314, Fig. 6]

$$\frac{\partial S(E_1, E - E_1)}{\partial E_1} - \frac{\partial S(E_1, E - E_1)}{\partial E_2} = 0, \tag{5.6.26}$$

or, using (5.6.5),

$$\frac{1}{T_1} - \frac{1}{T_2} = 0, \tag{5.6.27}$$

which means that the equilibrium value of E_1, $E_2 = E - E_1$ can be computed if we know the explicit form of the function $S(E_1, E_2, V, N)$ through (5.6.26).

We also see that:

$$dS = \left(\frac{1}{T_1} - \frac{1}{T_2}\right) dE_1, \tag{5.6.28}$$

and if $T_1 > T_2, dS > 0$ means $dE_1 < 0$. In other words, the energy of the warmer part decreases, or the heat flows from hot to cold, which is of course what we experience when S increases, i.e. in any spontaneous process.

Similarly, we may ask what will be the volumes V_1 and V_2 occupied by the two subsystems (with $V = V_1 + V_2$), if the separation between them is a movable piston and the thermal equilibrium has taken place ($T_1 = T_2$); keeping the number of particles N_1, N_2 of each subsystem fixed, we get from (5.6.5):

$$\frac{\partial S(V_1, V - V_1)}{\partial V_1} - \frac{\partial S(V_1, V - V_1)}{\partial V_2} = \frac{P_1}{T_1} - \frac{P_2}{T_2} = 0, \tag{5.6.29}$$

i.e., since $T_1 = T_2$, $P_1 = P_2$, which again corresponds to our experience.

Finally, we can determine the number of particles N_1 and N_2 in equilibrium if the separation between the two subsystems allows the exchange of particles and the thermal equilibrium has taken place ($T_1 = T_2$); use (5.6.17):

$$\frac{\partial S(N_1, N - N_1)}{\partial N_1} - \frac{\partial S(N_1, N - N_1)}{\partial N_2} = -\frac{\mu_1}{T_1} + \frac{\mu_2}{T_2} = 0, \tag{5.6.30}$$

i.e. $\mu_1 = \mu_2$, which does not really correspond to our intuitive experience, but gives a physical meaning to the parameter μ.

Reasoning as in (5.6.28), one can show that a piston will move in the direction of the region with the lowest pressure and that particles will flow towards the low μ region.

One may think of the temperature difference as a sort of "force" that pushes energy towards the low temperature region, of the pressure as a "force" expanding the volume towards the low pressure region and of the chemical potential as a "force" pushing the particles towards the low μ region. Of these three examples, only the pressure is a mechanical force (per unit area); for the others, the expression is used by analogy.

Actually, there is an interesting difference between extensive and intensive variables: the first ones, energy, volume and number of particles (or amount of matter) are not directly perceived by us but can be directly measured; the intensive variables, temperature, pressure and chemical potential on the other hand are measured indirectly but are directly perceived: we feel the heat or the pressure and even the chemical potential through our sense of taste, because the chemical potentials are in relation to the concentrations of different substances like the pressure is to the volume and the temperature to the energy.

5.7 An Example: The Ideal Gas

So, far we have not given any concrete example of what a fundamental relation may look like. Later we will explain how to compute it from first (i.e. microscopic) principles, but we would also like to have an example based on purely macroscopic considerations (otherwise, what would thermodynamic be good for?).

We will show that, starting from purely phenomenological laws, one can derive the fundamental relation in the case of ideal gases.

Let us start with the two empirical relations valid for ideal gases:

$$PV = NkT \tag{5.7.1}$$

which is an extension of (5.2.2) when the variable N, the number of particles in the gas, is introduced, and k is Boltzmann's constant (see Appendix 5.C). We also have

$$E = cNkT \tag{5.7.2}$$

where c is a constant depending on the gas. For monoatomic gas, we have $c = \frac{3}{2}$ (so that this is the value that we will use in Sect. 6.5), but for other gases we can have $c = \frac{5}{2}$ or $c = \frac{7}{2}$.[7]

In order to obtain the fundamental equation from (5.7.1), (5.7.2), we will need:

5.7.1 Mathematical Identities and the Gibbs–Duhem Relations

Let us start with the following identity:

Euler's relation for homogeneous functions of several variables.

Let $f : \mathbb{R}^n \to \mathbb{R}$ be a C^1 homogeneous function of degree one:

$$f(\lambda x_1, \ldots, \lambda x_n) = \lambda f(x_1, \ldots, x_n) \tag{5.7.3}$$

then differentiating with respect to λ and setting $\lambda = 1$, we get:

$$f(x_1, \ldots, x_n) = \sum_{i=1}^{n} x_i \frac{\partial f(x_1, \ldots, x_n)}{\partial x_i}. \tag{5.7.4}$$

Let us apply this to (5.6.14), where $E, S, V, N_1, \ldots, N_l$ are intensive functions, which means that $E(\lambda S, \lambda V, \lambda N_1, \ldots, \lambda N_l) = \lambda E(S, V, N_1, \ldots, N_l)$ (which is an extension of (5.6.1)). Using (5.6.3), (5.6.15), we get:

$$E(S, V, N_1, \ldots, N_l) = TS - PV + \sum_{i=1}^{l} \mu_i N_i. \tag{5.7.5}$$

taking the differential of (5.7.5), we obtain:

$$dE = TdS + SdT - PdV - VdP + \sum_{i=1}^{l} \mu_i dN_i + \sum_{i=1}^{l} N_i d\mu_i. \tag{5.7.6}$$

Subtracting from that the relation (5.6.16), we have:

$$SdT - VdP + \sum_{i=1}^{l} N_i d\mu_i = 0 \tag{5.7.7}$$

---

[7] This comes from the inclusion of vibrational or rotational modes of a molecule, see e.g. Amit and Verbin [12, Part IV, Chap. 1].

which is known as the *Gibbs–Duhem relation*.

We can proceed in a similar way with the homogenous function $S = S(E, V, N_1, \ldots, N_l)$ and obtain:

$$S(E, V, N_1, \ldots, N_l) = \frac{E}{T} + \frac{P}{T} V - \sum_{i=1}^{l} \frac{\mu_i}{T} N_i. \tag{5.7.8}$$

and, by differentiating and using (5.6.17), we get:

$$E \, d\left(\frac{1}{T}\right) + V \, d\left(\frac{P}{T}\right) - \sum_{i=1}^{l} N_i \, d\left(\frac{\mu_i}{T}\right) = 0, \tag{5.7.9}$$

another Gibbs–Duhem relation.

5.7.2 Derivation of the Fundamental Relation for an Ideal Gas

Let $e = \frac{E}{N}$, $v - \frac{V}{N}$, $s = \frac{S}{N}$, be the energy, volume and entropy per particle, and consider for simplicity, the case where the number of types of particles $l = 1$. We can rewrite (5.7.1), (5.7.2) as

$$\frac{P}{T} = k\frac{N}{V} = \frac{k}{v} \tag{5.7.10}$$

and

$$\frac{1}{T} = ck\frac{N}{E} = \frac{ck}{e}. \tag{5.7.11}$$

Let us write the Gibbs–Duhem relation (5.7.9), after dividing both sides by N, as:

$$d\left(\frac{\mu}{T}\right) = e \, d\left(\frac{1}{T}\right) + v \, d\left(\frac{P}{T}\right). \tag{5.7.12}$$

Now, inserting (5.7.10), (5.7.11) into (5.7.12); we get:

$$d\left(\frac{\mu}{T}\right) = -e\frac{ck}{e^2}de - v\frac{k}{v^2}dv = -ck\frac{de}{e} - k\frac{dv}{v},$$

which can be integrated as:

$$\frac{\mu}{T} - (\frac{\mu}{T})_0 = -ck \ln \frac{e}{e_0} - k \ln \frac{v}{v_0}, \tag{5.7.13}$$

where the index 0 refers to some (arbitrary) initial state.[8]

Now, inserting (5.7.10), (5.7.11) and (5.7.13) into (5.7.8), we get:

$$s(e, v) = \frac{ck}{e}e + \frac{k}{v}v - \frac{\mu}{T} = (c + 1)k - \left(\frac{\mu}{T}\right)_0 + ck \ln \frac{e}{e_0} + k \ln \frac{v}{v_0}, \quad (5.7.14)$$

or,

$$s(e, v) = (c + 1)k + k \ln e^c v - \left(\frac{\mu}{T}\right)_0 - k \ln e_0^c v_0 \quad (5.7.15)$$

This gives our final formula (using $e = \frac{E}{N}$, $v = \frac{V}{N}$):

$$S(E, V, N) = Nk \ln \frac{E^c V}{N^{c+1}} + Ns_0, \quad (5.7.16)$$

with

$$s_0 = (c + 1)k - \left(\frac{\mu}{T}\right)_0 - k \ln e_0^c v_0. \quad (5.7.17)$$

This gives an explicit example of a fundamental relation. We will leave s_0 undetermined here, as well as the constants C_0, C_1 below, and we will fix those constants when we compute the formula for the entropy of the ideal gas in Sect. 6.4.

Solving for E, we get

$$E(S, V, N) = N^{1+1/c} \frac{\exp(\frac{s-s_0}{ck})}{V^{1/c}}. \quad (5.7.18)$$

From this and (5.6.7) we get

$$T(S, V, N) = \frac{\partial E}{\partial S} = \frac{E(S, V, N)}{Nck},$$

which gives

$$E(T, V, N) = cNkT \quad (5.7.19)$$

and

$$S(T, V, N) = Nk \ln \frac{(ckT)^c V}{N} + Ns_0, \quad (5.7.20)$$

with s_0 given by (5.7.17).[9]

---

[8] For another derivation, see Callen [66, p. 68].

[9] In thermodynamics, the entropy is defined up to a constant, which depends on the initial state starting from which the form $\frac{\delta Q}{T}$ is integrated, see (5.5.6, 5.7.15).

From (5.6.3) and (5.7.18), we get:

$$P = -\frac{\partial E}{\partial V} = \frac{E}{cV} \tag{5.7.21}$$

which, combined with (5.7.19), gives (5.7.1).

5.7.3 The Fundamental Relation in Other Variables for an Ideal Gas

From (5.6.6) and (5.7.19), (5.7.20), we get the Helmhotz free energy:

$$F(T, V, N) = E(T, V, N) - T S(T, V, N) = cNkT - NkT \ln \frac{(ckT)^c V}{N} - T N s_0 \tag{5.7.22}$$

Finally, we can compute the grand potential:

$$\Phi(T, V, \mu) = E(T, V, \mu) - T S(T, V, \mu) - \mu N(T, V, \mu) = F(T, V, \mu) - \mu N(T, V, \mu).$$

From (5.6.18), we know that $\mu(T, V, N) = \frac{\partial F}{\partial N}$ and from (5.7.22), we get

$$\mu(T, V, N) = (c + 1)kT - kT \ln \frac{(ckT)^c V}{N} - T s_0 = -kT \ln \frac{(ckT)^c V}{N} + C_0 T, \tag{5.7.23}$$

with $C_0 = (c + 1)k - s_0 = (\frac{\mu}{T})_0 + k \ln e_0^c v_0$.
Inverting this, we get:

$$N(T, V, \mu) = C_1 (ckT)^c V \exp\left(\frac{\mu}{kT}\right), \tag{5.7.24}$$

where $C_1 = \exp(-\frac{C_0}{k})$. Combining this and (5.7.22), we have:

$$\Phi(T, V, \mu) = F(T, V, N(T, V, \mu)) - \mu N(T, V, \mu)$$
$$= cN(T, V, \mu)kT - N(T, V, \mu)kT \ln \frac{(ckT)^c V}{N(T, V, \mu)} - T N(T, V, \mu)s_0$$
$$+ N(T, V, \mu)kT \ln \frac{(ckT)^c V}{N(T, V, \mu)} - T N(T, V, \mu)C_0$$
$$= -kT N(T, V, \mu) = -kT C_1 (ckT)^c V \exp\left(\frac{\mu}{kT}\right), \tag{5.7.25}$$

since $C_0 = (c+1)k - s_0$. This gives:

$$\Phi(T, \mu) = \lim_{V \to \infty} \frac{\Phi(T, V, N)}{V} = -kTC_1(ckT)^c \exp\left(\frac{\mu}{kT}\right). \tag{5.7.26}$$

5.7.4   Adiabatic Transformations

By definition, for adiabatic transformations, we have $\delta Q = 0$. For an ideal gas, combining (5.3.9) and (5.7.2), we get:

$$\delta Q = 0 = dE + PdV = cNkdT + PdV$$

Using (5.7.1) we get $cNkdT + \frac{NkTdV}{V} = 0$ or $\frac{cdT}{T} + \frac{dV}{V} = 0$; Integrating this, we have:

$$T^c V = \text{Constant}$$

or $TV^{1/c} = \text{Constant}$, which is (5.4.3), with $\gamma = 1 + 1/c$.

5.7.5   Be Careful with Derivatives!

Students of thermodynamics (myself included in the past) get easily confused as to which derivative to take of which quantity in order to derive new formulas.

For example, using formulas like (5.6.3) in order to compute the pressure, it is essential to take the derivative of the function $E(S, V)$ or $E(S, V, N)$ with respect to V, i.e. to start from a fundamental relation. If we took instead the derivative of $E(T, V, N)$ with respect to V, we would get the wrong result. Take the ideal gas: if we take the derivative of the fundamental relation, as in (5.7.21), we get (5.7.1), but if we take the derivative with respect to V of $E(T, V, N)$, given by (5.7.2), we get 0!

5.8   What Is All This Good For?

Thermodynamics imposes restrictions on what processes are possible or not. But it is not limited to that: once we have a fundamental relation like (5.6.14) expressing the energy as a function of the extensive variables (the entropy, volume and number of particles), one obtains through differentiation expression for the intensive variables (temperature, pressure and chemical potential) as functions of those extensive variables, see (5.7.10), (5.7.11), (5.6.15).

One can also use Legendre transforms to express those quantities as functions of other variables, some extensive some intensive, see e.g. (5.6.18) or (5.6.22).

At the level of thermodynamics, fundamental relations have to be guessed or derived by integration from empirical formulas for ideal gases such as (5.7.1) and (5.7.2). It is only in the context of statistical mechanics that fundamental relations can be derived from first principles.

But one can obtain more relations by using a property of mixed derivatives of smooth functions on n variables: let $f : \mathbb{R}^n \to \mathbb{R}$ be twice continuously differentiable, then:

$$\frac{\partial^2 f}{\partial x_i \partial x_j} = \frac{\partial^2 f}{\partial x_j \partial x_i} \tag{5.8.1}$$

$\forall i, j = 1, \ldots, n$.

Starting from (5.6.16) and using (5.8.1), we get:

$$\left(\frac{\partial T}{\partial V} \right)_{S,N} = -\left(\frac{\partial P}{\partial S} \right)_{V,N}$$

$$\left(\frac{\partial T}{\partial N} \right)_{S,V} = \left(\frac{\partial \mu}{\partial S} \right)_{V,N}$$

$$\left(\frac{\partial P}{\partial N} \right)_{S,V} = -\left(\frac{\partial \mu}{\partial V} \right)_{S,N} \tag{5.8.2}$$

where on both sides, the function are functions of S, V, N, and the variables indicated as indices of the parentheses are kept constant.

These relations are called *Maxwell relations*. One can derive many more of them, by using (5.8.1) together with (5.6.18), (5.6.22) or the differential form of (5.7.7). For a complete list of Maxwell relations and some their applications, we refer to Callen [66, Chap. 7].

5.9 Summary

Thermodynamics is a strange science, widely used and highly reliable, but at the same time rather abstract and hard to understand intuitively. In fact, its understanding is only possible through the statistical mechanics discussed in the next chapter but that was developed after thermodynamics.

There are basically three laws of thermodynamics[10]: the zeroth one that simply says that equilibrium states exist; the latter are supposed to be reached if one keeps a system isolated long enough and these states are characterized by functional relations

---

[10] If we ignore the so-called third law which is less fundamental than the other ones.

between variables such as pressure, volume, density, temperature etc., see Sect. 5.2. Equilibrium thermodynamics studies the changes between those equilibrium states, in particular those that are possible and those that are not.

The first law is, strictly speaking, a consequence of mechanics: the total energy of an isolated system is conserved. Yet, in the context of thermodynamics, one distinguishes between two macroscopic aspects of changes in that total energy: exchanges of heat with the surroundings of the system and exchanges of work. This distinction is somewhat anthropomorphic, since work means useful work for us, but it can be given a perfectly quantitative and measurable definition, see Sect. 5.3.1. Since not all energy changes involve only exchanges of work, what is left in those changes is by definition heat, see (5.3.7).

The second law is based on the observation that one cannot freely use heat to produce work. This observation led to two precise and equivalent statements, those of Lord Kelvin and of Clausius see Sect. 5.4. Their equivalence is proven thanks to the Carnot cycle, see Sects. 5.4.1 and 5.4.2.

But the Carnot cycle has more implications, thanks to (5.4.7). It implies that the differential form $\frac{\delta Q}{T}$ is exact, namely that the exist a state function S, the entropy, satisfying $dS = \frac{\delta Q}{T}$. This definition is rather unintuitive, but one gets from (5.4.8) that, in a closed system, any transformation will lead to a non decrease of entropy, see Proposition 5.1.

The next step is to express the entropy in terms of the other extensive variables, the energy E, volume V and particle number N or numbers if there are several types of particles. That expression is called a fundamental relation, because all the thermodynamic information is contained in it and the formulas for the intensive variables such as the pressure, the temperature and the chemical potential can be obtained from the fundamental relation by taking derivatives. One can invert that relation and express E as a function of S, V, N and that constitutes also a fundamental relation, see Sect. 5.6.2.

But it is convenient to have fundamental relations involving other variables. That cannot be done by a simple change of variables, but requires a Legendre transformation. We consider two such transformations (and mentioned two others): one that takes the Legendre transform of $E(S, V, N)$ with respect to S and gives rise to the Helmoltz free energy $F(T, V, N)$ and a further Legendre transform with respect to the particle number(s) that leads to the grand potential $\Phi(T, V, \mu)$. Those new functions contain the same thermodynamic information as the fundamental relations and everything else can be derived from them by taking derivatives, see Sects. 5.6.3 and 5.6.4.

Since in a closed system, entropy tends to increase, one is tempted to characterize equilibrium by the principle of maximum entropy, see Sect. 5.6.5. This principle implies that the temperature, the pressure and the chemical potential behave in the expected way, see (5.6.27), (5.6.29), (5.6.30).

To give an example of the above ideas, one can obtain the fundamental relation $S(E, V, N)$ (5.7.16) for ideal gases: start from the empirical equations $PV = NkT$ and $E = cNkT$, and integrate them. This is important because this relation can be

derived from Boltzmann's formula for the entropy given in the next chapter. It is easy
to obtain, via Legendre transformations, the formulas for the Helmoltz free energy
and the grand potential for ideal gases.

5.10   Exercises

Some of the exercises in this section are taken from Amit and Verbin [12] and from
Callen [66].

5.1. Check (5.7.8), (5.7.9), (5.6.22).

5.2. Calculate the amount of heat transferred to an ideal gas of N molecules during:

(1) A process at constant temperature.

(2) A process at constant volume.

(3) A process at constant pressure.

Hint: use (5.3.9) and (5.7.2).

5.3. An ideal gas is at temperature T_1 and volume V_1 is taken through a process
at constant pressure to a state of higher temperature T_2. It is then taken via a
process at constant volume to a state of temperature T_1 and finally back to the
initial state in a process at constant temperature.

(1) Calculate the amount of heat transferred to the gas in the cycle.

(2) Same as (1) but in a reversed cycle.

(3) What would be the result if δQ was an exact differential?

(4) What is the work done by the gas during that cycle?

5.4. Starting from (5.7.20):

(1) Calculate the entropy increase of an ideal gas in a process at constant temperature.

(2) Calculate the entropy increase of an ideal gas in a process at constant volume.

(3) Is there a process in which the entropy remains constant?

5.5. Consider a system whose Helmholtz free energy is given by: $F(T, V) = -CVT^4$ where $C > 0$ is a constant (no particle number variable here).[11]

(1) Calculate the entropy of that system.

(2) Calculate the pressure of that system and its equation of state.

(3) Calculate the energy of that system.

(4) What is the chemical potential of that system?

(5) Calculate the equation of the adiabatics of that system.

5.6. Check the Maxwell relations (5.8.2) and derive those associated with the grand potential Φ.

5.7. Consider the following differential forms; identify those that are exact and, for them, find a function F so that $\omega = dF$.

(1) $\omega_1 = xy dx + y dy$.

(2) $\omega_2 = \sin x \cos y dx + \sin y \cos x dy$.

(3) $\omega_3 = x^2 dx + y^2 dy$.

(4) $\omega_4 = y^2 dx + x^2 dy$.

5.8. One mole of a monoatomic gas performs a Carnot cycle between the temperatures 200 degrees Kelvin and 100 degrees Kelvin while the volume on the upper isotherm goes from one liter to four liters. The volume on the lower isotherm goes from four liters to one liter.

How much work is performed during this cycle and how much heat is exchanged with the two sources?

5.9. What is the Legendre transform of $f(x) = x^\alpha$, $\alpha > 1$, $x > 0$? Same question for $f(x) = e^x$.

5.10. A system satisfies the equations:

$$e = \tfrac{1}{2} Pv$$

and

$$T^2 = a \frac{e^{\frac{3}{2}}}{v}.$$

---

[11] This is the formula for the free energy of a photon gas which, being part of quantum physics, will not be discussed in this book. See e.g. Amit and Verbin [12, Part IV, Chap. 4]. See exercise 6.15 for a calculation leading to the statistical mechanical formula for the energy of that system.

Find its fundamental equation of the form $S(E, V, N)$.

5.11. For each of the following fundamental equations, find the corresponding equations of state, namely $T = T(s, v)$ and $P = P(s, v)$, with $s = \frac{S}{N}$, $v = \frac{V}{N}$.

(1) $E = a\frac{S^3}{NV}$

(2) $E = a\frac{S^2}{N} - b\frac{V^2}{N}$

(3) $E = a\frac{S^2}{V}\exp(b\frac{S}{N})$

5.12. A system obeys the equations

$$E = aVT^4$$

$$P = \frac{E}{3V}$$

Find its fundamental equation $S = S(E, V)$.

Appendix

5.A   Differential Forms

Given a continuous function $f : \mathbb{R} \to \mathbb{R}$, one can integrate it: $\int_a^b f(x)dx$ and, if f happens to be the derivative of another function: $f(x) = F'(x)$, then we have the fundamental theorem of calculus:

$$F(b) - F(a) = \int_a^b f(x)dx. \tag{5.A.1}$$

Reciprocally, given a differentiable function $F(x)$, (5.A.1) always holds for $f(x) = F'(x)$.

A natural question is whether those formulas can be extended to functions $\mathbf{f} : \mathbb{R}^n \to \mathbb{R}^n$. If $\mathbf{f} = dF$ for a differentiable function $F : \mathbb{R}^n \to \mathbb{R}$, we will derive the analogue of (5.A.1) below, see (5.A.4). But does there exist functions $\mathbf{f}$ that are not of the form $\mathbf{f} = dF$?

We will consider here, for simplicity, $n = 2$, and write $\mathbf{f} : \mathbb{R}^2 \to \mathbb{R}^2$ as $\mathbf{f}(x, y) = (f_1(x, y), f_2(x, y))$. We associate a differential form to such a function:

$$\omega = f_1(x, y)dx + f_2(x, y)dy, \tag{5.A.2}$$

which can simply be taken as a convenient way to represent the function $\mathbf{f}$. Now, can one integrate ω from a point $(x_1, y_1) \in \mathbb{R}^2$ to another point $(x_2, y_2) \in \mathbb{R}^2$. In order to do that, we need to define a *path* in $\mathbb{R}^2$ joining these two points, which is simply a differentiable function:

$$\gamma : [0, 1] \to \mathbb{R}^2$$

such that $\gamma(0) = (x_1, y_1)$, $\gamma(1) = (x_2, y_2)$. Let us write, for $t \in [0, 1]$, $\gamma(t) = (x(t), y(t))$. Then, one can define the integral of ω over γ as:

$$\int_\gamma \omega = \int_0^1 \left(f_1(x(t), y(t)) \frac{dx(t)}{dt} + f_2(x(t), y(t)) \frac{dy(t)}{dt} \right) dt, \qquad (5.A.3)$$

which is an ordinary integral for a real valued function defined on $[0, 1]$. But one may ask:

1. Does the integral depend on the choice of the parametrization of the path, namely on the "speed" with which the path γ is travelled? In other words, if we define $t = t(\tau)$ a monotone map from $[0, 1]$ to $[0, 1]$ and let $\tilde{\gamma}(\tau) = \gamma(t(\tau))$, will the integral defined by (5.A.3) with $\tilde{\gamma}$ instead of γ be different from (5.A.3)?
2. Does the integral depend on the choice of the path γ or only on its endpoints?
3. Does a formula similar to (5.A.1) hold?

The answer to the first question is no and checking it is a simple exercise (just perform a change of variable $t = t(\tau)$ in (5.A.3)). The answer to the second question is yes, it may depend on the path.[12] To answer the third question, let us consider the following example; let $F : \mathbb{R}^2 \to \mathbb{R}$ be a smooth function and define $f_1(x, y) = \frac{\partial F}{\partial x}$, $f_2(x, y) = \frac{\partial F}{\partial y}$ or, in other words, $\omega = dF$, where dF is the differential of F. Beware of the notation: here ω is just a formal expression of the form (5.A.2), while dF is a real differential of the function F. But, when there exists a function F so that $\omega = dF$, one says that the form ω is *exact*.

Then inserting this in (5.A.3), we get:

$$\int_\gamma \omega = \int_\gamma dF = \int_0^1 \left(\frac{\partial F}{\partial x}(x(t), y(t)) \frac{dx(t)}{dt} + \frac{\partial F}{\partial y}(x(t), y(t)) \frac{dy(t)}{dt} \right) dt$$

$$= \int_0^1 \frac{dF}{dt}(x(t), y(t)) dt = F(1) - F(0) = F(x_2, y_2) - F(x_1, y_1),$$

$$(5.A.4)$$

which is indeed the analogue of (5.A.1).

We can ask whether there are examples of forms ω where the integral (5.A.3) is independent of the path γ, but that are not of the form $\omega = dF$. The answer to that

---

[12] Consider for example the form $\omega = ydx$, which obviously does not satisfy (5.A.6), and the square formed in $\mathbb{R}^2$ whose corners are $A = (0, 0)$, $B = (1, 0)$, $C = (1, 1)$, $D = (0, 1)$. It is easy to see, from (5.A.3), that $\int_A^B ydx = \int_B^C ydx = \int_A^D ydx = 0$, because either x is constant or $y = 0$ on these paths. But $\int_D^C ydx = 1$, so that $\int_{\gamma_1} ydx = 0 \neq \int_{\gamma_2} ydx = 1$, where γ_1 is the path $A \to B \to C$ and γ_2 the path $A \to D \to C$.

latter question is negative: suppose that we have an ω of the form (5.A.2) so that the integral in (5.A.3) does not depend on the path γ. Then we could define a function $F : \mathbb{R}^2 \to \mathbb{R}$ by the formula:

$$F(x, y) = \int_\gamma \omega \qquad\qquad (5.A.5)$$

where γ is any path with $\gamma(0) = (x_0, y_0)$ and $\gamma(1) = (x, y)$, where (x_0, y_0) is an arbitrarily chosen "initial" point. If we chose another "initial" point, say (x_1, y_1), then we would obtain another function $\tilde{F}$, but differing from F by a constant: $F(x, y) - \tilde{F}(x, y) = \int_{\tilde{\gamma}} \omega$, where $\tilde{\gamma}$ is any path with $\tilde{\gamma}(0) = (x_0, y_0)$, $\tilde{\gamma}(1) = (x_1, y_1)$. It is easy to check that, if F is defined by (5.A.5), then $dF = \omega$.

A more interesting question is whether one can recognize that ω is of the form dF for some $F : \mathbb{R}^2 \to \mathbb{R}$, without knowing F. First, observe that, if there is such an F, we must have

$$\frac{\partial f_1}{\partial y} = \frac{\partial f_2}{\partial x} \qquad\qquad (5.A.6)$$

since both terms are then equal to $\frac{\partial^2 F}{\partial x \partial y}(x, y)$.
Another necessary condition is that

$$\int_{cycle} \omega = 0$$

where a cycle is a closed curve in $\mathbb{R}^2$, since, by (5.A.4) the integral equals $F(x, y) - F(x, y) = 0$, for any point (x, y) belonging to the cycle.

The remarquable fact is that each of these necessary conditions is also sufficient for ω to be of the form dF for some $F : \mathbb{R}^2 \to \mathbb{R}$. We have the fundamental result[13]:

Proposition 5.2 *Given a ω of the form (5.A.2), the following properties are equivalent:*

1. *There exists $F : \mathbb{R}^2 \to \mathbb{R}$, so that $\omega = dF$, i.e. $f_1(x, y) = \frac{\partial F}{\partial x}$, $f_2(x, y) = \frac{\partial F}{\partial y}$.*
2. $\frac{\partial f_1}{\partial y} = \frac{\partial f_2}{\partial x}$, $\forall (x, y) \in \mathbb{R}^2$.
3. $\int_{cycle} \omega = 0$ *for any closed cycle in $\mathbb{R}^2$.*

To prove the Proposition, note that [3] implies [1], because [3] implies that the integral in (5.A.5) is independent of γ. Indeed, consider two paths γ_1, γ_2 going from (x_0, y_0) to (x, y) and consider the cycle composed of γ_1 followed by $\tilde{\gamma}_2$, which is the reversed of the path γ_2 (and thus going from (x, y) to (x_0, y_0)). By [3] the integral over that cycle vanishes and equals $\int_{\gamma_1} \omega + \int_{\tilde{\gamma}_2} \omega = \int_{\gamma_1} \omega - \int_{\gamma_2} \omega = 0$, which shows

---

[13] For simplicity we assume that ω is defined in $\mathbb{R}^2$, but the result holds also if ω is defined in any simply connected open subset of $\mathbb{R}^2$ (i.e. an open subset in which any simple closed curve can be shrunk to a point continuously in the subset or, intuitively, a subset with no holes in it).

that the integral in (5.A.5) is independent of γ, which implies [1]. We already saw that [1] implies [2]. To see that [2] implies [3], consider an "infinitesimal" cycle γ joining (x, y), $(x + dx, y)$, $(x + dx, y + dy)$ and $(x, y + dy)$. Writing down $\int_\gamma \omega$, we see that is approximately equal to the integral of $\frac{\partial f_2}{\partial x} - \frac{\partial f_1}{\partial y}$ over the interior of γ. Hence [2] implies that $\int_{cycle} \omega = 0$ holds for infinitesimal cycles. By filling the interior of any cycle by a finite number of "infinitesimal" ones adjacent to each other, one shows that $\int_{cycle} \omega = 0$ for any closed cycle.

Formula (5.3.3) is an example of a differential form that is not exact and so is, see (5.3.9), $\delta Q = dE + PdV$. Indeed the amount of work or heat exchanged when one passes from a state A to a state B depends on the path taken in the space of parameters of the equilibrium states.

The main application of this proposition here is the derivation of (5.5.4) from (5.5.3). Formulas (5.6.2), (5.6.4), (5.6.9), (5.6.16), (5.6.22) or (5.7.6) are exact differentials.

5.B   Legendre Transforms

5.B.1   Mathematical Definition

It is customary to describe a curve in $\mathbb{R}^2$ by the equation $y = f(x)$. But suppose that f is strictly convex, which means $f''(x) > 0$ or, equivalently, that $f'(x)$ is strictly increasing. In that case, the map $x \to f'(x)$ can be inverted and we can define a function $x(p)$ by:

$$f'(x(p)) = p, \tag{5.B.1}$$

namely $x(p)$ is the value of the x variable for which the slope of the curve $y = f(x)$ at the point $(x(p), f(x(p)))$ equals p. So, if we want, we can parametrize the curve $(x, f(x))$ by the slope rather than by the x coordinate and define $\tilde{f}(p) = f(x(p))$ (why would we want to do that? See below!).

One problem with that definition is that it is not invertible: given $\tilde{f}(p)$ we cannot recover the function f it came from. This is easy to see since all the functions of the form $f_c(x) = f(x) + c$, for c a constant will satisfy (5.B.1) and therefore $\tilde{f}_c(p)$ is independent of c.

A more intelligent transformation is the *Legendre transformation*, defined as:

$$f^*(p) = px(p) - f(x(p)), \tag{5.B.2}$$

where $x(p)$ is the unique (by assumption of convexity) solution of (5.B.1). It is easy to see that $-f^*(p) = f(x(p)) - px(p)$ is the intersection with the y axis of the tangent to the curve $(x, f(x))$ whose slope equals p. Now one can give some motivation for this transformation: one can either parametrize a convex function in the usual way $y = f(x)$ or as the envelope of a set of straight lines (that are therefore the tangents to the curve $(x, f(x))$) parametrized by their slope p and their intersection $-f^*(p)$ with the y axis. It is easy to check that, since f is convex, f^* is also convex. Indeed, from (5.B.2), we get $f^{*'}(p) = x(p) + px'(p) - f'(x(p))x'(p) = x(p)$ by (5.B.1). So, $f^{*''}(p) = x'(p)$, which is strictly positive since, from (5.B.1), we get $x'(p) = \frac{1}{f''(x(p))}$ and, by convexity of f, $f''(x(p)) > 0$.

The transformation (5.B.2) is invertible and is in fact an involution, meaning that the Legendre transform of $f^*(p)$ gives $f(x)$: $f(x) = f^{**}(x)$. To see this, it will be convenient to use another notation and define a function g as: $g(p) = x(p)$ (otherwise there may be some confusion below between the function $x(p)$ and the variable x). Then, using (5.B.2) twice:

$$f^{**}(x) = xp(x) - f^*(p(x)) = xp(x) - p(x)g(p(x)) + f(g(p(x))) \qquad (5.B.3)$$

where $p(x)$ is the unique solution of

$$f^{*'}(p(x)) = x.$$

From (5.B.2) we obtain (with $g(p) = x(p)$):

$$f^{*'}(p) = g(p) + pg'(p) - f'(g(p))g'(p),$$

but from (5.B.1), we get $f'(g(p)) = f'(x(p)) = p$, and thus, $f^{*'}(p) = g(p)$. Combining this with (5.B.1), we get $p(x) = g^{-1}(x)$, which means that $g(p(x)) = x$. Inserting this in (5.B.3), we have:

$$f^{**}(x) = xg^{-1}(x) - g^{-1}(x)g(p(x)) + f(g(p(x)))$$
$$= xg^{-1}(x) - g^{-1}(x)x + f(x) = f(x)$$

which proves that the Legendre transformation is an involution (and therefore invertible).

An alternative definition of the Legendre transformation is

$$f^*(p) = \sup_x(px - f(x)) \qquad (5.B.4)$$

since the supremum is attained when $p = f'(x)$, namely for $x = x(p)$ which is the solution of (5.B.1). However, the advantage of this definition is that it makes sense for functions that are simply convex ($f''(x) \geq 0$) and not necessarily strictly convex ($f''(x) > 0$) and the Legendre transformation (5.B.4) is still invertible.

One can also extend this definition to convex functions of n variables, $f : \mathbb{R}^n \to \mathbb{R}$:

$$f^*(\mathbf{p}) = \sup_x(\mathbf{p} \cdot \mathbf{x} - f(\mathbf{x}))$$

where $\mathbf{p} = (p_1, \ldots, p_n)$, $\mathbf{x} = (x_1, \ldots, x_n)$ and $\cdot$ is the scalar product. Convex here means that the matrix of second derivatives $(\frac{\partial^2 f}{\partial x_i \partial_j})^n_{i,j=1}$ is positive definite. If it is strictly positive definite, then the equation:

$$\nabla f(\mathbf{x}) = \mathbf{p}$$

has a unique solution $\mathbf{x}(\mathbf{p})$ and $f^*(\mathbf{p}) = \mathbf{p} \cdot \mathbf{x}(\mathbf{p}) - f(\mathbf{x}(\mathbf{p}))$.

5.B.2 Physical Applications

There are two major applications of Legendre transformations in physics: one in mechanics and the other one in thermodynamics, used here. In mechanics, one associates to any mechanical systems with conservative forces (see (3.2.2)) and generalized coordinates[14] $\mathbf{q} \in \mathbb{R}^n$ a Lagrangian function

$$L(\mathbf{q}, \dot{\mathbf{q}}) = \frac{1}{2}(\dot{\mathbf{q}} \cdot A(\mathbf{q})\dot{\mathbf{q}}) - V(\mathbf{q}) \tag{5.B.5}$$

where $\dot{\mathbf{q}}$ denotes generalized velocities (the time derivatives of the generalized coordinates) and $A(\mathbf{q})$ is, $\forall \mathbf{q}$, a positive definite matrix depending on the generalized coordinates (but which is constant in Cartesian coordinates).

Since the function $L(\mathbf{q}, \dot{\mathbf{q}})$ is strictly convex with respect to the variables $\dot{\mathbf{q}}$, one can perform a Legendre transformation with respect to those variables:

$$H(\mathbf{q}, \mathbf{p}) = \dot{\mathbf{q}}(\mathbf{q}, \mathbf{p}) \cdot \mathbf{p} - L(\mathbf{q}, \dot{\mathbf{q}}) \tag{5.B.6}$$

where $\dot{\mathbf{q}}(\mathbf{q}, \mathbf{p})$ is obtained from solving

$$\mathbf{p} = \frac{\partial L}{\partial \dot{\mathbf{q}}}(\mathbf{q}, \dot{\mathbf{q}}).$$

From (5.B.5), we obtain $\dot{\mathbf{q}}(\mathbf{q}, \mathbf{p}) = A^{-1}(\mathbf{q})\mathbf{p}$ and inserting this in (5.B.6), we get:

$$H(\mathbf{q}, \mathbf{p}) = \frac{1}{2}(\mathbf{p} \cdot A^{-1}(\mathbf{q})\mathbf{p}) + V(\mathbf{q}).$$

The simplest example of a matrix A is a diagonal one $A_{ij} = m_i \delta_{ij}$, with m_i the masses of the particles, which gives rise to the Hamiltonian (3.3.1).

Unfortunately, the applications of Legendre transformations in thermodynamics are made with the opposite sign than the one used in mathematics and in mechanics.

---

[14] In Chap. 3 and in the rest of this book, we consider only Cartesian coordinates. But it is often convenient to introduce more general systems of coordinates, like polar, spherical or cylindrical coordinates.

Indeed, we have, see (5.6.6), (5.6.7), $F = E - TS$, while if we adhered to definition (5.B.2) we would write $F = TS - E$. With that modified definition, the Legendre transformation is still invertible, but with a change of sign, since we have in (5.6.10) $E = F + TS$.

5.C Physical Units and Boltzmann's Constant

Work or energy has the dimension of a force times a length, $[E] = [F][L]$, where $[\cdot]$ denotes the dimension of a quantity. Pressure has the dimension of a force divided by a surface, $[P] = \frac{[F]}{[L]^2}$, so that $[P][V] = [E]$, as we saw in (5.3.1).

The dimension of temperature is a bit more tricky: given (5.5.4), (5.5.6), it is natural to give the same dimensions to the energy and the temperature and thus to consider the entropy to be dimensionless.

Relation (5.4.7) gives a way to measure (in principle) ratios of temperatures. Besides, temperatures cannot go below $-273, 16$ degree Celsius. Counting temperatures in degree Celsius but defining $-273, 16$ degree Celsius as the absolute 0 gives a scale of temperatures in *Kelvin degrees*.

Then Boltzmann's constant k becomes a conversion constant between units of the same quantity (like converting centimeters into meters); k equals 1.380649×10^{-23} joules/kelvin. This is a very small number, but in (5.7.1), (5.7.2), for example, as well as in many other formulas, k is multiplied by the number of particles, which is of the order of Avogadro's number $N_A \approx 10^{23}$.

The entropy is also given in joules/kelvin, but we have to remember that both joules and kelvin denote units of energy.

Chapter 6
Equilibrium Statistical Mechanics

In this chapter, we will present Boltzmann's ideas concerning the microscopic basis of equilibrium thermodynamics. In Chap. 8, we will extend those ideas to the approach to equilibrium. We rely on the works of Lebowitz, Goldstein, Tumulka, Zanghì, and others [67, 149–155, 212–214, 217–221, 223–225, 314, 333].

6.1 Microstates and Macrostates

As we said in the Introduction, the whole of statistical mechanics consists in the derivation of macroscopic laws from microscopic ones. We saw in Chap. 3 what are the microscopic laws, in a classical framework. In Chap. 5 we introduced the macroscopic laws in equilibrium.

We will start by defining macroscopic laws more generally and explain what "derive" means.

The *microstates* of a classical mechanical system are elements of $\mathbb{R}^{6N}$, and we wrote in Sect. 3.3 a vector $\mathbf{x} \in \mathbb{R}^{6N}$ as a pair $\mathbf{x} = (\mathbf{q}, \mathbf{p})$, with $\mathbf{q} = (\vec{q}_1, \vec{q}_2, \ldots, \vec{q}_N) \in \mathbb{R}^{3N}$, $\mathbf{p} = (\vec{p}_1, \vec{p}_2, \ldots, \vec{p}_N) \in \mathbb{R}^{3N}$ and defined the time evolution of $\mathbf{x}(t) = T^t(\mathbf{x}(0))$ through Hamilton's equations (3.3.3), (3.3.4), (3.3.10).

If one works with closed systems, as one usually does, one considers a closed and bounded subset $\Omega \subset \mathbb{R}^{6N}$ (the spatial part is bounded because the system is enclosed in a box and its velocities are bounded because the total energy of the system is finite[1]) which we will refer to as the *phase space* of the system.

Our definition of the macrostate will be more vague: we are all familiar with macroscopic quantities like the volume, the pressure, the temperature or the density of physical systems. These are quantities that we have a direct access to, without

---

[1] This follows from the formula (3.3.1) for the energy, assuming the potential function to be bounded from below.

© Springer Nature Switzerland AG 2022
J. Bricmont, *Making Sense of Statistical Mechanics*, Undergraduate Lecture
Notes in Physics, https://doi.org/10.1007/978-3-030-91794-4_6

knowing anything about the microscopic composition of matter. In fact, physicists started by studying such quantities long before the had any idea about atoms and people like the 19th century Austrian physicist Ernst Mach studied them while denying the existence of atoms.

As long as we study terrestrial phenomena, those macroscopic quantities characterize systems composed of a number of atoms which is very large but too large. The objects studied here are, to use Lars Onsager phrase "large compared to atoms, but small compared to the sun", which means in practice that we can neglect gravitational forces.

To define abstractly a *macrostate*, we introduce a map from the phase space of the system into $\mathbb{R}^L$:

$$M : \Omega \to \mathbb{R}^L, \tag{6.1.1}$$

where $L << N$. Those variables could be the energy or the density but we will give more examples of such variables in a moment. By definition, $M(\mathbf{x})$ is the macrostate associated to the microstate $\mathbf{x}$. The important point is that the macrostate is a *function* of the microstate. For example, if we know $\mathbf{x} = (\mathbf{q}, \mathbf{p})$, then we know the total energy of the system given by (3.3.11). Obviously, we also know the total density of the system. The condition $L << N$ means that we have far fewer macroscopic variables than microscopic ones; otherwise there would be no point in speaking of a macroscopic description.

We already introduced the distinction between microstates and macrostates in the simple example of coin tossing in Sect. 2.3.1, where 0 denoted 'head' and 1, 'tail'; we called the 'space' of results of any single tossing, $\{0, 1\}$, the 'individual phase space' and the space of all possible results of N tossings, $\Omega = \{0, 1\}^N$, was called the 'total phase space'. The variables N_0, N_1 that count the number of heads (0) or tails (1) in N tossings were the *macroscopic* variables.

Now, let us introduce some less trivial examples of macrostates. Let $\Delta(\vec{u})$ be the cubic cell of size δ^3 centered around $\vec{u} \in (\delta\mathbb{Z})^3$ and let $\mathbf{q} = (\vec{q}_1, \ldots, \vec{q}_N) \in \mathbb{R}^{3N}$, be an element of the 'configuration space' of the system, i.e. the space consisting of the positions for all the particles. Define

$$N_{\vec{u}}(\mathbf{q}) = |\{i \,|\, \vec{q}_i \in \Delta(\vec{u}), \; i \in \{1, \ldots, N\}\}|, \tag{6.1.2}$$

the number of particles in $\Delta(\vec{u})$.

Assume that the particles are enclosed in a box Λ which is a disjoint union of cubic cells of size δ^3, label $\vec{u}_1, \ldots, \vec{u}_L$ the centers of the cubic cells with $\Delta(\vec{u}) \subset \Lambda$ and let

$$M_1(\mathbf{x}) = (n_i(\mathbf{q}))_{i=1}^L \equiv \left(\frac{N_{\vec{u}_i}(\mathbf{q})}{N} \right)_{i=1}^L, \tag{6.1.3}$$

where N is the total number of particles in Λ. The set of numbers $(n_i(\mathbf{q}))_{i=1}^L$ is similar to the histogram defined in (2.3.10) or the empirical distribution defined in (2.3.14).

The map $M_1(\mathbf{x})$ takes values in $\mathbb{R}^L$, with L the number of cubic cells of size δ^3 in Λ, or $L = \frac{|\Lambda|}{\delta^3}$.

To give another example of a macrostate, let $\mathbf{v} = (\vec{v}_1, \ldots, \vec{v}_N) \in \mathbb{R}^{3N}$, with each $\vec{v}_i \in \mathbb{R}^3$, and let:

$$N_{\vec{u}}(\mathbf{v}) = |\{i \,|\, \vec{v}_i \in \Delta(\vec{u}), \ i \in \{1, \ldots, N\}\}|, \tag{6.1.4}$$

which counts, for each cubic cell $\Delta(\vec{u})$, of size δ^3, centered around $\vec{u} \in (\delta\mathbb{Z})^3$, and for any given set of velocities of all the particles, the number of particles whose velocities lie in $\Delta(\vec{u})$. Labelling the centers of the cubic cells as before $\vec{u}_1, \ldots, \vec{u}_L$, we define the *macrostate* $M_2(\mathbf{x})$ by the set of variables:

$$M_2(\mathbf{x}) = (n_i(\mathbf{v}))_{i=1}^L = \left(\frac{N_{\vec{u}_i}(\mathbf{v})}{N} \right)_{i=1}^L \tag{6.1.5}$$

Here the number L of macroscopic variables $n_i(\mathbf{v})$ may seem to be infinite, since $\vec{u} \in (\delta\mathbb{Z})^3$, but the number of non-zero values of $N_{\vec{u}}(\mathbf{v})$ is finite, if we assume that the total energy of the system is bounded, since then the velocity of each particle is also bounded.

We may do the same thing with both the positions and the momenta. Assuming the total energy of the system to be bounded, we can introduce a box $\tilde{\Lambda} \subset \mathbb{R}^6$ such that the pair $(\vec{q}_i, \vec{p}_i)$ of each particle of the system belongs to $\tilde{\Lambda}$. We can again introduce a partition of $\tilde{\Lambda}$ into L cubic cells $\Delta(\vec{u})$ of size δ^6 centered around $\vec{u} \in (\delta\mathbb{Z})^6$.

Then, we define, as before:

$$N_{\vec{u}}(\mathbf{q}, \mathbf{p}) = |\{i \,|\, (\vec{q}_i, \vec{p}_i) \in \Delta(\vec{u}), \ i \in \{1, \ldots, N\}\}|, \tag{6.1.6}$$

and define the *macrostate* $M_3(\mathbf{x})$ by:

$$M_3(\mathbf{x}) = (n_i(\mathbf{x}))_{i=1}^L = \left(\frac{N_{\vec{u}_i}(\mathbf{q}, \mathbf{p})}{N} \right)_{i=1}^L. \tag{6.1.7}$$

Here, the number L equals the number of cubic cells of size δ^6 in $\tilde{\Lambda}$, or $L = \frac{|\tilde{\Lambda}|}{\delta^6}$.

In the limit $N \to \infty$ and $L \to \infty$, $|\Delta(\vec{u})| \to 0$, the function $M_1(\mathbf{x})$ will, in some sense, converge to a continuous function $\rho(x)$ which is the *local density* of the system. In the same limit, the function $M_2(\mathbf{x})$ will converge to a distribution of velocities (which, in equilibrium, is a Gaussian function, the Maxwellian, see (6.2.20)). Finally, in that limit, the function $M_3(\mathbf{x})$ will converge to Boltzmann's f function (see Sect. 8.6).

When we consider those limits, there seems to be a contradiction with the condition $L << N$ introduced in the definition (6.1.1) of the macrostate, since a function requires in principle an infinite number of parameters to be defined (think for example of its expansion in a Fourier series), while, in physical systems, N is always finite. But this objection can be answered by noticing that we do not need infinitely many parameters to characterize those functions. If they are not too irregular, in practice a finite (and not too large) number of Fourier coefficients (or of coefficients of an expansion of the function in another basis) will suffice.

Note that, in the physics literature, the phase space denoted Ω here is often written Γ and called the Γ-space while the space $\mathbb{R}^6$ (or a subset of that space), whose partition defines the macrostate $M_3(\mathbf{x})$ is called the μ-space.

Given a macrostate map M and value $M(0)$ of the macrostate, one introduces the set of microstates $\mathbf{x}$ corresponding to the same value $M(0)$ of the macrostate:

$$\Omega(M(0)) = M^{-1}(M(0)), \tag{6.1.8}$$

where M^{-1} is the inverse map of M. Thus $\Omega(M(0)) \subset \Omega$ is the set of microstates $\mathbf{x}$ such that $M(\mathbf{x}) = M(0)$.

The macrostate can evolve in time (in equilibrium situations it will not change but we will consider time changes later) and its time evolution $M(0) \to M(t)$ is *induced* by the one of the microstate $\mathbf{x}(0) \to \mathbf{x}(t) = T^t\mathbf{x}(0)$:

$$M(0) = M(\mathbf{x}(0)) \to M(t) = M(\mathbf{x(t)}). \tag{6.1.9}$$

In statistical mechanics, one would like to consider only macrostates whose time evolution is *autonomous*, which means that the evolution of $M(t)$ is the same for all $\mathbf{x}(t) \in \Omega(M(t))$. Or, in formulas, if we start at some initial state 0, and some macrostate $M(0)$, $M(t) = M(\mathbf{x}(t)) = M(T^t(\mathbf{x}(0)))$ is constant as a function of $\mathbf{x}(0) \in \Omega(M(0))$.

The reason for this limitation is obvious: since we do not have access to the microstate $\mathbf{x}(0)$, if the evolution was not autonomous, we would have no way to predict the future behavior of $M(0)$ if all we know about the system is its initial macrostate.

If the evolution of $M(t)$ is autonomous, then one can say that the evolution of $M(t)$, which is called a *macroscopic law*, has been *reduced to* or *derived from* the microscopic one $\mathbf{x}(0) \to \mathbf{x}(t)$, in a straightforward way (see footnote 22 in Chap 2).

We will see in Chap. 8 that evolutions of macroscopic quantities are never strictly speaking autonomous, but can be almost autonomous. This will lead to quite a lot of subtleties, but in this chapter, we will study equilibrium macrostates, which are time-independent (but which have their own subtleties).

There is one final remark to add to our definition of macrostate, which is motivated by the following observation: suppose we divide the box Λ into two equal sub-boxes Λ_1, Λ_2, and that we put $\frac{N}{2} - 1$ particles in Λ_1 and $\frac{N}{2} + 1$ particles in Λ_2. Then, according to our definitions, the macrostate defined by the number of particles in Λ_1 and Λ_2 would be different from the one where we would have $\frac{N}{2}$ particles both in Λ_1 and Λ_2.

But obviously there would be no relevant physical differences between the two situations. What we need to do is to coarse grain the set of values taken by the map M in order to obtain physically different values (see e.g. Tumulka [314, Sect. 7.1.2] for a discussion of this idea). For example, we could divide the interval $[0, 1]$ into sub-intervals $I_l, l = 1, \ldots, K$ of equal length $\eta = \frac{1}{K}$, and modify definition (6.1.3) into:

$$M_1(\mathbf{x}) = (l_1(\mathbf{x}), l_2(\mathbf{x}), \ldots l_L(\mathbf{x})), \tag{6.1.10}$$

where $l_1(\mathbf{x}), l_2(\mathbf{x}), \ldots l_L(\mathbf{x})$ are defined by: $n_1(\mathbf{q}) \in I_{l_1}(\mathbf{x}), n_2(\mathbf{q}) \in I_{l_2}(\mathbf{x}), \ldots n_L(\mathbf{q}) \in I_{l_L}(\mathbf{x})$. That makes our definition of the macro-state less precise, since it now depends of the choice of the length η of the intervals I_l, but somewhat more realistic physically. Now, the map $M_1(\mathbf{x})$ takes values in a subset of $\mathbb{R}^L$: $\{1, \ldots, K\}^L$.

Of course, with this redefinition, there are still sets of values of $(n_i(\mathbf{q}))_{i=1}^L$ that are very close to each other but correspond to different values of $M_1(\mathbf{x})$ in (6.1.10) (think of numbers near the boundaries of some interval I_l), but there are far fewer such sets than with the definition (6.1.3). We can extend this coarse graining to (6.1.5) and (6.1.7) and we will always assume below that we use this redefinition.

6.2 Dominance of the Equilibrium Macrostate

The most important property of our map M, which will allow us to understand the notion of equilibrium is that *this map is many to one in a way that depends on value taken by M.*

To explain this, think again of the simple example of N coin tossing: a microstate is a sequence of results e.g. $(H, T, H, H, \ldots, T)$ and Ω is the set of such sequences; we have $|\Omega| = 2^N$.

The macrostate is $M = (N_0, N_1)$, with N_0 = number of heads, N_1 = number of tails.

If $N_0 = N$, it corresponds to a unique microstate $(H, H, H, H, \ldots, H)$

But if $N_0 = \frac{N}{2}$ then there are approximately $\frac{2^N}{\sqrt{N}}$ microstates giving rise to that value of M (see (6.2.7) below).

In (2.3.4), we gave the number $|M^{-1}(N_0)|$ of microstates corresponding to a given value of M, i.e. of N_0:

$$|M^{-1}(N_0)| = \frac{N!}{N_0! N_1!} = \frac{N!}{N_0!(N - N_0)!} \tag{6.2.1}$$

It is interesting to consider that number in the large N limit; using Stirling's formula ($\ln n! \approx n \ln n - n$ see Appendix 6.A.2), we have

$$\ln |M^{-1}(N_0)| \approx +N \ln N - N_0 \ln N_0 - N_1 \ln N_1 - N + N_0 + N_1$$

$$= -N_0 \ln \frac{N_0}{N} - N_1 \ln \frac{N_1}{N}$$

since $N_0 + N_1 = N$.

If we take the limit of that quantity divided by N as $N \to \infty$, with $\frac{N_0}{N} = n_0$ fixed (and $\frac{N_1}{N} = 1 - n_0$), we get:

Fig. 6.1 The curve $S(x) = -x \ln x - (1-x) \ln(1-x)$, $0 \le x \le 1$

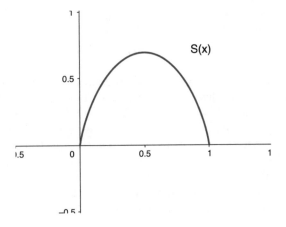

$$\lim_{N \to \infty} \frac{\ln |M^{-1}(N_0)|}{N} = -n_0 \ln n_0 - (1-n_0) \ln(1-n_0) \tag{6.2.2}$$

Let us call the right hand side of (6.2.2) $s(n_0)$. The function $s(x)$ is defined on $[0, 1]$ and has the following properties (see Fig. 6.1):

1. $s(x)$ is positive and symmetric around $\frac{1}{2}$ ($s(\frac{1}{2} - y) = s(\frac{1}{2} + y)$ for $0 \le y \le \frac{1}{2}$).
2. $s(x)$ is strictly concave ($s''(x) = -\frac{1}{x} - \frac{1}{1-x} < 0$).
3. $s(0) = s(1) = 0$ (with the convention $0 \ln 0 = 0$) and $s(x)$ reaches its maximum at $x = \frac{1}{2}$, with $s(\frac{1}{2}) = \ln 2$, $s'(\frac{1}{2}) = 0$, $s''(\frac{1}{2}) = -4$.

So, being somewhat sloppy with the interchange of limits, we get:

$$|M^{-1}(N_0)| \approx \exp(N s(n_0)) \tag{6.2.3}$$

Let us divide the set Ω into the set Ω_{eq} of (almost) "equilibrium" microstates (called like that by analogy with what we will do later):

$$\Omega_{eq} \equiv \{\omega \in \Omega \mid |n_0 - \frac{1}{2}| = |n_1 - \frac{1}{2}| \le \epsilon\} \tag{6.2.4}$$

for some $\epsilon > 0$ and the rest, the "non-equilibrium" microstates[2]: $\Omega_{neq} = \Omega \setminus \Omega_{eq}$. Let us estimate the relative size of the two sets: $\frac{|\Omega_{neq}|}{|\Omega_{eq}|}$. Since there are at most N possible values of n_0 with $|n_0 - \frac{1}{2}| > \epsilon$, we get from (6.2.3): $|\Omega_{neq}| \le N \exp(N(s(\frac{1}{2} - \epsilon))$, since, by the properties of $s(x)$, we have $s(n_0) \le s(\frac{1}{2} - \epsilon)$ for $|n_0 - \frac{1}{2}| > \epsilon$.

On the other hand, we can bound from below $|\Omega_{eq}|$ by $|\{\omega \in \Omega \mid |n_0 - \frac{1}{2}| \le \frac{\epsilon}{2}\}|$, which, by (6.2.3) and the properties of $s(x)$ satisfies: $|\{\omega \in \Omega \mid |n_0 - \frac{1}{2}| \le \frac{\epsilon}{2}\}| \ge \exp(N s(\frac{1}{2} - \frac{\epsilon}{2}))$ (since $s(n_0) \ge s(\frac{1}{2} - \frac{\epsilon}{2})$ for $|n_0 - \frac{1}{2}| \le \frac{\epsilon}{2}$). So, we get:

---

[2] Ω_{eq} coincides with what we called $G_N(\epsilon)$ in (2.3.2).

$$\frac{|\Omega_{neq}|}{|\Omega_{eq}|} \leq N \exp\left[N\left(s\left(\frac{1}{2} - \epsilon\right) - s\left(\frac{1}{2} - \frac{\epsilon}{2}\right)\right)\right] \tag{6.2.5}$$

This is a very rough estimate, whose only virtue is that it is easy to prove. Now, for x small, we have, since $s''(\frac{1}{2}) = -4$, by expanding $s(\frac{1}{2} - x)$ to third order in x, we have, for $|x| \leq \epsilon$, $s(\frac{1}{2} - x) = s(\frac{1}{2}) - 2x^2(1 + \mathcal{O}(\epsilon))$, so that (6.2.5) gives:

$$\frac{|\Omega_{neq}|}{|\Omega_{eq}|} \leq N \exp[-\mathcal{O}(N\epsilon^2)]. \tag{6.2.6}$$

If we take $N \approx 10^{23}$ and $\epsilon \approx 10^{-6}$, which is rather small, we still get that the ratio $\frac{|\Omega_{neq}|}{|\Omega_{eq}|}$ is less than $\exp(-\mathcal{O}(10^{11}))$, namely minuscule.

Of course, one could remove the factor N in (6.2.6) by changing the constant in $\mathcal{O}(N\epsilon^2)$; we include it here because it makes that bound easier to understand (see also (6.2.11) below).

We have done this calculation in some detail because the same scheme will occur repeatedly. Obviously this calculation holds for any system composed of elements that can be in two states: heads or tails here, but also for a lattice with empty or occupied cells or with up or down spins at each site, see Chap. 9.

Note that we did not use the word "probability" here; we simply *counted* the number of microstates corresponding to different macrostates and found that the overwhelming majority of them are those where the macrostate is close to the equilibrium value $n_0 = \frac{1}{2}$.

One must also remark that if we counted the number of microstates corresponding to the macrostate having *exactly* the equilibrium value $n_0 = \frac{1}{2}$, then, it would be the largest number corresponding to any given macrostate, but it would not be the overwhelming majority compared to all the other macrostates put together.[3]

But, from a physical point of view, it makes sense to group together macrostates that are close to the equilibrium value, as we did at the end of Sect. 6.2, since no experiment or observation will distinguish between those macrostates if ϵ in (6.2.4) is small enough.[4]

---

[3] In fact, one gets, from Stirling's formula (6.A.3), that

$$\frac{|\{\omega \in \Omega \mid n_0 = \frac{1}{2}\}|}{|\Omega|} = \mathcal{O}\left(\frac{1}{\sqrt{N}}\right);$$

Indeed, use (6.2.1), with $N_0 = \frac{N}{2}$, assuming N to be even. Then, since $|\Omega| = 2^N$, (6.A.3) gives:

$$\frac{|\{\omega \in \Omega \mid n_0 = \frac{1}{2}\}|}{|\Omega|} = \frac{N^N \exp(-N)\sqrt{2\pi N}}{2^N \left(\frac{N}{2}\right)^N \exp(-N)(\sqrt{\pi N})^2} = \mathcal{O}\left(\frac{1}{\sqrt{N}}\right). \tag{6.2.7}$$

[4] In [211] Lavis criticizes the approach outlined here, in particular the way I presented it in [53], by emphasizing the necessity of grouping macrostates close to the equilibrium value, but nobody ever denied that and the law of large numbers always speaks of values close to the average one,

This generalizes in a straightforward way to N particles each of which can be in L possible states (with L fixed, independently of N). The set of microstates is $\Omega = \{1, \ldots, L\}^N$, and we can take as macrostates the fractions of particles in each state: $\mathbf{n} = (n_1, \ldots, n_L)$, where $n_i = \frac{N_i}{N}$, with N_i being the number of particles is state i. Following the previous computations, which were done for $L = 2$, we have from the multinomial formula[5]:

$$|M^{-1}(\mathbf{n})| = \frac{N!}{\prod_{i=1}^{L} N_i!} \tag{6.2.8}$$

and we get:

$$\lim_{N \to \infty} \frac{\ln |M^{-1}(\mathbf{n})|}{N} = -\sum_{i=1}^{L} n_i \ln n_i \equiv s(\mathbf{n}) \tag{6.2.9}$$

The function $s(\mathbf{x})$ is defined on $[0, 1]^L$ and has the following properties:

1. $s(\mathbf{x})$ is positive and symmetric under permutations of its arguments.
2. $s(\mathbf{x})$ is strictly concave over $[0, 1]^L$.
3. $s(\mathbf{x})$ reaches its maximum[6] at $\mathbf{x}_0 = (\frac{1}{L}, \ldots, \frac{1}{L})$, with $s(\mathbf{x}_0) = \ln L$.

So, the macrostate corresponding to the largest number of microstates correspond to a uniform distribution: $n_i = \frac{1}{L}, \forall i = 1, \ldots, L$.

We can again divide the set Ω into the set Ω_{eq} of (almost) equilibrium microstates:

$$\Omega_{eq} \equiv \{\mathbf{n} \in \Omega \mid |\mathbf{n} - \mathbf{x}_0| \leq \epsilon\} \tag{6.2.10}$$

for some $\epsilon > 0$ and the rest, the non-equilibrium microstates: $\Omega_{neq} = \Omega \setminus \Omega_{eq}$.

Estimates similar to the ones leading to (6.2.6) (left as an exercise) yield again:

$$\frac{|\Omega_{neq}|}{|\Omega_{eq}|} \leq N^{L-1} \exp[-\mathcal{O}(N\epsilon^2)], \tag{6.2.11}$$

where N^{L-1} is an upper bound on the number of possible values of n_i, $i = 1, \ldots, L$ (with $\sum_{i=1}^{L} n_i = 1$).

This calculation applies to the distribution of particles in the L sub-boxes of the box Λ, see (6.1.3), with $n_i = \frac{N_{\bar{u}_i}(\mathbf{q})}{N}$, $i = 1, \ldots, L$.

---

not exactly equal to it, see (2.3.2), (2.3.7), (2.3.11). See Lazarovici and Reichert [212, Sect. 5.3] for more discussion of Lavis' views.

[5] To get this formula, list all the permutations of the N particles; put the N_1 first particles in state 1, the next N_2 particles in state 2 etc. Thus we get $N!$ permutations the N particles to be divided by the number $\prod_{i=1}^{L} N_i!$ of permutations of the N_i particles in each group $i = 1, \ldots, L$, since permutations in those groups do not change the final distribution of the particles among states.

[6] This can be shown by using Lagrange's multipliers, as one does below, but with only the constraint (6.2.14) and not the constraint (6.2.15).

Consider now a system of N particles each of which can be in L possible states, but with an "energy" variable e_i, $i = 1, \ldots, L$, associated to each such state, and with the sum of the energies being fixed: $E = \sum_{\alpha=1}^{N} e_{i(\alpha)}$.

The set of microstates is $\Omega = \{1, \ldots, L\}^N$, and we can take as macrostates the fractions of particles in each state: $\mathbf{n} = (n_1, \ldots, n_L)$, where $n_i = \frac{N_i}{N}$ with N_i being the number of particles in state i. In terms of the macrostate, the total energy equals: $E = \sum_{i=1}^{L} N_i e_i$ and $\frac{E}{N} = \sum_{i=1}^{L} n_i e_i$.

We need to maximize $|M^{-1}(\mathbf{n})| = \frac{N!}{\prod_{i=1}^{L} N_i!}$ for N large, but under the constraints $\sum_{i=1}^{L} n_i = 1$ and $\sum_{i=1}^{L} n_i e_i = E$. Using (6.2.9), we see that we must maximize the function $s(\mathbf{n}) = -\sum_{i=1}^{L} n_i \ln n_i$ under the above constraints. Using the method of Lagrange's multipliers, we maximize[7]:

$$-\sum_{i=1}^{L} n_i \ln n_i - \beta \left(\sum_{i=1}^{L} n_i e_i - \frac{E}{N} \right) - \lambda \left(\sum_{i=1}^{L} n_i - 1 \right), \tag{6.2.12}$$

where $\lambda \left(\sum_{i=1}^{L} n_i - 1 \right)$ is included because of the constraint $\sum_{i=1}^{L} n_i = 1$.
Setting the derivative with respect to n_i equal to 0 for each i, we get:

$$-\ln n_i - 1 - \beta e_i - \lambda = 0,$$

or

$$n_i = \exp(-\beta e_i - 1 - \lambda), \tag{6.2.13}$$

where the constants β and λ are determined by the constraints:

$$\sum_{i=1}^{L} n_i = \sum_{i=1}^{L} \exp(-\beta e_i - 1 - \lambda) = 1 \tag{6.2.14}$$

and

$$\sum_{i=1}^{L} n_i e_i = \sum_{i=1}^{L} \exp(-\beta e_i - 1 - \lambda) e_i = \frac{E}{N}. \tag{6.2.15}$$

From (6.2.14) we get

$$\exp(-1 - \lambda) = \frac{1}{\sum_{i=1}^{L} \exp(-\beta e_i)}$$

Thus, we have from (6.2.13):

---

[7] We denote β one of the multipliers because it will be shown later to coincide with the $\beta = \frac{1}{kT}$ of thermodynamics.

$$n_i = \frac{\exp(-\beta e_i)}{\sum_{i=1}^{L} \exp(-\beta e_i)}, \tag{6.2.16}$$

and (6.2.15) can be rewritten as:

$$\frac{\sum_{i=1}^{L} \exp(-\beta e_i) e_i}{\sum_{i=1}^{L} \exp(-\beta e_i)} = \frac{E}{N}. \tag{6.2.17}$$

Formula (6.2.17) defines $e = e(\beta)$. Let us compute the derivative of $e(\beta) = \frac{E}{N}$ with respect to β:

$$-\frac{\partial e(\beta)}{\partial \beta} = <e^2> - <e>^2, \tag{6.2.18}$$

where we use the notation $<f> = \frac{\sum_{i=1}^{L} \exp(-\beta e_i) f(i)}{\sum_{i=1}^{L} \exp(-\beta e_i)}$ and $e(i) = e_i$. Formula (6.2.18) follows from the formula for the derivative of the ratio of functions in (6.2.17).

Since the variance $<e^2> - <e>^2 > 0$ (unless e_i is constant in i) by (2.A.10), $e(\beta)$ is decreasing in β and we can thus invert (6.2.17); this inversion defines implicitly $\beta = \beta(e)$ as a function of $e = \frac{E}{N}$.

It is important to note that, although we use the notation β in anticipation for its later identification with $\frac{1}{kT}$, in Sect. 6.6, where T is the thermodynamic temperature defined in (5.6.5), for the moment β is just a Lagrange multiplier (which is of course a function of the variable e).

Define $\mathbf{n}_{eq} = (n_i)_{i=1}^{L}$, with the n_i's given by (6.2.16), namely the equilibrium distribution of the n_i's, we have that (6.2.19) for $\mathbf{n} = \mathbf{n}_{eq}$ equals:

$$s(\mathbf{n}_{eq}) = -\frac{\sum_{i=1}^{L} \exp(-\beta e_i)(-\beta e_i - \ln(\sum_{i=1}^{L} \exp(-\beta e_i)))}{\sum_{i=1}^{L} \exp(-\beta e_i)}$$

$$= \beta <e> + \ln \sum_{i=1}^{L} \exp(-\beta e_i), \tag{6.2.19}$$

with $<e> = \frac{E}{N}$.

If one divides the set of microstates into equilibrium and non-equilibrium ones, as in (6.2.10):

$$\Omega_{eq} \equiv \{\mathbf{n} \in \Omega \mid |\mathbf{n} - \mathbf{n}_{eq}| \leq \epsilon\}$$

for some $\epsilon > 0$ and the rest, $\Omega_{neq} = \Omega \setminus \Omega_{eq}$, one can again derive an estimate of the form (6.2.11).

This computation can also be applied to the distribution of the velocities of free particles in a box, at least approximately. We use the macroscopic variables (6.1.5), with the approximation that the momenta $\vec{p}_i = m\vec{v}_i$ is constant in each cubic cell $\Delta(\vec{u}_1), \ldots, \Delta(\vec{u}_L)$. We have the constraint that the total energy is fixed: $\sum_{i=1}^{N} \frac{|\vec{p}_i|^2}{2m} = E$. Applying the calculation leading to (6.2.16), we obtain the Maxwellian or Maxwell–Boltzmann distribution of velocities:

Fig. 6.2 A partition of the phase space Ω (represented by the entire square) into regions $\Omega_0, \Omega_1, \Omega_2, \dots$ corresponding to microstates that are macroscopically indistinguishable from one another, i.e. that give rise to the same value of M. The region Ω_{eq} labelled "thermal equilibrium" corresponds to the value of M corresponding to the overwhelming majority of microstates. This figure is inspired by a similar picture in Chap. 7 of Penrose [255]

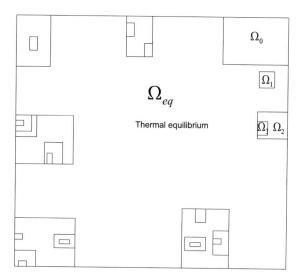

$$n_i = \frac{\exp(-\frac{\beta|\vec{p}_i|^2}{2m})}{\sum_{i=1}^{L} \exp(-\frac{\beta|\vec{p}_i|^2}{2m})}. \qquad (6.2.20)$$

We will rederive this distribution in Sect. 6.6.3.1, see (6.6.16).

We can summarize the calculations done here, through the following picture of the phase space.

In Fig. 6.2, each region in Ω corresponds to the set of microstates giving rise to the same value of M and we denote these regions by $\Omega_0, \Omega_1, \Omega_2, \dots$. Let us stress that Fig. 6.2 is highly "abstract" since the phase space Ω represented there by a two-dimensional square is in reality a subset of a space of dimension of order 10^{23}.

Moreover, the function M usually takes a continuum of values and in Fig. 6.2 we do as if those values were discrete. But we did that approximation in (6.1.10) and it simplifies our illustration.

In the example of coin tossing, the region Ω_{eq} labelled equilibrium in Fig. 6.2 corresponds to having approximately as many heads and tails, i.e. the set of coin tosses defined by (6.2.4). The region Ω_0 may correspond to having approximately one third heads and two third tails, Ω_1 may correspond to having approximately one quarter heads and three quarters tails, etc.

In the example of the gas in the box, the region Ω_{eq} labelled "thermal equilibrium" in Fig. 6.2 corresponds to an approximate uniform distribution of the particles in the box and an approximate Maxwellian distribution of their velocities (6.2.20).

The region Ω_0 in Fig. 6.2 might correspond to all the particles being in one half of the box, another region might correspond to all the particles being in the other half, yet another region, say Ω_1, might correspond to all the particles being in an even smaller part of the box etc.

Of course, the regions $\Omega_0, \Omega_1, \Omega_2 \ldots$ in Fig. 6.2 are not drawn to scale: if the size of the region in the phase space where all the particles are concentrated in one part of the box is 2^{-N} smaller than the one where the particles are uniformly distributed in Λ, and N is of the order of the Avogadro's number, $N \approx 10^{23}$, that region where all the particles are concentrated in one part of the box could not be visible at all if those regions were drawn to scale.

The thermal equilibrium region Ω_{eq} is almost equal to the entire phase space Ω, as shown by estimate like (6.2.6), (6.2.11) and all the non-equilibrium regions put together (all the particles concentrated in one part of the box, or the distribution of the velocities being different from the Maxwellian one) occupy only a tiny fraction of Ω.

6.3 Typicality

In each of the examples above, the configurations in Ω_{eq} are *typical*, to use the term defined Sect. 2.3.1. In the situations considered in statistical mechanics, we always have a large parameter N, which, for mathematical convenience, we let tend to ∞. The typicality of the configurations in Ω_{eq} is expressed by bounds of the form (6.2.6, 6.2.11).

As we discussed in Sect. 2.6 typical events do not need to be explained, since, by definition of "probable" or by Cournot's principle, typical events are what we expect to happen. Thus, equilibrium states do not need to be explained. What need to be explained are non-equilibrium states and the evolution from non-equilibrium states towards equilibrium ones (see Chap. 8).

6.4 Entropy in Equilibrium

Let us start with a general and fundamental definition:

Definition 6.1 To every macrostate function M we can assign a Boltzmann entropy $S_B(\mathbf{x})$, which is a function of the microstate:

$$S_B(\mathbf{x}) = k \ln |\Omega_M(\mathbf{x})|, \tag{6.4.1}$$

where $|E|$ is either the number of elements in the set E, if we are in a discrete situation with $|E| < \infty$, or the Lebesgue measure of E in the continuous case, where $\Omega_M(\mathbf{x})$ is a subset of the phase space $\mathbb{R}^{6N}$; here k is Boltzmann's constant (see Appendix 5.C).[8]

---

[8] That formula is inscribed on Boltzmann's tomb in Vienna, with Ω replaced by W, W for Wahrscheinlichkeit (probability in German), but it was first written in that form by Max Planck, who however credited the idea to Boltzmann.

Remark 6.2 It should be emphasized that this function is defined on the set of microstates Ω but its definition depends on the choice of the map M, and therefore is sometimes considered to have a "subjective" aspect (see Sect. 8.9 for a detailed discussion).

Remark 6.3 Most of equilibrium statistical mechanics consists in justifying formula (6.4.1) and in deriving its consequences. We define

$$S_{eq}(\mathbf{x}) = k \ln |\Omega_M(\mathbf{x})| = k \ln |\Omega_{eq}| \approx k \ln |\Omega|, \tag{6.4.2}$$

$\forall \mathbf{x} \in \Omega_{eq}$, where Ω_{eq} is the region labelled "thermal equilibrium" in Fig. 6.2 (by definition, $|\Omega_M(\mathbf{x})|$ is constant for $\mathbf{x} \in \Omega_{eq}$). Although the "correct" conceptual definition is $k \ln |\Omega_{eq}|$, we will use later the approximate formula $k \ln |\Omega|$, which is justified by the dominance of the equilibrium macrostate shown in Sect. 6.2.

Remark 6.4 The reader might worry that the size $|\Omega_M(\mathbf{x})|$, in the continuous situation, is not dimensionless: its numerical value depends on the units that we use to measure volumes in phase space. However, changing those units will change the value $|\Omega_M(\mathbf{x})|$ by a multiplicative factor, hence change the value of $S_{eq}(\mathbf{x})$ by an additive constant; but, as we saw in Chap. 5, physical quantities such as the pressure and the temperature are defined by derivatives of the entropy, see (5.6.5) and are unaffected by this additive constant. Since the multiplicative constant will be exponential of N, the additive constant will be linear in N and that will affect the value of the chemical potential μ, see (5.6.17). But the latter is defined as the derivative of the energy with respect to the number N of particles, see (5.6.15), and the energy is also defined up to an additive constant (that can be linear in N).

In many physics textbooks, this arbitrariness in the choice of units is "solved" by dividing $|\Omega_M(\mathbf{x})|$ by a power of Planck's constant h and appealing to quantum mechanics, where the set of states is discrete, to justify that normalization. However, here we want to formulate classical statistical mechanics in a self-consistent way.

The same remark about the arbitrariness in the choice of units holds for the other thermodynamic potentials, see Sect. 6.5, in particular (6.5.1), (6.5.5) and (6.5.8), (6.5.12), but without more consequences.

Note also that, when we will introduce probability distributions or ensembles in Sect. 6.6, we will always deal with ratios of phase space volumes in the microcanonical ensemble or by similar ratios in the other ensembles, that are thus unaffected by this multiplicative factor.

Before justifying (6.4.2) in Sect. 6.7, let us collect the explicit formulas for (6.4.2) that were computed before.

1. For a system of particles being in two possible states, see (6.2.2):

$$S(n_o) = -kN(n_0 \ln n_0 - (1 - n_0) \ln(1 - n_0)),$$

whose maximum is reached for $n_0 = \frac{1}{2}$, with $S(\frac{1}{2}) = k \ln 2$.

2. For a system of particles being in L possible states, see (6.2.9):

$$S(\mathbf{n}) = -kN \left(\sum_{i=1}^{L} n_i \ln n_i \right),$$

whose maximum is reached for $\mathbf{n}_{eq} = (n_i)_{i=1}^{L}$, with $n_i = \frac{1}{L}$, $\forall i = 1, \ldots, L$, and $S(\mathbf{n}_{eq}) = k \ln L$. In particular, this applies the fraction of the number of particles in sub-boxes Λ_i of a box Λ, see (6.1.3).

3. For a system of particles being in L possible states, each of which has an energy e_i, with the total energy $\frac{E}{N} = \sum_{i=1}^{L} n_i e_i$ fixed, we get for the maximum value of $S(\mathbf{n})$, see (6.2.19):

$$S(\mathbf{n}_{eq}) = k(\beta E + N \ln \sum_{i=1}^{L} \exp(-\beta e_i)),$$

4. Entropy of an ideal gas

We will consider now the equilibrium situation for a simple monoatomic ideal gas. Here, the macrostate will be defined by the following variables: the energy E, the volume V, the number of particles N, the temperature T, the pressure P and the chemical potential μ.

A gas is ideal or non interacting if $V_{ij} = 0$ in (3.2.5) and $V_i(\vec{q}_i) = 0$ for $\vec{q}_i \in \Lambda$, where Λ is a bounded subset of $\mathbb{R}^3$, and $V_i(\vec{q}_i) = \infty$ for $\vec{q}_i \notin \Lambda$, which means that the particles do not interact with each other but are confined in the box Λ. The phase space volume seems to be given by $|\Omega(E, V, N)|$, the Lebesgue measure of $\Omega(E, V, N)$, which is the set of microstates $(\mathbf{q}, \mathbf{p}) \in \mathbb{R}^{6N}$ such that the particles are in the box Λ with $|\Lambda| = V$ and with total energy $E = \sum_{i=1}^{N} \frac{|\vec{p}_i|^2}{2m}$. That last constraint means that the vector $\mathbf{p}$ lies on the surface of a sphere of radius $\sqrt{2mE}$ in a space of dimension $3N$. For simplicity, we will set $2m = 1$.

But before computing $|\Omega(E, V, N)|$ we must observe that there is some overcounting. To understand this, consider a system made of two identical particles (by that, we mean that all their known intrinsic physical properties such as their mass or charge are the same) with positions and momenta given by $(\vec{q}_1, \vec{p}_1)$, $(\vec{q}_2, \vec{p}_2)$.

Now interchange the two particles: take particle 1 and put it in position $\vec{q}_2$ with momentum $\vec{p}_2$ and vice-versa. Obviously nothing changes to the physical situation of the world, since we have the same distribution of mass and charge as before this interchange, and it is only through those distributions that the particles interact with each other and with the rest of the world. In other words, our *labelling* of the particles is devoid of physical meaning.

This was clearly stated by Nino Zanghì[9]:

---

[9] This comes from his course notes on Statistical Physics, available on his webpage http://www.ge. infn.it/~zanghi/. It is quoted by Tumulka [314, Sect. 5.4], who adds: the clarification of the status of identical particles, both classical and quantum, is mainly due to Leinaass and Myrheim [228].

The textbooks [of classical mechanics and statistical mechanics] tend to underline that the proper description of identical particles can be achieved only within the framework of quantum mechanics. The standard argument is something like this: *Particles are identical if they cannot be distinguished by means of measurements. So, if particles have the same mass, charge, etc., they could be distinguished only by their location in space, as it is the case in classical mechanics. However, in quantum mechanics, particles do not have trajectories. Therefore they cannot be distinguished if they have the same mass, charge, etc.. Thus the notion of identical particles is purely a quantum one without any classical equivalent.* The conclusion is faulty and the argument is wrong.

The notion of identical particles was already recognized by Gibbs. In order to correctly calculate the entropy change in a process of mixing to identical fluids or gases (at the same temperature, etc.), Gibbs postulated that states differing only by permutations of identical particles should not be counted as distinct.

Thus, we should divide $|\Omega(E, V, N)|$ by the number $N!$ of permutations of those N particles, and define

$$S(E, V, N) \equiv S_{eq}(E, V, N) = k \ln \frac{|\Omega(E, V, N)|}{N!} \tag{6.4.3}$$

Since the area of a sphere of radius R in $\mathbb{R}^n$ is given by $S_{n-1} = c_n R^{n-1}$ with $c_n = 2\frac{\pi^{\frac{n}{2}}}{\Gamma(\frac{n}{2})}$, where Γ is the gamma function,[10] we have

$$|\Omega(E, V, N)| = V^N c_{3N} E^{\frac{3N-1}{2}} \tag{6.4.5}$$

and thus, using Stirling's formula, see (6.A.2), which is valid also for the gamma function, $\ln N! \approx N \ln N - N$ and $\ln \Gamma\left(\frac{3N}{2}\right) \approx \frac{3N}{2} \ln\left(\frac{3N}{2}\right) - \frac{3N}{2}$, the entropy $S(E, V, N)$ defined in (6.4.3) equals:

$$S(E, V, N) \tag{6.4.6}$$
$$\approx k\left(N \ln V - N \ln N + N + \frac{3N}{2} \ln E - \frac{3N}{2} \ln\left(\frac{3N}{2}\right) + \frac{3N}{2} + \frac{3N}{2} \ln \pi\right)$$
$$= kN\left(\ln v + \frac{3}{2} \ln\left(\frac{2\pi}{3}e\right) + \frac{5}{2}\right) \tag{6.4.7}$$

where $v = \frac{V}{N}$, $e = \frac{E}{N}$ and we neglect terms $o(N)$. We can also introduce the entropy per particle:

---

[10] The gamma function is a generalization of the factorial: for $z \in \mathbb{C}, \Re z > 0$,

$$\Gamma(z) = \int_0^\infty x^{z-1} e^{-x} dx \tag{6.4.4}$$

and for $z = n, n \in \mathbb{N}$, we have $\Gamma(n) = (n-1)!$, see Appendix 6.A.2.

$$s(e, v) = \lim_{N \to \infty} \frac{S(E, V, N)}{N} = k\left(\ln v + \frac{3}{2} \ln \left(\frac{2\pi}{3} e \right) + \frac{5}{2} \right), \qquad (6.4.8)$$

where the limit is taken with $e = \frac{E}{N}$, $v = \frac{V}{N}$ fixed.

Since $s(e, v)$ gives the correct thermodynamic formula for the entropy of an ideal gas in equilibrium (see (5.7.15), (5.7.16), (5.7.17)), with $c = \frac{3}{2}$ and where the constant $s_0 = (c + 1)k + \frac{3}{2}k \ln(\frac{2\pi}{3}) = \frac{5}{2}k + \frac{3}{2}k \ln(\frac{2\pi}{3})$, which fixes the value of $-(\frac{\mu}{T})_0 - k \ln e_0^c v_0 = \frac{3}{2}k \ln(\frac{2\pi}{3})$ in (5.7.17). We obtain, by taking partial derivatives:

$$\frac{\partial S(E, V, N)}{\partial E} = \frac{\partial s(e, v)}{\partial e} = \frac{3Nk}{2E} = \frac{1}{T}$$

and thus $E = \frac{3NkT}{2}$, which is (5.7.2). We have also

$$\frac{\partial S(E, V, N)}{\partial V} = \frac{Nk}{V} = \frac{P}{T}$$

which gives $PV = NkT$, which is (5.7.1).

6.5 Other Equilibrium Potentials

If we accept the definition (6.4.2), (6.4.3) of the equilibrium entropy (which will be justified in Sect. 6.7), then we can obtain formulas for the other thermodynamic potential in terms of microscopic variables.

We will start with the Helmholtz free energy and proceed formally in our derivation. Consider the *canonical partition function*:

$$Z_c(T, V, N) = \frac{1}{N!} \int_\Omega \exp(-\beta E(\mathbf{x})) d\mathbf{x} \qquad (6.5.1)$$

with $\beta = \frac{1}{kT}$ and the symbol $\int \cdot d\mathbf{x}$ means either an integral over $\Lambda^{3N} \times \mathbb{R}^{3N}$ with $d\mathbf{x} = d\mathbf{q} d\mathbf{p}$ or a sum over all discrete states of the system, with N fixed and without the factor $\frac{1}{N!}$.

Remember that, although we write $\beta = \frac{1}{kT}$, both β and T here are just parameters that have not yet been identified with the corresponding thermodynamic quantities (this identification will be done in Sect. 6.7).

Now, rewrite (6.5.1) in the continuous case, using $|\Omega(E, V, N)| = \int \delta(E(\mathbf{x}) - E) d\mathbf{x}$, and with $Z = Z_c$,

$$Z(T, V, N) = \int_0^\infty \exp(-\beta E) \int \frac{\delta(E(\mathbf{x}) - E)}{N!} d\mathbf{x}$$

$$= \int_0^\infty \exp(-\beta E) \frac{|\Omega(E, V, N)|}{N!} dE \qquad (6.5.2)$$

$$= \int_0^\infty \exp\left(-\beta E + \frac{S(E, V, N)}{k}\right) dE.$$

Writing $-\beta E + \frac{S(E,V,N)}{k} = N(-\beta e + \frac{s(e,v)}{k})$ and using Laplace's method for $N \to \infty$ leads to the following approximation[11]:

$$Z(T, V, N) \approx \exp\left(N \sup_E \left(-\beta \frac{E}{N} + \frac{S(E, V, N)}{kN}\right)\right), \qquad (6.5.3)$$

in the sense that

$$\lim_{N \to \infty} \frac{\ln Z(T, V, N)}{N} = \sup_e \left(-\beta e + \frac{s(e, v)}{k}\right), \qquad (6.5.4)$$

with $s(e, v)$ defined in (6.4.8).

Since $s(e, v)$ given in (6.4.8) is a concave function of e, the supremum (which is a maximum) is obtained by setting $\beta k = \frac{\partial s(e,v)}{\partial e}$ or $\frac{\partial s(e,v)}{\partial e} = \frac{1}{T}$, which is (5.6.5) and we have, at that supremum, $-\beta e + \frac{s(e,v)}{k} = -\beta e(T, v) + \frac{s(e(T,v),v)}{k}$, with the function $e(T, v)$ obtained from (5.6.8).

Going back to the definition (5.6.6) of Helmoltz free energy, in terms of Legendre's transform of $s(e, v)$ with respect to e, $f(T, v) = e(T, v) - Ts(e(T, v), v)$ we get:

$$f(T, v) = -kT \lim_{N \to \infty} \frac{\ln Z(T, V, N)}{N} = e(T, v) - Ts(e(T, v), v), \qquad (6.5.5)$$

which gives the formula for the Helmoltz free energy in terms of microscopic variables.

From (6.5.5), we also obtain the formula for the entropy as a function of (T, V, N):

$$S(T, V, N) = k(\ln Z(T, V, N) + \beta E(T, V, N))$$

$$= k\left(\ln Z(T, V, N) - \beta \frac{\partial \ln Z(T, V, N)}{\partial \beta}\right), \qquad (6.5.6)$$

which will be used when we will connect the microscopic formula for the entropy and the thermodynamic one (5.6.4).

For an ideal gas, we have, setting the mass of the particle $m = \frac{1}{2}$ (as we did in the derivation of (6.4.8)),

---

[11] See Appendix 6.A.1 for a discussion of that method and the justification of (6.5.4) below.

$$Z(T, V, N) = \frac{V^N (\sqrt{\pi k T})^{3N}}{N!},$$

since in (6.5.1) we have N integrals over the position variables in Λ giving rise to the factor V^N, and $3N$ integrals of the form $\int_{\mathbb{R}} \exp(-\beta p^2) dp = \sqrt{\pi k T}$. Using (6.5.5), this gives

$$f(T, v) = -kT \left(\ln v + \frac{3}{2} \ln(\pi k T) + 1 \right), \tag{6.5.7}$$

which coincides with (5.7.22) with $c = \frac{3}{2}$, with the constant $s_0 = \frac{5}{2}k + \frac{3}{2}k \ln(\frac{2\pi}{3})$ fixed as in (6.4.8). It is easy to check that $f(T, v)$ given by (6.5.7) is the Legendre transform of $s(e, v)$ given by (6.4.8), by using $e = \frac{3}{2}kT$ and so, $\frac{3}{2} \ln(\frac{2\pi}{3} e) = \frac{3}{2} \ln(\pi k T)$.

A calculation similar to (6.5.3), (6.5.4), (6.5.5) can be done for the grand potential defined in (5.6.21). Define the *grand-canonical partition function*:

$$\mathcal{Z}_g(T, V, \mu) = \sum_{N=0}^{\infty} \exp(\beta \mu N) Z(T, V, N)$$

$$= \sum_{N=0}^{\infty} \exp(\beta \mu N) \int_{\Omega} \exp(-\beta E(\mathbf{x})) \frac{d\mathbf{q} d\mathbf{p}}{N!} = \sum_{N=0}^{\infty} z^N \int_{\Omega} \exp(-\beta E(\mathbf{x})) \frac{d\mathbf{q} d\mathbf{p}}{N!}$$

$$\tag{6.5.8}$$

with $z = \exp(\beta \mu)$ and $Z(T, V, N)$ defined in (6.5.1).

Then we get

$$\mathcal{Z}_g(T, V, \mu) \approx \exp \left(V \sup_N \left(\beta \mu \frac{N}{V} + \frac{\ln Z(T, N, V)}{V} \right) \right), \tag{6.5.9}$$

whose rigorous meaning is:

$$\lim_{V \to \infty} \frac{\ln \mathcal{Z}_g(T, V, \mu)}{V} = \beta \sup_n (\mu n - f(T, n) n), \tag{6.5.10}$$

with $n = \frac{N}{V}$ fixed and $f(T, n) = -kT \lim_{N \to \infty} \frac{\ln Z(T, N, V)}{N}$, where we write $\frac{\ln Z(T, N, V)}{V} \approx -\beta f(T, n) n$ in order to pass from (6.5.9) to (6.5.10).

To prove (6.5.9), one has to show that the infinite sum over N can be bounded by the supremum in (6.5.9), which requires some estimates that we will not do here. But note that, intuitively, since N grows (on average) linearly with V, one expects the sum to grow also linearly with V (if the contributions coming from very large N are negligible) and thus not to contribute to the limit in (6.5.9).

Thus, the grand potential, being the Legendre transform with respect to N of $F(T, V, N)$, see (5.6.21) and (5.B.4) for the expression of the Legendre transform, equals

$$\Phi(T, V, \mu) = -kT \ln \mathcal{Z}_g(T, V, \mu), \qquad (6.5.11)$$

whose limit $V \to \infty$ defines

$$\Phi(T, \mu) = -kT \lim_{V \to \infty} \frac{\ln \mathcal{Z}_g(T, V, \mu)}{V} = f(T, \mu)n - \mu n, \qquad (6.5.12)$$

where n is expressed in terms of μ.

Formula (6.5.12) can be written as $-kT \ln(\mathcal{Z}_g(T, V, \mu)) \approx V\Phi(T, \mu) = \Phi(T, V, \mu)$, which means that, since, by (5.6.22), the pressure $P = -\frac{\partial \Phi(T, V, \mu)}{\partial V}$:

$$P = kT \frac{\partial \ln \mathcal{Z}_g(T, V, \mu)}{\partial V} \approx kT \lim_{V \to \infty} \frac{\ln \mathcal{Z}_g(T, V, \mu)}{V} = kT\Phi(T, \mu). \quad (6.5.13)$$

For an ideal gas, we have $\mathcal{Z}_g(T, V, \mu) = \sum_{N=0}^{\infty} \frac{\exp(\beta\mu N) V^N (\sqrt{\pi kT})^{3N}}{N!}$ which, summing over N, equals $\exp(e^{\beta\mu} V (\sqrt{\pi kT})^3)$. Using (6.5.12), we get:

$$\Phi(T, \mu) = -kT \lim_{V \to \infty} \frac{\ln \mathcal{Z}_g(T, V, \mu)}{V} = -kT(\pi kT)^{\frac{3}{2}} e^{\beta\mu},$$

which coincides with the formula (5.7.26) for the limit $\Phi(T, \mu) = \lim_{V \to \infty} \frac{\Phi(T, V, N)}{V}$ defining the grand potential $\Phi(T, V\mu)$, with $c = \frac{3}{2}$, and with the constants $C_0 = (c+1)k - s_0 = \frac{5}{2}k - s_0 = -\frac{3}{2}k \ln \frac{2\pi}{3}$ and $C_1 = \exp(-\frac{C_0}{k}) = (\frac{2\pi}{3})^{\frac{3}{2}}$, see (5.7.23), (5.7.24), determined by $s_0 = \frac{5}{2}k + \frac{3}{2}k \ln(\frac{2\pi}{3})$, i.e. by its value in (6.4.8).

6.6 The Equilibrium Ensembles

The most traditional approach to statistical mechanics, which is also the most effective from a computational point of view, is via "ensembles" which is the word physicists use to refer, in this context, to probability distributions (or measures) over the set of microstates. In what follows we will use the expression "probability distributions."

6.6.1 Definition of the Ensembles

There are three basic such probability distributions: micro-canonical, canonical and grand-canonical. Since the latter two can be derived from the micro-canonical distribution, and since they are all, in some sense, equivalent to each other, we will concentrate on the micro-canonical distribution, when discussing the meaning and the justification of the use of ensembles in Sects. 6.6.4 and 6.8.

But let us first define those distributions as purely mathematical objects:

The *micro-canonical distribution* $d\nu_m$ is a probability distribution on $\Omega(E, V, N)$ (indexed by (E, V, N)), given, in the continuous case, by:

$$d\nu_m(\mathbf{x}) = \frac{\mu_{\text{Liouville}}(\mathbf{x})d\mathbf{x}}{|\Omega(E, V, N)|} \qquad (6.6.1)$$

where $\mu_{\text{Liouville}}$ has been defined in (3.4.11); and, in the discrete case by

$$\nu_m(\mathbf{x}) = \frac{1}{|\Omega|}. \qquad (6.6.2)$$

Let us call $\frac{|\Omega(E,V,N)|}{N!}$ the *micro-canonical partition function*.

The *canonical distribution* $d\nu_c$ is a probability distribution on $\cup_E\Omega(E, V, N)$, indexed by (T, V, N), and given, in the continuous case (where $\cup_E\Omega(E, V, N) = \mathbb{R}^{6N}$), by:

$$d\nu_c(\mathbf{x}) = \frac{\exp(-\beta E(\mathbf{x}))d\mathbf{x}}{Z_c}, \qquad (6.6.3)$$

and, in the discrete case by:

$$\nu_c(\mathbf{x}) = \frac{\exp(-\beta E(\mathbf{x}))}{Z_c} \qquad (6.6.4)$$

with Z_c defined in (6.5.1). Note that, in (6.6.1), (6.6.3) the factor $N!$ appears both in the numerator and the denominator.

Finally the *grand-canonical distribution* $d\nu_g$ is a probability distribution on $\cup_{E,N}\Omega(E, V, N)$, indexed by (T, V, μ), and given, in the continuous case, by:

$$d\nu_g(\mathbf{x}) = \frac{1}{N!}\frac{\exp(\beta\mu N(\mathbf{x}) - \beta E(\mathbf{x}))d\mathbf{x}}{\mathcal{Z}_g}, \qquad (6.6.5)$$

and, in the discrete case by:

$$\nu_g(\mathbf{x}) = \frac{\exp(\beta\mu N(\mathbf{x}) - \beta E(\mathbf{x}))}{\mathcal{Z}_g}. \qquad (6.6.6)$$

with $\mathcal{Z}_g$ defined in (6.5.8) in the continuous case and by the sum over $\mathbf{x}$ of the numerator of (6.6.6) in the discrete case.

6.6.2 The Gibbs Entropy

There is another definition of the entropy that is, in equilibrium, equivalent to the Boltzmann entropy given by (6.4.2). To each measure absolutely continuous with respect to the Lebesgue measure $d\nu(\mathbf{x}) = \rho(\mathbf{x})d\mathbf{x}$, with $\int_\Omega \rho(\mathbf{x}))d\mathbf{x} = 1$ or, in the discrete case, $\sum_{\mathbf{x}} \nu(\mathbf{x}) = 1$, one can associate an entropy[12]:

Definition 6.5 The Gibbs entropy $S_G(\mathbf{x})$ of a probability measure $d\nu(\mathbf{x}) = \rho(\mathbf{x})d\mathbf{x}$ is defined by:

$$S_G(\nu) = -k \int_\Omega \ln \rho(\mathbf{x})\rho(\mathbf{x})d\mathbf{x} = -k \int_\Omega \ln \frac{d\nu(\mathbf{x})}{d\mathbf{x}} d\nu(\mathbf{x}), \qquad (6.6.7)$$

or, in the discrete case,

$$S_G(\nu) = -k \sum_{\mathbf{x}} \ln \nu(\mathbf{x})\nu(\mathbf{x}). \qquad (6.6.8)$$

If we take $\nu = \nu_c$, defined by (6.6.3) or (6.6.4) in the formulas above, we get:

$$S_G(\nu_c) = -k(-\beta \int_\Omega E(\mathbf{x})\rho(\mathbf{x})d\mathbf{x}) - \ln Z_c), \qquad (6.6.9)$$

or,

$$T S_G(\nu_c) = \int_\Omega E(\mathbf{x})\rho(\mathbf{x})d\mathbf{x} + kT \ln Z_c = < E > + kT \ln Z_c, \qquad (6.6.10)$$

If we divide both side by N and let $N \to \infty$, and compare with (6.5.6), we see that

$$\lim_{N\to\infty} \frac{S_G(\nu_c)}{N} = s(e(T, v), v).$$

This means that *the Gibbs and the Boltzmann entropies coincide in equilibrium*, when $N \to \infty$. One can easily verify that this holds also in the discrete case.

6.6.3 Equivalence of Ensembles

There is a sense in which all ensembles are equivalent; they describe the same systems but through different variables: (E, V, N) for the micro-canonical ensemble, (T, V, N) for the canonical one, and (T, V, μ) for the grand-canonical ensemble.

But we have to specify what "equivalent" means; we already saw in Sect. 6.5 that the partition functions associated to each ensemble lead, after taking logarithms and multiplying them by some T-dependent factor, to the basic thermodynamic functions, that are related to each other through Legendre's transformations. So, these ensembles are equivalent at the level of the thermodynamic functions since those functions are "equivalent" because of the invertibility of the Legendre's transformations.

Let us now see in which sense the ensembles are equivalent as probability distributions.

---

[12] We shall see in Chap. 7 that this definition coincides with the one of the Shannon entropy.

6.6.3.1   From the Microcanonical Distribution to the Canonical and
Grand Canonical Ones

We will first show that the micro-canonical probability distribution "implies" the
other ones in the sense that the latter are the probability distributions "induced" on
subsystems by the micro-canonical distribution or, using the probability language of
Appendix 2.A.6, that the canonical and grand-canonical distributions are the marginal
distributions of the micro-canonical probability distribution on subsystems of larger
systems. We will deal only with the continuous case, since both cases are similar.

Consider a box Λ, with $|\Lambda| = V$, containing a set of N particles whose total
energy is E. And consider a subsystem having $\tilde{N}$ particles in a box $\tilde{\Lambda} \subset \Lambda$, with
$|\tilde{\Lambda}| = \tilde{V}$, whose configuration is $\mathbf{x}$, with $E(\mathbf{x}) = \tilde{E}$. We will assume that $\tilde{N}$, $\tilde{V}$ and
$\tilde{E}$ are fixed while we let $N, V, E \to \infty$.

The probability density of that particular state with respect to the micro-canonical
distribution on $\Omega(E, V, N)$ is proportional to:

$$\frac{|\Omega(E - \tilde{E}, V - \tilde{V}, N - \tilde{N})|}{|\Omega(E, V, N)|}, \tag{6.6.11}$$

since $|\Omega(E - \tilde{E}, V - \tilde{V}, N - \tilde{N})|$ is the size of the set of states in $\Omega(E, V, N)$
whose configuration inside $\tilde{\Lambda}$ is fixed and has and energy $\tilde{E}$ and a number of particles
$\tilde{N}$.

So, if we consider the relative probabilities of two configurations $\mathbf{x}_1$, $\mathbf{x}_2$ of the
subsystem of energies $E(\mathbf{x}_1) = \tilde{E}_1$ and $E(\mathbf{x}_2) = \tilde{E}_2$, we get:

$$\frac{|\Omega(E - \tilde{E}_1, V - \tilde{V}, N - \tilde{N})|}{|\Omega(E - \tilde{E}_2, V - \tilde{V}, N - \tilde{N})|}, \tag{6.6.12}$$

The ratio (6.6.12) can be written, using (6.4.3), (6.4.8), as:

$$\exp(k^{-1}(S(E - \tilde{E}_1, V - \tilde{V}, N - \tilde{N}) - S(E - \tilde{E}_2, V - \tilde{V}, N - \tilde{N}) \tag{6.6.13}$$
$$\approx \exp(k^{-1}N(s(e - \tilde{e}_1, v - \tilde{v}, 1 - \tilde{n}) - s(e - \tilde{e}_2, v - \tilde{v}, 1 - \tilde{n})),$$

with the lower case letters referring to the variables in capital letters divided by N.

Using the conditions $\tilde{E}_1 \ll E$, $\tilde{E}_2 \ll E$, $\tilde{V} \ll V$, $\tilde{N} \ll N$ (since $\tilde{E}_i, i =$
$1, 2, \tilde{V}, \tilde{N}$ are fixed while $E, V, N \to \infty$ with $\frac{E}{N} = e$, $\frac{V}{N} = v$ fixed), and thus $\tilde{e}_i \ll$
$e, i = 1, 2, \tilde{v} \ll v, \tilde{n} \ll 1$, we can expand (6.6.14) to second order in a Taylor
series:

$$s(e - \tilde{e}_1, v - \tilde{v}, 1 - \tilde{n}) - s(e - \tilde{e}_2, v - \tilde{v}, 1 - \tilde{n})$$
$$= -\frac{\partial s(e, v)}{\partial e}(\tilde{e}_1 - \tilde{e}_2) + \mathcal{O}(\tilde{e}_1^2 + \tilde{e}_2^2 + +(|\tilde{e}_1| + |\tilde{e}_2|)(|\tilde{v}| + |\tilde{n}|)). \tag{6.6.14}$$

We have $N(\tilde{e}_1 - \tilde{e}_2) = (\tilde{E}_1 - \tilde{E}_2)$, $N(\tilde{e}_1^2 + \tilde{e}_2^2) = \mathcal{O}(\frac{1}{N})$ and $N(|\tilde{e}_1 + |\tilde{e}_2|)(|\tilde{v}| + |\tilde{n}|)) = \mathcal{O}(\frac{1}{N})$. So, neglecting those terms in (6.6.14), and using $\frac{\partial s(e,v)}{\partial e} = \frac{1}{T}$, with $T = T(E, V, N)$, which follows from (5.6.5) (remember that T is intensive: $T(E, V, N) = T(e, v, 1)$) if we accept the identification of (6.4.2) with the thermodynamic entropy, which will be justified in Sect. 6.7, we get that:

$$\frac{|\Omega(E - E(\mathbf{x}_1), V - \tilde{V}, N - \tilde{N})|}{|\Omega(E - E(\mathbf{x}_2), V - \tilde{V}, N - \tilde{N})|} \approx \exp[-\frac{1}{kT}(E(\mathbf{x}_1) - E(\mathbf{x}_2))], \quad (6.6.15)$$

which leads to:

$$d\nu_c(\mathbf{x}) = \frac{\exp(-\beta E(\mathbf{x}))d\mathbf{x}}{Z_c}, \quad (6.6.16)$$

where Z_c, given by (6.5.1) is a normalization constant and this "proves" (assuming that one justifies the approximations done here) that the canonical distribution (6.6.3) is the marginal distribution, see (2.A.20),[13] induced on a subsystem having $\tilde{N}$ particles in a box $\tilde{\Lambda} \subset \Lambda$, with $|\tilde{\Lambda}| = \tilde{V}$ by the micro-canonical distribution on $\Omega(E, V, N)$.

Note that if the energy is given only by the kinetic energy term (3.3.2), $E(\mathbf{x}) = \sum_{i=1}^{N} \frac{\|\vec{p}_i\|^2}{2m_i}$, the canonical distribution is the Maxwell-Boltzmann distribution "derived" approximately by maximizing the entropy in (6.2.20).

In many derivations of the canonical distribution, one imagines a sub-system in contact with a "heat bath" with which it can exchange energy.

Our derivation makes no such assumption; it is simply a mathematical fact that the induced or marginal probability distribution of the micro-canonical distribution on a sub-system of energy $\tilde{E}$ is the canonical distribution. This holds for a system of non-interacting particles, for which the notion of interactions with a heat bath does not make sense (some books invoke such interactions in order to derive the canonical distribution on a subsystem and then neglect them).

In fact, the idea that we need an interaction with a heat bath in order to derive the canonical distribution on a subsystem is implicitly a dynamical argument that assumes that the sub-system is initially out of equilibrium and converges towards equilibrium through its interactions with the heat bath. The important point here is that our derivation makes no reference to non-equilibrium states.

A similar calculation for a subsystem in a box $\tilde{\Lambda} \subset \Lambda$, with $|\tilde{\Lambda}| = \tilde{V}$ whose configuration is $\mathbf{x}$, with $E(\mathbf{x}) = \tilde{E}$ and the number of particle $N(\mathbf{x}) = \tilde{N}$ yields the grand-canonical distribution (exercise).

---

[13] To see the connection with (2.A.20), take $E(\mathbf{x})$ as the random variable F, $\mu = \nu_m$, the micro-canonical probability distribution, and ν the canonical distribution ν_c given by (6.6.16).

6.6.3.2 From the Canonical and Grand Canonical Distributions to the Microcanonical One

What about implications in the other direction? If we start, say, with the canonical distribution, can we recover the micro-canonical one?

The answer is yes, and is a consequence of the law of large numbers. To simplify matter, we will consider again systems of independent particles. First observe that, if one considers a function F depending on a finite number of variables:

$$F = F(\vec{q}_1, \vec{q}_2, \ldots, \vec{q}_L, \vec{p}_1, \vec{p}_2, \ldots, \vec{p}_L),$$

we have:

$$\lim_{N \to \infty} \int F \, d\nu_m(\mathbf{x}) = \lim_{N \to \infty} \int F \, d\nu_c(\mathbf{x}). \tag{6.6.17}$$

Indeed, on the one hand, the probability distribution induced, when $N \to \infty$, by $\nu_m(\mathbf{x})$ on the L variables

$$(\vec{q}_1, \vec{q}_2, \ldots, \vec{q}_L, \vec{p}_1, \vec{p}_2, \ldots, \vec{p}_L)$$

is the canonical one, with $\beta = \beta(e)$, as we saw in the previous subsection.

On the other hand, for independent particles, the probability distribution induced on those variables by the canonical distribution $\nu_c(\mathbf{x})$ is just the canonical distribution restricted to those particles since the canonical distribution is a product measure for independent particles:

$$\nu = \prod_{i=1}^{\infty} \nu_i(\vec{q}_i, \vec{p}_1), \tag{6.6.18}$$

with $\nu_i(\vec{q}_i, \vec{p}_1) = \frac{\exp(-\beta E(\vec{q}_i, \vec{p}_1))}{Z}$.

But there is a much deeper result. Consider a function F which is an average of functions of the position and momentum of one particle

$$F = \frac{1}{L} \sum_{i=1}^{L} f(\vec{q}_i, \vec{p}_i) \tag{6.6.19}$$

with $f : \mathbb{R}^6 \to \mathbb{R}$.

Let $<f> = \lim_{N \to \infty} \int f(\vec{q}_i, \vec{p}_i) d\nu_m(\mathbf{x}) = \lim_{N \to \infty} \int f(\vec{q}_i, \vec{p}_i) d\nu_c(\mathbf{x})$, by (6.6.17) ($\int f(\vec{q}_i, \vec{p}_i) d\nu_c(\mathbf{x})$ is independent of i).

Then, we have the following consequence of the law of large numbers of Sect. 2.4:

Proposition 6.6 $\forall \epsilon > 0$,

$$\lim_{L \to \infty} \lim_{N \to \infty} \nu_m(|F - <f>| > \epsilon) = \lim_{L \to \infty} \lim_{N \to \infty} \nu_c(|F - <f>| > \epsilon) = 0. \tag{6.6.20}$$

To prove (6.6.20), we first observe that, for $N \to \infty$, the marginal distribution of ν_m on the L variables $(\vec{q}_1, \vec{q}_2, \ldots, \vec{q}_L, \vec{p}_1, \vec{p}_2, \ldots, \vec{p}_L)$ is the canonical one. And for the second equality, use the law of large numbers (2.3.8) for the canonical distribution, i.e. for the product measure (6.6.18).

One could also get a "strong" version of that statement namely that $F \to < f >$ with probability one with respect to ν_m or with respect to ν_c taking first limits $N \to \infty$ and then $L \to \infty$ (see (2.3.9)).

This means that average quantities like F in (6.6.19) tend to a constant value in the large N and large L limits.

In particular, if one is interested in the average energy, i.e. (6.6.19) with $f(\vec{q}_i, \vec{p}_i) = E(\vec{q}_i, \vec{p}_i)$, one gets that the set of microstates with average energy $e = \frac{1}{L} \sum_{i=1}^{L} E(\vec{q}_i, \vec{p}_i)$ (with $e = e(T)$) becomes, in those limits, a set of probability one. So, in those limits, the canonical distribution becomes concentrated on configurations whose average energy is the same as in the corresponding micro-canonical distribution.

But it is much more general than that: *all* functions of the form (6.6.19) tend to *the same* constant value with probability one, both with respect to the micro-canonical distribution and with respect to the canonical one. That is a (strong) sense in which both "ensembles" are equivalent.

And one can extend this to a similar equivalence result with respect to the grand-canonical distribution.

This result extends also to interacting particles under fairly general assumptions on the nature of the interactions, but proving it would go beyond the scope of this book. See Tumulla [314, Sect. 7.3.5] for a summary of results and and references to the original results, including Fisher [130], Lanford [205], Ruelle [280], Aizenman, Goldstein and Lebowitz [2].

6.6.4 The Meaning of Ensembles

In statistical mechanics, one often tries to compute or to estimate the average or expected value $< f >$ for various functions of one or several variables (but in finite number). We will see examples of such computations when we will study phase transitions in Chap. 9.

However, we are not fundamentally interested in those averages. Indeed, what could they mean? The way ensembles are sometimes introduced is as a collection of "identical" systems over which one computes "ensemble averages". But then it would seem that we are interested in averages of quantities depending on one or few variables, where the average is taken over a large collection of identical systems.

But what we measure or observe are not such quantities, which by definition (if they depend on one or few variables) are microscopic and not accessible to us. We are interested only in the value of *macroscopic quantities* such as F in (6.6.19) for large L. Moreover, we are *not* interested in averages of such quantities over a large number of "identical" systems but in *typical* values of those quantities, i.e. in values that are the same for most microstates of the ensemble or, in other words, that take the same value for almost all microstates, with respect to the equilibrium distributions.

And that is why Proposition 6.6 states exactly what we need: the macroscopic quantities take a constant value over the vast majority of microstates (where, in the continuum case, "majority" means of probability close to one relatively to the equilibrium distributions). That value is, by definition, the equilibrium one.

Thus, there is a big conceptual difference between the averages $< F >$ and $< f >$. In $< F >$ we average a function that is essentially constant, hence the average is equal to that constant value (unless the function F takes very different values on the small set of microstates where its value differs from the equilibrium one). The function f on the other hand depends very much on which microstate it is evaluated.

However, from a computational point of view, it is easier to study, compute or estimate $< f >$ than $< F >$, because f depends only on few microscopic variables and both the canonical and the grand-canonical distributions give an explicit formula for $< f >$.

But, by (6.6.20), we know that $< F > \approx < f >$. So, although the two averages are conceptually radically different and what we are really interested in is $< F >$ or rather, the constant value taken by the function F that coincides with this average, from a practical point of view, people working in statistical mechanics study $< f >$. The situation can be summarized by:

Proposition 6.7 *Fundamental Formula*

$$\text{Equilibrium value of } F \approx < F > \approx < f > . \tag{6.6.21}$$

Although this identity is just a restatement of (6.6.20), we will call it the "fundamental formula". Indeed, it allows us to study its right hand side in order to know its left hand side. If we forget the conceptual distinction between the two, one easily gets confused by the idea that averages like $< f >$ amounts to averaging a (variable) function of the microstate over a large number of "identical systems", while this averaging is merely a tool to compute values of almost constant functions like F.

In his defense of a Gibbsian approach, i.e. an approach based on ensembles, David Wallace [326, Sect. 4] objects to the Boltzmannian approach that transport coefficients (such as thermal conductivity) can be computed using the two-time correlation function $C(t) = < X(t)X(0) >$ where $X(t)$ is the position of some particle at time t, and the average $< \cdot >$ is taken with respect to the micro-canonical distribution. Wallace wrote: "since C(t) is an explicitly probabilistic quantity, it is not even defined on the Boltzmannian approach." To which Goldstein, Lebowitz, Tumulka and Zanghì [155, Sect. 5] reply: "Actually, that is not correct. The individualist[14] will be happy as soon as it is shown that for most phase points in [the micro-canonical energy surface],[15] the rate of heat conduction is practically constant and can be computed from $C(t)$ in the way considered." We will see other examples of macro-

---

[14] Meaning the Boltzmannian.

[15] My paraphrase.

scopic quantities that are computed by taking averages of local quantities over some probability distribution in Sect. 9.2.

6.7 Justification of the Entropy Formula (6.4.2)

Finally, we must justify the entropy formula (6.4.2) and, to do that, we must show that the entropy function, so defined, also satisfies the relation that defines it in thermodynamics[16]:

$$dS = \frac{dQ}{T} \equiv \frac{dE + dW}{T} \tag{6.7.1}$$

This will also show that the function $\beta(e)$ defined implicitly in (6.2.17) equals $\frac{1}{kT(E)}$, for the variable $T = T(E)$ in (6.7.1).

Actually, instead of proving directly that (6.4.2) satisfies (6.7.1), we will prove (6.7.1) for the formula (6.5.6), equivalent to (6.4.2).

To prove (6.7.1), we first need a formula for dW. Let the energy of every microstate depend on some external parameter which we will denote by λ; this could be the volume of the box in which the system is enclosed, or an electric or magnetic field acting on it, or some other variable.

So, $\forall \mathbf{x} \in \Omega$, we have $E(\mathbf{x}) = E(\mathbf{x}, \lambda)$. The parameter λ could be a multicomponent vector (in which case expressions like $\frac{\partial f}{\partial \lambda} d\lambda$ below should be replaced by $\sum_i \frac{\partial f}{\partial \lambda_i} d\lambda_i$), but to keep the notation simple, we use a scalar variable λ here.

Now the (infinitesimal) work done by an individual microstate when λ is varied is given by

$$dW(\mathbf{x}, \lambda) = -\frac{\partial E(\mathbf{x}, \lambda)}{\partial \lambda} d\lambda, \tag{6.7.2}$$

where we put minus sign because, if λ is the volume of the box in which the system is enclosed, then a positive work done by the system corresponds to an increase of λ but a decrease of its total energy (see (5.3.1), (5.3.2)).

But since E is a macroscopic quantity, its value is constant over most microstates and thus is the same as its average in any of our equilibrium distributions. It will be convenient to use the average with respect to the canonical distribution, since we want to identify the microscopic entropy defined by (6.5.6) with the thermodynamic one defined by (6.7.1):

$$\int d\nu_c(\mathbf{x}) dW(\mathbf{x}, \lambda) = -\frac{\int d\mathbf{x} \frac{\partial E(\mathbf{x}, \lambda)}{\partial \lambda} \exp(-\beta E(\mathbf{x}, \lambda))}{\int d\mathbf{x} \exp(-\beta E(\mathbf{x}, \lambda))} d\lambda. \tag{6.7.3}$$

Observe that the right hand side of (6.7.3) equals $\frac{1}{\beta} \frac{\partial \ln Z}{\partial \lambda} d\lambda$, with $Z = Z_c$, so that:

---

[16] We will also give an intuitive version of the entropy formula in Appendix 6.C.

$$dW(\mathbf{x}, \lambda) = \frac{1}{\beta} \frac{\partial \ln Z}{\partial \lambda} d\lambda \tag{6.7.4}$$

for $\mathbf{x} \in \Omega_{eq}$.

A concrete example of (6.7.4) is given by the formula for the pressure[17]:

$$P = \frac{1}{\beta} \frac{\partial \ln Z}{\partial V},$$

where P is the pressure and $P dV$ the infinitesimal work done by the gas.

Now let us compute the differential dE, starting from the formula $E = -\frac{\partial \ln Z}{\delta \beta}$. We have:

$$dE = -\frac{\partial^2 \ln Z}{\partial \beta \partial \lambda} d\lambda - \frac{\partial^2 \ln Z}{\partial^2 \beta} d\beta. \tag{6.7.5}$$

Combining (6.7.4) and (6.7.5), we get:

$$dE + dW = \left(\frac{1}{\beta} \frac{\partial \ln Z}{\partial \lambda} - \frac{\partial^2 \ln Z}{\partial \beta \partial \lambda} \right) d\lambda - \frac{\partial^2 \ln Z}{\partial^2 \beta} d\beta,$$

or,

$$\beta(dE + dW) = \left(\frac{\partial \ln Z}{\partial \lambda} - \beta \frac{\partial^2 \ln Z}{\partial \beta \partial \lambda} \right) d\lambda - \beta \frac{\partial^2 \ln Z}{\partial^2 \beta} d\beta. \tag{6.7.6}$$

Compute $k^{-1} dS$ from the formula (6.5.6) $k^{-1} S = \ln Z - \beta \frac{\partial \ln Z}{\partial \beta}$:

$$
\begin{aligned}
k^{-1} dS &= \left(\frac{\partial \ln Z}{\partial \lambda} - \beta \frac{\partial^2 \ln Z}{\partial \beta \partial \lambda} \right) d\lambda + \left(\frac{\partial \ln Z}{\partial \beta} - \frac{\partial \ln Z}{\partial \beta} - \beta \frac{\partial^2 \ln Z}{\partial^2 \beta} \right) d\beta \\
&= \left(\frac{\partial \ln Z}{\partial \lambda} - \beta \frac{\partial^2 \ln Z}{\partial \beta \partial \lambda} \right) d\lambda - \beta \frac{\partial^2 \ln Z}{\partial^2 \beta} d\beta. \tag{6.7.7}
\end{aligned}
$$

Comparing (6.7.7) and (6.7.6) proves (6.7.1) and shows also that $\beta = \beta(e)$ defined implicitly in (6.2.17) equals $\frac{1}{kT(E)}$, with $T(E)$ as in (6.7.1), or (5.6.5), i.e. the thermodynamic temperature.

6.8 What Justifies the Microcanonical Distribution?

Our justification of the equilibrium statistical mechanical formalism is based on three ideas:

---

[17] For a detailed mechanical derivation of that formula, see e.g. Amit and Verbin [12, p. 231].

1. The equilibrium values of macroscopic quantities are typical with respect to the equilibrium distributions.
2. All equilibrium distributions are equivalent in the sense that they give, with probability almost equal to 1, equal values to the macroscopic quantities.
3. The fundamental formula (6.6.21) gives us a practical way to compute the equilibrium values of macroscopic quantities.

But a skeptical mind could still ask: "why are you satisfied by showing that the values of macroscopic quantities are typical with respect to the equilibrium distributions?" "What is so special about those distributions that being typical with respect to them has an explanatory value or counts as a justification for the equilibrium statistical formalism"?

It is enough to answer that question for the micro-canonical distribution, since the other distributions are equivalent to that one. And since the micro-canonical distribution is just the Lebesgue measure on phase space restricted to the constant energy surfaces, it is enough to discuss the Lebesgue measure.

Before getting into that discussion, one must notice that all configurations are typical with respect to some probability distribution, if only with respect to the delta measure concentrated on that configuration. That example may seem silly, and it is, but it shows that something has to be said to justify the choice of the probability distribution with respect to which the values of macroscopic quantities are typical.

There are (at least) four types of justifications for the role played by the Lebesgue measure in the foundations of statistical mechanics.

1. *"Justification" by the consequences*: let us simply accept the role of the Lebesgue measure as a basic principle of physics, an "axiom" if you wish, just like Newton's laws or Schrödinger's equation.[18] After all, there is no deeper justification for the fundamental laws of physics than the fact that their empirical consequences coincide with our observations. One can sometimes give some sort of hand waving arguments to "derive" those laws, but those arguments appeal to some other assumptions that have no other justification than their consequences.

If we accept that the role of the micro-canonical distribution is fundamental in equilibrium statistical mechanics, then one can derive all the empirical predictions of that field, not only the thermodynamic formalism, but also the theory of phase transitions, see Chap. 9. So, that is the justification of the introduction of that distribution, period.

This "justification" is the simplest of all, hence somewhat attractive, until one realizes that it is no justification at all. One just postulates what ought to be justified. Yet, even if the fundamental laws of physics can only be justified through their consequences, one might think that any probability distribution on the set of microstates considered here, is not as fundamental or as "objective" as those laws and therefore that its introduction might demand an argument.

---

[18] See Albert [9, Chap. 3] for an elaboration of this argument and criticisms of the other justifications given below.

2. *Justification by stationarity.* One knows, from Liouville's theorem, that the
   Lebesgue measure (or Liouville's measure, see Sect. 3.4.1) is invariant under
   the Hamiltonian flow. One could then argue that choosing this measure is natural
   because that choice does not depend on the moment in time when it is made.
   Notice first that we have, in general, no proof that the Lebesgue measure is the
   only invariant one (even among absolutely continuous ones) under the Hamiltonian flow. Proving that would amount to proving ergodicity of the Hamiltonian
   flow and that is very hard to do except in special cases (see item [5] in Sect. 4.3.1).
   Moreover, as we saw in Sect. 4.6.2, there exists non-Hamiltonian systems, for
   which the Lebesgue measure is *not* invariant under the dynamical evolution and
   that do possess an invariant (Sinai–Bowen–Ruelle) measure ν whose support is a
   set $\mathcal{A}$ of zero Lebesgue measure. Moreover, one can, in certain cases, prove that,
   for almost every initial condition $\mathbf{x}(0)$ with respect to the Lebesgue measure on a
   set of non-zero Lebesgue measure containing $\mathcal{A}$, the time evolution $\mathbf{x}(0) \rightarrow \mathbf{x}(t)$
   drives the trajectory towards $\mathcal{A}$ and the statistics of the time spent by the trajectory
   close to subsets of $\mathcal{A}$ is proportional to the ν-measure of those subsets.[19]
   The fact that one is interested, for those systems, in trajectories whose initial
   conditions are typical with respect to the Lebesgue measure, which is not invariant
   under the dynamical evolution, shows that one cannot take stationarity as a general
   principle justifying the use of the Lebesgue measure.[20]
   Besides, we shall introduce in Sect. 8.8 another "natural" time evolution of measures, under which the Liouville measure is not invariant. Indeed, if measures are
   not physical quantities, it is not obvious why they should evolve according to the
   microscopic dynamics.

3. *Justification by the indifference principle.* Let us first consider the situation where
   the set of microstates is discrete and finite. Then, when we speak of typicality, we
   simply refer to the vast majority of microstates, as in the example of coin tossing.
   Another way to say this is that the "measure of typicality" that we use is simply the
   counting measure on the set of microstates. And this is the measure that follows
   from the indifference principle: if we do not have any reason to think that one
   state is more likely to occur than another, then we assign an equal probability to
   all of them, i.e. we use the counting measure.
   One can then argue that the Lebesgue measure is the natural extension of the
   counting measure to the continuum situation and is thus justified on the basis of
   the indifference principle. Indeed, at the end of appendix 2.A.2, we noticed that the
   Lebesgue measure is equivalent, using the binary expansion of real numbers, to
   a product measure on sequences of symbols 0 and 1 with equal weight $\frac{1}{2}$ for each
   symbol. As long as we look at finite strings of such symbols, that product measure
   is indeed a counting measure. Thus, it is natural to consider the extension of this

---

[19] This happens not only in the examples of Sect. 4.6.2 but also for more general "chaotic" dynamical
systems for which ν is a Sinai-Ruelle-Bowen measure and the set $\mathcal{A}$ is a "strange attractor", see e.g.
Bowen [45], Bowen and Ruelle [46], Eckmann and Ruelle [120], Ruelle [281, 282], Sinai [295].

[20] See Allori [11, Sect. 3.4.2] for a criticism of this observation.

measure to infinite sequences, which is the Lebesgue measure, as an extension of the counting measure. See Pitkowsky [258] for an elaboration of this argument.

4. *Justification by naturalness.* This is somewhat similar to the previous justification but does not rely so much on the indifference principle. One can argue that the Lebesgue measure is "natural" because it is the only measure which is invariant under spatial translations (and rotations, although one does not have to assume that in order to prove uniqueness), see Appendix 2.A.2.

   For the chaotic dynamical systems mentioned in point 2 above, it is this naturalness property of the Lebesgue measure that leads us to consider as relevant the initial conditions that are typical with respect to that measure and not those that are typical with respect to the invariant measure.

5. *Typicality measure vs probability measure.* Finally, one line of thought is that we do not need the "full" micro-canonical measure, but only a measure of typicality, the latter giving only a size "close to 1" to certain sets and a size "close to 0" to their complements, see e.g. Goldstein [149, p. 53], or Allori [11, p. 14], while a usual probability measure will give a size to all (Borel) sets, size which can be any number between 0 and 1.

   But this is not clear: indeed, we will see in Chap. 8 that one defines the time-dependent Boltzmann entropy as the logarithm of the size of a set, whose size changes continuously in time (see (8.1.6), which is a generalization of (6.4.1)); hence that size cannot be limited to values close to 1 or close to 0.

   Moreover, because of the extension theorem 2.4, constructing a "full" probability measure is not that complicated: it is enough to give a size to every interval in $\mathbb{R}$ or rectangle in $\mathbb{R}^n$, which for the Lebesgue measure is rather natural, and, next, give a size to unions of disjoint intervals in $\mathbb{R}$ or disjoint rectangles in $\mathbb{R}^n$, equal to the sum of the sizes of those intervals or rectangles, which is also quite natural. Then, by the extension theorem, the size of all Borel sets is automatically defined. The same is true for product spaces, where it is enough to give a size to union of disjoint cylinder sets of the form (2.A.5), and then use the extension theorem to assign a size to all sets in the cylindrical σ-algebra. So that, if one wants to give a size to "big" sets it is not obvious how to do so in a natural way without defining also, at least implicitly, a size for all Borel sets.[21]

The reader is invited to choose his or her favorite justification. My own preference is the fourth one, but other choices can be defended.

A final question to be answered is whether the probabilities considered here are subjective or objective in the sense of Chap. 2. They are certainly not to be understood in a frequentist way, since we are not interested in the statistics of what would happen to many "identical" macroscopic systems but in regularities that occur in almost all of

---

[21] Roderich Tumulka has pointed out to me (private communication) that non additive "measures" (that is, positive valued maps defined on families of sets that do not satisfy (2.2.1) and thus are not real measures but that may satisfy other properties than additivity), could be relevant for certain approaches to quantum mechanics and that a notion of typicality, distinct from the one of probability, may be relevant for statistical mechanics on unbounded phase spaces on which there is no natural probability measure.

them. In fact the misleading presentation of ensembles as ways to take averages over many identical systems has a frequentist flavor, but, as we explained in Sect. 6.6.4, this is the wrong way to justify the introduction of ensembles.

Should then the probabilities in our three ensembles be understood as subjective probabilities? Yes, except that we are mostly interested in events whose probability is close to 0 or 1, namely in typical events (but see point [5] above—we do not consider only those sets when we study approach to equilibrium). Then this subjective notion of probability can be related to our intuitive notion of explanation as we argued in Sect. 2.6.

6.9   Summary

There are two basic ideas in this chapter: first, the notion of macrostate as opposed to microstate, and the dominance of the equilibrium macrostate, and then Boltzmann's formula for the entropy.

The notion of macrostate is introduced in Sect. 6.1; it is defined by a map from the set of microstates to a few macroscopic parameters. One of its main properties is that this map is many to one in a way that depends enormously on the value of the macrostate, i.e. some values of the macrostate correspond to many more microstates than others; in particular, one (approximate) value of the macrostate corresponds to the vast majority of microstates and that value is called the equilibrium value, see Sect. 6.2 and Fig. 6.2.

The next basic idea is Boltzmann's definition of the entropy (6.4.1), which is a function of the microstate defined by or parametrized by the choice of the macrostate. In Sect. 6.4, we compute that function in various simple situations and in particular for an ideal gas (6.4.7), where we recover the thermodynamic formula (5.7.16).

Once we have a microscopic formula for the entropy, it is easy to obtain the corresponding formulas for the Helmoltz free energy and for the grand potential, see Sect. 6.5.

But statistical mechanics allows us to go far beyond formulas for the thermo-dynamic functions by introducing equilibrium "ensembles" i.e. probability distri-butions of the set of microstates see Sect. 6.6. The basic such distribution is the micro-canonical one, which is uniform over the set of microstates. The justification for that probability distribution is discussed in Sect. 6.8, but if one accepts it, one can introduce other, and sometimes more convenient, probability distributions, the canonical and the grand-canonical ones, see Sect. 6.6.1. The main point of Sect. 6.6.3 is that all these distributions are equivalent in the sense that they give the same value to all the macroscopic quantities in the thermodynamic limit of large volumes and large number of particles. In fact, a stronger result holds, which is also a conse-quence of the law of large numbers: macroscopic observables take constant values almost everywhere with respect to the micro-canonical, canonical or grand-canonical distributions, see Proposition 6.6.

Once one considers ensembles, one can define a new entropy, the Gibbs one: unlike Boltzmann's entropy which is a function of the microstate, the Gibbs entropy is defined on probability distributions, but coincides with the Boltzmann entropy in equilibrium, see Sect. 6.6.2 (for non equilibrium situations, see Chap. 8).

In Sect. 6.6.4, we discussed the meaning of these ensembles. We stressed that one is not interested, from a physical point of view, in averages of physical quantities with respect to probability distributions, but with the values of those quantities that hold typically or almost everywhere, which is only true for macroscopic quantities. But the fundamental formula (6.6.21) allows us to compute those typical values by taking averages of microscopic quantities with respect to some ensemble. This is a crucial distinction between what one does in practice (taking averages of microscopic quantities with respect to some ensemble) and what one is interested in theory (typical values of macroscopic quantities), which is sometimes overlooked when one discusses the physical relevance of ensembles.

Finally, in Sect. 6.7, we justify Boltzmann's definition of the entropy (6.4.1) by showing that it coincides with the thermodynamic definition (6.7.1). For that, we need to give an appropriate definition of thermodynamic work (6.7.2), which is also a macroscopic quantity. Then, justifying (6.4.1) it reduced to a computation. We give another, more intuitive, derivation of the entropy formula in Appendix 6.C.

Finally, in Sect. 6.8 we discussed several conceptual justifications of the micro-canonical probability distribution.

6.10 Exercises

Some of the exercises in this section are taken from Callen [66], from Sator and Pavloff [284], and from Thompson [308].

6.1. Prove the estimate (6.2.11).

6.2. By reasoning as in Sect. 6.6.3.1, show that the micro-canonical distribution on Ω^Λ induces the grand-canonical distribution on a subsystem in a box $\tilde\Lambda \subset \Lambda$, with $|\tilde\Lambda| = \tilde V$ whose configuration is $\mathbf{x}$, with $E(\mathbf{x}) = \tilde E$ and the number of particle $N(\mathbf{x}) = \tilde N$.

6.3. Consider a set of variables $s_i = \pm 1, i = 1, \ldots, N$ and associate to each configuration $\mathbf{s} = (s_i)_{i=1}^N$ an energy $E(\mathbf{s}) = \sum_{i=1}^N s_i$. Compute explicitly the function

$$S(E, N) = k \ln |\Omega(E, N)|$$

with $|\Omega(E, N)| = |\{\mathbf{s} = (s_i)_{i=1}^N \mid E(\mathbf{s}) = E\}|$. Compute also

$$s(e) = \lim_{\substack{N, E \to \infty \\ \frac{E}{N} = e}} \frac{S(E, N)}{N}.$$

6.4. Let

$$Z(\beta, N) = \sum_{\mathbf{s}} \exp(-\beta E(\mathbf{s}))$$

with $E(\mathbf{s})$ as in exercise 6.3. Compute

$$f(T) = -kT \lim_{N \to \infty} \frac{\ln Z(\beta, N)}{N}$$

6.5. Check that $s(e)$ in exercise 6.3 and $f(T)$ in exercise 6.4 are Legendre transforms of one another.

6.6. Consider a set of variables $n_i \in \mathbb{N}, i = 1, \ldots, N$ and associates to each configuration $\mathbf{n} = (n_i)_{i=1}^N$ an energy $E(\mathbf{n}) = \sum_{i=1}^N n_i$. Compute explicitly the function

$$S(E, N) = k \ln |\Omega(E, N)|$$

with $\Omega(E, N) = \{\mathbf{n} = (n_i)_{i=1}^N \mid E(\mathbf{n}) = E\}$. Compute also

$$s(e) = \lim_{\substack{N, E \to \infty \\ \frac{E}{N} = e}} \frac{S(E, N)}{N}.$$

Hint: to compute $|\Omega(E, N)|$, we have to distribute E units of energy in N boxes; the units of energy are indistinguishable but the boxes are distinguishable. This is like putting E indistinguishable marbles in N distinguishable boxes. Consider E marbles and $N - 1$ sticks. Arrange the $E + N - 1$ objects in some order and put all the marbles before the first stick in the first box, all the marbles after the first stick and before the second stick in the second box etc. Then compute the number of such arrangements, but taking into account the fact that permuting the boxes or the marbles within each box does not change the final distribution of the marbles in the boxes.

6.7. Let

$$Z(\beta, N) = \sum_{\mathbf{n}} \exp(-\beta E(\mathbf{n}))$$

with $\mathbf{n}$ and $E(\mathbf{n})$ as in exercise 6.6. Compute

$$f(T) = -kT \lim_{N \to \infty} \frac{\ln Z(\beta, N)}{N}$$

6.8. Check that $s(e)$ in exercise 6.6 and $f(T)$ in exercise 6.7 are Legendre transforms of one another.

6.9. A crystal is composed of N atoms, some of which can be detached from the crystal, at an energy cost ϵ, called defects. Let n be the number of those defects. Let E_0 be the energy of the crystal without defects.
Compute the entropy of that system as a function of N and n for N, n and $N - n$ large.

Compute the temperature of that system and the number $n(T)$ of defects as a function of the temperature, for N, n and $N - n$ large.

Hint: Let n be the number of those defects. Let E_0 be the energy of the crystal without defects. The total energy is then: $E = E_0 + n\epsilon$

6.10. A system is composed of N molecules of H_2 that can be in four possible states: one, called "para", of energy 0 and three, called "ortho", of energy ϵ. Let n_i the number of molecules in state $i = 1, 2, 3, 4$.

Compute the entropy of that system as a function of N and n_1, for N, n_1 and $N - n_1$ large.

Compute the temperature of that system and the energy $E(T)$ as a function of T, for N, n_1 and $N - n_1$ large. What is the value of $\lim_{T \to 0} E(T)$ and $\lim_{T \to \infty} E(T)$ and in which fraction of molecules are in the para or the ortho state in those limits?

6.11. One considers N classical harmonic oscillators, of coordinates $(q_i, p_i)_{i=1}^N$ and individual energies $E_i = p_i^2 + q_i^2$. The total energy of the system is $E = \sum_{i=1}^N E_i$.

Compute $\Omega(E, N)$, for N, E, large and show that one recovers the result of exercise 6.6 for $s(e)$ when $e \to \infty$ (which corresponds to high temperatures).[22]

6.12. Consider a Hamiltonian of the form (3.3.1) with $V_{ij} = 0$ in (3.2.5),i.e. with only one-body potentials.

Give a formula for the canonical and grand-canonical partition functions and deduce from the latter that the equation $PV = NkT$ still holds.

6.13. Consider a modification of exercise 6.6 with $E(\mathbf{n}) = \sum_{i=1}^L n_i \omega_i$, where L is the number of vibrational modes of a crystal.[23] Give an expression for the free energy of that system in terms of a sum over these modes $(\omega_i)_{i=1}^L$.

6.14. Suppose that, in exercise 6.13, the sum over these modes can be approximated[24] by an integral with density $CV\omega^2 d\omega$, for a constant C, and running from 0 to ω_M, where the maximum value of ω is determined by

$$CV \int_0^{\omega_M} \omega^2 d\omega = 3N$$

(because the number of vibrational modes in the crystal equals three times the number N of atoms). Express ω_M as a function of N and V.

Express the free energy and the average energy as integrals over ω.

6.15. For photons, we have a sum over modes as in exercise 6.13, approximated by an integral as exercise 6.14, but the density in the approximating integral is $\frac{V}{\pi^2 c^3} \omega^2 d\omega$, where c is the velocity of light and the integral runs from 0 to ∞.

---

[22] The energy of exercise 6.6 is, with a suitable choice of constants, the energy of N quantum harmonic oscillators; the limit when $e \to \infty$ for quantum harmonic oscillators coincides with the classical value obtained here.

[23] This is the basis of Debye's model of crystals, see footnote 7.9.

[24] See Callen [66, Sect. 16-7] for the justification of this approximation.

Using the formula[25]:

$$\int_0^\infty \frac{x^3}{e^x - 1} dx = \frac{\pi^4}{15},$$

derive the exact dependence of $E(T)$. The result is the famous Stefan-Boltzmann law.

Appendix

6.A Asymptotics

6.A.1 Laplace's Method

Consider integrals of the form

$$\int_0^\infty \exp(\lambda f(x)) dx \qquad (6.A.1)$$

for λ large and positive. Assume that $f(x) \to -\infty$ as $x \to \infty$, sufficiently last so that the integral converges for all $\lambda > 0$. Assume also that the function f has a unique maximum at $x = x_0$, $0 < x_0 < \infty$. Write that integral as:

$$\exp(\lambda f(x_0))$$

$$\left(\int_{|x-x_0| \le \epsilon} \exp(\lambda(f(x) - f(x_0))) dx + \int_{|x-x_0| > \epsilon} \exp(\lambda(f(x) - f(x_0))) dx \right),$$

for some $\epsilon > 0$. Since, $\forall x$, $|x - x_0| > \epsilon$, $f(x) - f(x_0) < \delta$ for some $\delta > 0$ (since x_0 is the unique maximum of f), and the integral in (6.A.1) converges, we have that the second integral in (6.A.1) is bounded by $\exp(-c\lambda)$, for some $c > 0$.

When $|x - x_0| \le \epsilon$, we approximate $f(x) - f(x_0) \approx -\frac{a(x-x_0)^2}{2}$, with $-a = f''(x_0)$; since x_0 is a strict maximum of f, we have $f'(x_0) = 0$ and $a > 0$. Changing variables $y = \sqrt{\lambda a}(x - x_0)$ in $\int_{|x-x_0| \le \epsilon} \exp(-\lambda \frac{a(x-x_0)^2}{2}) dx$, we approximate that latter integral by $\frac{1}{\sqrt{\lambda a}} \int_{\mathbb{R}} \exp(-\frac{y^2}{2}) dy = \sqrt{\frac{2\pi}{\lambda a}}$ (we replace $\pm \lambda \epsilon$ by $\pm \infty$ in the limits of integration).

---

[25] Which the reader may derive by expanding $\frac{1}{e^x-1} = \sum_{n=1}^\infty e^{-nx}$ and using $\int_0^\infty x^3 e^{-x} dx = 3!$ and $\sum_{n=1}^\infty \frac{1}{n^4} = \frac{\pi^4}{90}$.

Altogether, we get the approximation:

$$\int_0^\infty \exp(\lambda f(x))dx \approx \exp(\lambda f(x_0))\sqrt{\frac{2\pi}{\lambda a}} = \exp(\lambda f(x_0))\sqrt{\frac{2\pi}{\lambda |f''(x_0)|}}, \quad (6.A.2)$$

as $\lambda \to \infty$. With more work, one can obtain much more detailed estimates than (6.A.2), see e.g. Erdelyi [123].

6.A.2 Stirling's Formula

Laplace's method can be used to justify the approximation to the factorial function $N! \approx N^N e^{-N}$ as $N \to \infty$. First write:

$$N! = \int_0^\infty x^N \exp(-x)dx$$

which can be proven recursively from $\int_0^\infty x^n \exp(-x)dx = n \int_0^\infty x^{n-1} \exp(-x)dx$ (which follows from integration by parts) and $\int_0^\infty \exp(-x)dx = 1$. Let $x = Ny$; we get:

$$\int_0^\infty x^N \exp(-x)dx = N^{N+1} \int_0^\infty y^N \exp(-Ny)dy = N^{N+1} \int_0^\infty \exp(N(\ln y - y))dy$$

The function $f(y) = \ln y - y$ has a unique maximum at $y = 1$ with $f''(1) = -1$, so that applying (6.A.2) gives:

$$N! \approx N^{N+1} \exp(-N)\sqrt{\frac{2\pi}{N}} = N^N \exp(-N)\sqrt{2\pi N} \quad (6.A.3)$$

One can check that the same approximation holds for the Gamma function (6.4.4).

6.B "Derivation" of Formula (6.5.4)

In order to apply Laplace's method to the integral (6.5.2), we need to verify that the function $f(e) = -\beta e + \frac{s(e,v)}{k}$ has a unique maximum, which follows from the concavity of the function $s(e, v)$, and check that the integral converges. That follows from the fact that $f(e) \approx -\beta e$, as $e \to \infty$, since $s(e, v) \approx \ln e$ as $e \to \infty$, see (6.4.7) (the latter computation is valid only for an ideal gas but the logarithmic growth of $s(e, v)$ as a function of e when $e \to \infty$ holds more generally).

6.C   An Intuitive Formula for the Entropy

An intuitive expression for the variation of entropy is most easily expressed in the situation of discrete variables.

Consider, as we did after (6.2.11), a system of N particles, each of which can be in L possible states, but with an "energy" variable $e_i(\lambda)$, $i = 1, \ldots, L$, associated to each such state and depending on an external parameter λ, with the sum of the energies being fixed: $E(\mathbf{x}, \lambda) = \sum_{i=1}^{N} e(x_i, \lambda)$.

The set of microstates is $\Omega = \{1, \ldots, L\}^N$, and we can take as macrostates the fractions of particles in each state: $\mathbf{n} = (n_1, \ldots, n_L)$, where $n_i = \frac{N_i}{N}$ with N_i being the number of particles in state i. In terms of the macrostate, the total energy equals: $E(\mathbf{x}, \lambda) = \sum_{i=1}^{L} N_i e_i(\lambda)$ and $\frac{E(\mathbf{x}, \lambda)}{N} = \sum_{i=1}^{L} n_i e_i(\lambda)$.

In the canonical ensemble, we get:

$$E = \sum_{\mathbf{x}} E(\mathbf{x}, \lambda) \frac{\exp(-\beta E(\mathbf{x}, \lambda))}{Z} = N \sum_{i=1}^{L} e_i(\lambda) n_i(\beta, \lambda)$$

with $E = \sum_{i=1}^{N} e(x_i, \lambda)$ and $n_i(\beta, \lambda)$ the fraction of states with energy $e(x_i, \lambda) = e_i(\lambda)$, given by $n_i(\beta, \lambda) = \frac{\exp(-\beta e_i(\lambda))}{Z}$. We get:

$$dE = N\left(\sum_{i=1}^{L} de_i(\lambda) n_i(\beta, \lambda) + \sum_{i=1}^{L} e_i(\lambda) dn_i(\beta, \lambda) \right). \qquad (6.C.1)$$

In the sum, the first term equals, using (6.7.2), (6.7.3),

$$N \sum_{i=1}^{L} \frac{\partial e_i}{\partial \lambda} n_i(\beta, \lambda) d\lambda = \sum_{\mathbf{x}} \frac{\partial E(\mathbf{x}, \lambda)}{\partial \lambda} \frac{\exp(-\beta E(\mathbf{x}, \lambda))}{Z} d\lambda = -dW.$$

Therefore, we get from (6.7.1):

$$T ds = \frac{T}{N} dS = \sum_{i=1}^{L} e_i(\lambda) dn_i(\beta, \lambda). \qquad (6.C.2)$$

So we see that the variation of entropy is given (up to the factor T) by the variation of the occupation numbers of the different energy levels, while the work is given by the variation of the individual energies. If no work is done (λ is constant) then we can still have a variation of the total energy through heat transfer and that translates microscopically into a variation of the occupation numbers n_i's: the occupation numbers of higher energy levels can increase or decrease (depending on the direction of the heat transfer) and that leads to a change of entropy.

One can also obtain directly (6.C.2) from the formula (6.7.7)

$$T ds = \frac{1}{N} \left(kT \frac{\partial \ln Z}{\partial \lambda} - \frac{\partial^2 \ln Z}{\partial \beta \partial \lambda} \right) d\lambda - \frac{\partial^2 \ln Z}{\partial^2 \beta} d\beta$$

and $n_i(\beta, \lambda) = \frac{\exp(-\beta e_i(\lambda))}{Z}$:
  We have:

$$\frac{1}{N} kT \frac{\partial \ln Z}{\partial \lambda} = - < \frac{\partial e(\lambda)}{\partial \lambda} >$$

and

$$\frac{1}{N} \frac{\partial^2 \ln Z}{\partial \beta \partial \lambda} = - < \frac{\partial e(\lambda)}{\partial \lambda} > + \beta \left(< \frac{\partial e(\lambda)}{\partial \lambda} e(\lambda) > - < \frac{\partial e(\lambda)}{\partial \lambda} > < e(\lambda) > \right).$$

Also,

$$\frac{1}{N} \frac{\partial^2 \ln Z}{\partial^2 \beta} = < e^2(\lambda) > - < e(\lambda) >^2 .$$

So,

$$
\begin{aligned}
T ds = & - < \frac{\partial e(\lambda)}{\partial \lambda} > d\lambda + < \frac{\partial e(\lambda)}{\partial \lambda} > d\lambda \\
& - \beta \left(< \frac{\partial e(\lambda)}{\partial \lambda} e(\lambda) > - < \frac{\partial e(\lambda)}{\partial \lambda} > < e(\lambda) > \right) d\lambda \\
& - (< e(\lambda)^2 > - < e(\lambda) >^2) d\beta \\
= & -\beta \left(< \frac{\partial e(\lambda)}{\partial \lambda} e(\lambda) > - < \frac{\partial e(\lambda)}{\partial \lambda} > < e(\lambda) > \right) d\lambda \\
& - (< e(\lambda)^2 > - < e(\lambda) >^2) d\beta
\end{aligned}
\tag{6.C.3}
$$

Let us compute the right hand side of (6.C.2)

$$\sum_{i=1}^{L} e_i(\lambda) dn_i(\beta, \lambda) = \sum_{i=1}^{L} e_i(\lambda) \frac{\partial n_i(\beta, \lambda)}{\partial \beta} d\beta + \sum_{i=1}^{L} e_i(\lambda) \frac{\partial n_i(\beta, \lambda)}{\partial \lambda} d\lambda \quad (6.C.4)$$

Computing each term in (6.C.4), we have:

$$\sum_{i=1}^{L} e_i(\lambda) \frac{\partial n_i(\beta, \lambda)}{\partial \beta} = -(< e(\lambda)^2 > - < e(\lambda) >^2),$$

and

$$\sum_{i=1}^{L} e_i(\lambda) \frac{\partial n_i(\beta, \lambda)}{\partial \lambda} = -\beta \left(< \frac{\partial e(\lambda)}{\partial \lambda} e(\lambda) > - < \frac{\partial e(\lambda)}{\partial \lambda} >< e(\lambda) > \right).$$

Inserting this in (6.C.4) and comparing with (6.C.3) proves (6.C.2).

Chapter 7
Information-Theoretic and Predictive Statistical Mechanics

7.1 Introduction

There is an approach to statistical mechanics different from the one proposed in Chap. 6 which is quite popular and is probably the most popular one. Although we do not share the "philosophy" of that approach (and we will explain why in Sect. 7.8), its popularity alone would be a sufficient reason for it to be included in this book. A better reason for that inclusion is that the formalism of this approach is very simple and elegant.

The basic focus of the information-theoretic approach is on probabilities or measures, rather than on microstates (and their corresponding macrostates). Limiting ourselves to equilibrium situations, the basic question is: how do we assign a probability distribution over microstates to a given system in equilibrium? Because of the words "we assign", one would naturally think that the meaning of probabilities here is the subjective one, which is sometimes openly stated by adherents of the information-theoretic approach (for example by Jaynes [180, 183] and Grandy [159]), but not always.

We shall first discuss probability distributions on finite sets and turn to continuous distributions in Sect. 7.5.

7.2 The Shannon Entropy

In Sect. 2.2.1, we introduced the probability distribution on N events corresponding to the principle of indifference, (2.2.2):

$$p_i = \frac{1}{N},$$

$\forall i = 1, \ldots, N.$

© Springer Nature Switzerland AG 2022
J. Bricmont, *Making Sense of Statistical Mechanics*, Undergraduate Lecture
Notes in Physics, https://doi.org/10.1007/978-3-030-91794-4_7

We introduced its generalization given by the principle maximum entropy, using the *Shannon entropy* of a probability distribution on N events, $\mathbf{p} = (p_i)_{i=1}^N$, $p_i \geq 0$, $\sum_{i=1}^N p_i = 1$, defined by:

$$S(\mathbf{p}) = - \sum_{i=1}^N p_i \log p_i. \tag{7.2.1}$$

where the logarithms will be taken in base 2 in this chapter, unless indicated otherwise (and noted log rather than ln), as is customary in information theory, instead of base e as used elsewhere in this book, since base e is more common in statistical mechanics; but if we multiplied the sum in (7.2.1) by an arbitrary constant k, the choice of that constant is equivalent to a change of the logarithm base. The minus sign in (7.2.1) guarantees that $S(\mathbf{p}) \geq 0$, since $p_i \leq 1$.

Note that this definition coincides, up to the choice of the base of logarithms, with the one of the Gibbs entropy in the discrete case, see (6.6.8) and definition 6.5.

One immediately observes that if $\mathbf{p}$ is such that $p_i = 1$ for some $i \in \{1, \dots, N\}$ and thus $p_j = 0$, $\forall j \neq i$, we have $S(\mathbf{p}) = 0$ (with the convention $0 \log 0 = \lim_{x \to 0} x \log x = 0$). On the other hand, if $p_i = \frac{1}{N}$, $\forall i = 1, \dots, N$, $S(\mathbf{p}) = \log N$. We also have:

$$0 \leq S(\mathbf{p}) \leq \log N, \tag{7.2.2}$$

which follows from the concavity of the log function and Jensen's inequality (see Appendix 7.B): $\sum_{i=1}^N \alpha_i F(x_i) \leq F(\sum_{i=1}^N \alpha_i x_i)$, with F concave, $\alpha_i \geq 0$ and $\sum_{i=1}^N \alpha_i = 1$; we take here $\alpha_i = p_i$, $x_i = \frac{1}{p_i}$, and $F(x) = \log x$.

This suggests that $S(\mathbf{p})$ measures how "spread out" $\mathbf{p}$ is or how much "uncertainty" there is in the probability distribution $\mathbf{p}$: if $p_i = 1$ for some $i \in \{1, \dots, N\}$ and $p_j = 0$, $\forall j \neq i$, the probability distribution is as concentrated as it can be (no spreading), there is no uncertainty in a "random" event with such a probability distribution, and $S(\mathbf{p}) = 0$. On the other hand, $S(\mathbf{p}) = \log N$ is maximal for the uniform distribution, which is the most spread out distribution and the one for which the uncertainty of a "random" event with such a probability distribution is maximal. That uniform distribution is of course the one associated to the indifference principle.

Another way to justify (7.2.1) is to view it as the average information gained when one performs an experiment with N possible outcomes, with probabilities $\mathbf{p} = (p_i)_{i=1}^N$. To see that, associate a number $I(j)$ representing the information gained when performing that sort of experiment and obtaining result j. Intuitively, one expects that if one repeats twice that experiment and one obtains results j_1 and j_2, the information associated to that combination of results $I(j_1 \cup j_2)$ must equal the sum $I(j_1) + I(j_2)$. But if $I(j)$ is a function of the probability p_j of result j, and if the different experiments are independent (so that the probability of results j_1 and j_2 in two successive experiments is $p_{j_1} p_{j_2}$), we have $I(p_{j_1} p_{j_2}) = I(p_{j_1}) + I(p_{j_2})$, which implies[1] that $I(j) = -k \log p_j$, with $k > 0$ if we want I to be positive. Taking the expectation value of $I(j)$ gives (7.2.1).

It turns out that one can prove that formula (7.2.1) is unique, if we want a function of $\mathbf{p}$ that measures the amount of dispersion or of "uncertainty" of the probability distribution $\mathbf{p}$. Indeed such a function should have the following properties:

1. $S(\mathbf{p})$ is a continuous function on $\mathbb{R}^N$.
2. The map $N \to S(\frac{1}{N}, \ldots, \frac{1}{N})$ is increasing in N.
3. For each N, and each $\mathbf{p} = (p_i)_{i=1}^N$, let $q_1, \ldots, q_M$ be defined by:

$$q_1 = p_1 + \cdots + p_{k_1} \ldots$$
$$q_j = p_{k_{j-1}+1} + \cdots + p_{k_j} \ldots$$
$$q_M = p_{k_{M-1}+1} + \cdots + p_{k_M},$$

with $M \leq N$, $1 \leq k_1 \leq k_2 \leq \ldots k_M = N$, and $\sum_{j=1}^M q_j = 1$. Then:

$$S(\mathbf{p}) = S(q_1, \ldots, q_M) + \sum_{j=1}^M q_j S\left(\frac{p_{k_{j-1}+1}}{q_j}, \ldots, \frac{p_{k_j}}{q_j}\right) \tag{7.2.3}$$

Then we have the following

Theorem 7.1 *Theorem: Let $S(\mathbf{p})$ be a function from $[0,1]^N \to R_+$ satisfying [1]-[3] above. Then, $S(\mathbf{p}) = -k \sum_{i=1}^N p_i \log p_i$ for some $k > 0$.*

Remark 7.2 Assumption [1] is necessary only if we want to allow for the possibility of irrational values of some p_i. When all the p_i's are rational, it is not needed to establish the uniqueness of the entropy formula (7.2.1).

Remark 7.3 Assumption [2] is natural if $S(\mathbf{p})$ measures the amount of spreading or of uncertainty in the probability distribution $\mathbf{p}$: the larger N is, the more uncertainty there is in the random event with the uniform probability distribution $p_i = \frac{1}{N}$, $\forall i = 1, \ldots, N$.

Remark 7.4 Assumption [3] is a consistency requirement: if we divide the events with probabilities p_i, $i = 1, \ldots, N$ into M subsets with $i \in \{k_{j-1} + 1, \ldots k_j\}$, $j = 1, \ldots, M$, and $k_0 = 0$, then the uncertainty associated to $\mathbf{p}$ is equal to the uncertainty associated to the distribution $q_1, \ldots, q_M$, plus the average of the uncertainties in each subset. The average is taken with respect to the distribution $q_1, \ldots, q_M$, and the uncertainty in each subset is the one of the induced probability distribution $(\frac{p_{k_{j-1}+1}}{q_j}, \ldots, \frac{p_{k_j}}{q_j})$ in the subset with label j.

Proof Let,[2]

---

[1] Let $f(x)$, $x \in \mathbb{R}_+$ be differentiable and satisfy $f(xy) = f(x) + f(y)$, $\forall x, y \in \mathbb{R}_+$. then differentiating with respect to y and setting $y = 1$, we get: $f'(x) = \frac{f'(1)}{x}$ and thus $f(x) = C \log x$ for some C (this results holds for more general functions than the differentiable ones)..

[2] This proof follows Jaynes [183, Sect. 11.3]. See also Khinchin [192].

$$A(N) = S\left(\frac{1}{N}, \ldots, \frac{1}{N}\right),$$

$p_i = \frac{1}{N}, \forall i = 1, \ldots, N$ and $q_j = \frac{n_j}{N}$, where $\sum_{j=1}^M n_j = N$. Then, from (7.2.3), we get:

$$S\left(\frac{1}{N}, \ldots, \frac{1}{N}\right) = S(q_1, \ldots, q_M) + \sum_{j=1}^M q_j S(\frac{1}{n_j}, \ldots, \frac{1}{n_j}), \qquad (7.2.4)$$

which implies, for $A(N)$,

$$A(N) = S(q_1, \ldots, q_M) + \sum_{j=1}^M q_j A(n_j). \qquad (7.2.5)$$

Now, let $N = ML$ and let $n_j = L, \forall j = 1, \ldots, M$. then we get from (7.2.4), where $q_j = \frac{L}{N} = \frac{1}{M}$:

$$A(ML) = S\left(\frac{1}{M}, \ldots, \frac{1}{M}\right) + \sum_{j=1}^M \frac{1}{M} S\left(\frac{1}{L}, \ldots, \frac{1}{L}\right) = A(M) + A(L). \quad (7.2.6)$$

As we saw in footnote 1, if this identity held for all positive real numbers (instead of only for positive integers), it would imply $A(N) = k \log N$ for some $k > 0$. The proof that this follows also from (7.2.6) restricted to integers is given in the Appendix.

Now, rewrite (7.2.5) with $q_j = \frac{n_j}{N}$ as:

$$\begin{aligned} S\left(\frac{n_1}{N}, \ldots, \frac{n_M}{N}\right) &= A(N) - \sum_{j=1}^M \frac{n_j}{N} A(n_j) \\ &= k\left(\log N - \sum_{j=1}^M \frac{n_j}{N} \log n_j\right) = -k\left(\sum_{j=1}^M \frac{n_j}{N} \log \frac{n_j}{N}\right). \end{aligned} \qquad (7.2.7)$$

We know that $k > 0$ because of [2] ($A(N)$ is increasing) and using property [1], namely the continuity of the map S, finishes the proof of the theorem.   □

7.3   The Maximum Entropy Principle

Since $S(\mathbf{p})$ measures how spread out the probability distribution is, one can use it to extend the indifference principle. This extension is called:

The maximum entropy principle.

Given some information about a system, assign to that system, among all the probability distributions that incorporate that information, the one having the largest entropy.

Of course, "incorporating some information" may seem vague and we'll be more specific in a moment. But the rationale behind this principle is to avoid including in our probability distribution information that we do not have. So, we want to choose the probability distribution which is as "indifferent" as possible; where indifferent is synonymous with "spread out" or "contains a maximum amount of uncertainty" or "contains a minimum amount of information".

It is easy to see why this principle extends the indifference principle: if we have no information, what follows from (7.2.2) and the calculation preceding it, is that the probability distribution with maximum entropy is the uniform one.

How does one incorporate information in our choice of probability distributions? There is no general answer to that question, but a class of such problems that are easy to solve is when our knowledge includes some average value of a function $F : \{1, \ldots, N\} \to \mathbb{R}$ defined on the set of states to which we want to assign a probability distribution; suppose that we know the value of $< F >$ defined as:

$$< F >= \sum_{i=1}^{N} F(i) p_i \tag{7.3.1}$$

The standard way to find maximum of (7.2.1) under the constraint (7.3.1) as well as $\sum_{i=1}^{N} p_i = 1$ is via Lagrange's multipliers, as we did in (6.2.12), namely maximizing:

$$-\sum_{i=1}^{N} p_i \ln p_i - \lambda \sum_{i=1}^{N} F(i) p_i - \lambda_0 \sum_{i=1}^{N} p_i, \tag{7.3.2}$$

where, in order to connect the computation done here with the rest of this book, we replace the base 2 logarithm in (7.2.1) by a base e logarithm. The maximum of (7.3.2) is found by setting equal to 0 the derivatives of (7.3.1) with respect to each $p_j, j = 1, \ldots, N$:

$$- \ln p_j - 1 - \lambda F(j) - \lambda_0 = 0$$

which implies:

$$p_j = C \exp(-\lambda F(j)) \tag{7.3.3}$$

where the constant is chosen so that $\sum_{i=1}^{N} p_i = 1$, i.e. $C = Z^{-1}$, with

$$Z = \sum_{i=1}^{N} \exp(-\lambda F(i)) \tag{7.3.4}$$

So, we get:

$$p_i = \frac{\exp(-\lambda F(j))}{Z}, \tag{7.3.5}$$

where the value of λ is given implicitly by:

$$< F >= -\frac{\partial \ln Z}{\partial \lambda} = \sum_{i=1}^{N} p_i F(i) = \frac{\sum_{i=1}^{N} F(i) \exp(-\lambda F(i))}{Z}. \tag{7.3.6}$$

However, this calculation does not *prove* that the distribution (7.3.3) is the only one that maximizes (7.2.1). But this is nevertheless true:

Proposition 7.5 *The probability distribution that maximizes the entropy (7.2.1) among the probability distributions satisfying (7.3.1) is given by (7.3.5), with Z and λ defined by (7.3.4), (7.3.6).*

Proof Let $\phi(x) = x \ln x - x + 1$, for $x > 0$. It is easy to see that $\phi(x)$ is strictly convex ($\phi''(x) = \frac{1}{x}$) with a minimum at $x = 1$, where $\phi(1) = 0$, so $\phi(x) > 0$, for $x > 0$.

Let $\mathbf{p}$ be given by (7.3.5) and $\mathbf{q}$ be another distribution. We have $\phi(\frac{q_i}{p_i}) \geq 0$, which implies:

$$- q_i \ln q_i + q_i \ln p_i \leq p_i - q_i. \tag{7.3.7}$$

Summing (7.3.7) over i gives:

$$S(\mathbf{q}) = - \sum_{i=1}^{N} q_i \ln q_i \leq - \sum_{i=1}^{N} q_i \ln p_i, \tag{7.3.8}$$

which is called *Gibbs' inequality*. But, by (7.3.5), $\ln p_i = - \ln Z - \lambda F(i)$, and so:

$$- \sum_{i=1}^{N} q_i \ln p_i = \ln Z + \lambda \sum_{i=1}^{N} F(i) q_i = \ln Z + \lambda < F > .$$

But we have also

$$S(\mathbf{p}) = - \sum_{i=1}^{N} p_i \ln p_i = \ln Z + \lambda \sum_{i=1}^{N} F(i) p_i = \ln Z + \lambda < F >,$$

and thus, (7.3.8) implies:

$$S(\mathbf{q}) \leq S(\mathbf{p}),$$

which proves the proposition. □

It is easy to see that all this extends to the situation where the function whose average value is known takes values in $\mathbb{R}^m$: $\mathbf{F} : \{1, \ldots, N\} \to \mathbb{R}^m$. We get, instead of (7.3.5):

$$p_i = \frac{\exp\left(-\sum_{l=1}^{m} \lambda_l F_l(i)\right)}{Z},\tag{7.3.9}$$

with

$$Z = \sum_{i=1}^{N} \exp\left(-\sum_{l=1}^{m} \lambda_l F_l(i)\right)\tag{7.3.10}$$

and where the values $\lambda_l, l = 1, \ldots, m$, are given implicitly by:

$$< F_n > = -\frac{\partial \ln Z}{\partial \lambda_n} = \sum_{i=1}^{N} F_n(i) p_i = \frac{\sum_{i=1}^{N} F_n(i) \exp(-\sum_{l=1}^{m} \lambda_l F_l(i))}{Z}$$

$\forall n = 1, \ldots m$.

The entropy (7.2.1), with ln instead of log, of the distribution maximizing that entropy is, for $\mathbf{p}$ defined by (7.3.9):

$$S(\mathbf{p}) = \ln Z + \sum_{l=1}^{m} \lambda_l < F_l > .\tag{7.3.11}$$

7.4 The Ensembles

Once we accept the maximum entropy principle, it is easy to derive the various ensembles introduced in Sect. 6.6. Let us start, for simplicity, with finite sets of states: $i = 1, \ldots, N$. Assume that each state has an energy e_i and a number of particles n_i. We consider three possibilities:

1. The numbers $(e_i)_{i=1}^{N}$ and $(n_i)_{i=1}^{N}$ are given and there is no constraint of the form (7.3.1).
2. The numbers $(n_i)_{i=1}^{N}$ are given as well as the average energy $< E >= \sum_{i=1}^{N} e_i p_i$. Thus, we have a constraint of the form (7.3.1) with $F(i) = e_i$.
3. The average energy $< E >= \sum_{i=1}^{N} e_i p_i$ and the average number of particles $< N >= \sum_{i=1}^{N} n_i p_i$ are given. We have two constraints of the form (7.3.1).

The maximum entropy principle and (7.3.5, 7.3.9) gives the corresponding probability distributions or ensembles maximizing (7.2.1) with the given contraints:

1. If e_i and n_i are given and there are no constraints the maximum entropy probability distribution is the uniform one:

$$p_i = \frac{1}{N} \quad \forall i = 1, \ldots, N,\tag{7.4.1}$$

which corresponds to the micro-canonical distribution.

2. We get, with $F(i) = e_i$,

$$p_i = \frac{\exp(-\lambda e_i)}{Z_c}, \tag{7.4.2}$$

with

$$Z_c = \sum_{i=1}^{N} \exp(-\lambda e_i), \tag{7.4.3}$$

which corresponds to the canonical distribution (with $\lambda = \beta$).
3. With two constraints of the form (7.3.1), $< E >= \sum_{i=1}^{N} e_i p_i$ and $< N >= \sum_{i=1}^{N} n_i p_i$, we get from (7.3.9):

$$p_i = \frac{\exp(-\lambda_1 e_i - \lambda_2 n_i)}{Z_g}, \tag{7.4.4}$$

with

$$Z_g = \sum_{i=1}^{N} \exp(-\lambda_1 e_i - \lambda_2 n_i), \tag{7.4.5}$$

which corresponds to the grand-canonical distribution (with $\lambda_1 = \beta, \lambda_2 = -\beta\mu$).

Note that in (7.4.3) and (7.4.5) there is no division by $N!$ because our states are distinguished by a label $i = 1, \ldots, N$ (for example, the label could refer to a particular location on a lattice).

Using (7.3.11), we get for the entropy in the various ensembles:

1. For **p** given by (7.4.1):

$$S(\mathbf{p}) = \ln N \tag{7.4.6}$$

2. For **p** given by (7.4.2):

$$S(\mathbf{p}) = (\ln Z_c + \beta < E >) \tag{7.4.7}$$

with Z_c given by (7.4.3).
3. For **p** given by (7.4.4):

$$S(\mathbf{p}) = (\ln Z_g + \beta < E > -\beta\mu < N >) \tag{7.4.8}$$

with Z_g given by (7.4.5).

7.5 Continuous Distributions and Relative Entropy

The natural extension of the notion of entropy to a continuous distribution $p(x) = f(x)dx, x \in \mathbb{R}, f(x) \geq 0, \int_{\mathbb{R}} f(x)dx = 1$, may be thought to be:

$$S(p) = -\int_{\mathbb{R}} f(x) \log f(x) dx \qquad (7.5.1)$$

But this function is not necessarily positive: for example, if $f(x)$ is uniformly distributed on an interval $[a, b]$, then $S(p)$ may be negative (exercise: check that).

Another defect of (7.5.1) is that it is not necessarily the limit of $S(\tilde{p}(n))$ for a sequence of distributions $\tilde{p}(n)$ that are discrete approximations of f. Indeed, suppose that f has support in $[0, 1]$, with $\int_0^1 f(x) dx = 1$. Divide $[0, 1]$ into n intervals $I_j = \left[\frac{j}{n}, \frac{j+1}{n} \right]$, $j = 0, \ldots, n-1$ and let $\tilde{p}_j = \frac{1}{n} f(x_j)$, where $x_j \in I_j$ is chosen so that $f(x_j) = \frac{1}{|I_j|} \int_{I_j} f(x) dx = n \int_{I_j} f(x) dx$, so that $\sum_{j=0}^{n-1} \tilde{p}_j = \int_0^1 f(x) dx = 1$. Then:

$$S(\tilde{p}(n)) = -\sum_{j=0}^{n-1} \tilde{p}_j \log \tilde{p}_j = -\frac{1}{n} \sum_{j=0}^{n-1} f(x_j) \log \left(\frac{f(x_j)}{n} \right) = \qquad (7.5.2)$$

$$-\frac{1}{n} \left[\sum_{j=0}^{n-1} f(x_j) \log f(x_j)) - \log n \sum_{j=0}^{n-1} f(x_j) \right] \not\to_{n \to \infty} -\int_0^1 f(x) \log f(x) dx,$$

because the term $\frac{1}{n} \log n \sum_{j=0}^{n-1} f(x_j) \to_{n \to \infty} \infty$.

This is one of the reasons why one introduces, given two probability distributions $dp(x) = f(x) dx$ and $dq(x) = g(x) dx$, the *relative entropy* of p with respect to q:

$$S(p\|q) = \int_{\mathbb{R}} f(x) \log \frac{f(x)}{g(x)} dx = \mathbb{E}_p \left(\log \frac{f(x)}{g(x)} \right) = -\mathbb{E}_p \left(\log \frac{g(x)}{f(x)} \right) \qquad (7.5.3)$$

Note that $S(p\|q)$ is finite only if the support of $f(x)$ is contained in the support of $g(x)$.

The relative entropy satisfies the following properties (identical to what holds in the discrete case):

1. $S(p\|q) \geq 0$.
2. $S(p\|q) = 0$ iff $p = q$.

This follows easily from Jensen's inequality (see Appendix 7.B), the expression on the right hand side of (7.5.3), and the convexity of the function $-\log$:

$$-\mathbb{E}_p \left(\log \frac{g(x)}{f(x)} \right) \geq -\log \mathbb{E}_p \left(\frac{g(x)}{f(x)} \right)$$

$$= -\log \int_{\mathbb{R}} \frac{g(x)}{f(x)} f(x) dx = -\log \int_{\mathbb{R}} g(x) dx = 0,$$

where the equality holds only if $\frac{g(x)}{f(x)} = 1$ almost everywhere.

Because of these properties, $S(p\|q)$ is also called *the Kullback–Leibler divergence* between p and q since it measures how much p and q differ; it is a sort of distance between these two probability distributions (however, it is not a real distance, since it is neither symmetric nor satisfies the triangle inequality).

For distributions $\mathbf{p}, \mathbf{q}$ on a finite set of N elements, we have:

$$S(\mathbf{p}||\mathbf{q}) = \sum_{i=1}^{N} \log \frac{p_i}{q_i} p_i$$

If $\mathbf{q}$ is the uniform distribution that gives a probability $\frac{1}{N}$ to each element, we have:

$$S(\mathbf{p}||\mathbf{q}) = \sum_{i=1}^{N} \log \frac{p_i}{\frac{1}{N}} p_i = \log N - S(\mathbf{p}),$$

with $S(\mathbf{p})$ given by (7.2.1), so that both the relative entropy $S(\mathbf{p}||\mathbf{q})$ and the Shannon entropy $S(\mathbf{p})$ can be positive.

More generally, if we have two probability measures, μ and ν on Ω, with μ absolutely continuous with respect to ν (see Appendix 2.A.6), $d\mu = F d\nu$, one defines:

$$S(\mu||\nu) = \int_{\Omega} \log F(x) d\mu = \mathbb{E}_{\mu}(\log F(x)) = -\mathbb{E}_{\mu}\left(\log \frac{1}{F(x)}\right).$$

Again, we have $S(\mu||\nu) \geq 0$, with equality iff $\mu = \nu$.

7.6 "Derivation" of the Second Law

There is an easy way to derive the second law of thermodynamics (see Jaynes [177] or Goldstein and Lebowitz [150]) from the maximum entropy principle.

Let Ω be the total phase space of an isolated physical system and $\Omega_0 \subset \Omega$ a subset of Ω in which the system is initially because of some constraint, like a wall separating two different gases or forcing the gas to be in one-half of a box.

The entropy of the initial probability distribution $\mathbf{p}_0$ is:

$$S(\mathbf{p}_0) = \max_{\mathbf{q}} S(\mathbf{q})$$

where the maximum is taken over all distributions $\mathbf{q}$ whose support lies Ω_0 (and that may also satisfy some contraints of the form (7.3.1)).

If we remove the constraint and let the physical system reach a new equilibrium, we get for the final entropy:

$$S(\mathbf{p}_f) = \max_{\mathbf{q}} S(\mathbf{q})$$

where the maximum is now taken over all distributions $\mathbf{q}$ satisfying the same constraints of the form (7.3.1) as initially but whose support lies Ω. Then, since $\Omega_0 \subset \Omega$, we have trivially:

$$S(\mathbf{p}_0) \leq S(\mathbf{p}_f)$$

since the maximum in the right hand side is taken over a larger subset than in the left hand side.

This "derives" the second law, but without any consideration of a time evolution between the two equilibrium states, the initial and the final ones.[3] Indeed, in the information theoretic approach, one insists that the maximum entropy formula applies only to equilibrium states.[4]

This is one of the reasons why we call this approach "predictive": it will predict in what direction the entropy evolves in between two equilibrium states, and also predicts, in principle, the value of the final entropy.

7.7 Shannon's Entropy and Communication

The entropy formula (7.2.1) is due originally to Shannon in the context of the technology of communication; Shannon was trying to answer the following question: what is the most efficient way to transmit a message over a given channel? This had nothing to do with thermodynamics or statistical physics, although the formal analogy between Shannon's formula (7.2.1) and Boltzmann and Gibbs entropies generated and enormous literature attempting to base statistical mechanics on the notion of "information" and of Shannon's entropy.

In this section, which is sort of an aside in this book, we will give two applications of Shannon's notion of entropy to the transmission of messages: the Shannon entropy allows us to estimate the number of the most probable messages and it also gives upper and lower bounds on the length of messages that can be transmitted efficiently.

The origin of the use of the word "entropy" by Shannon is due to von Neumann, according to Myron Tribus:

> In 1961 one of us (Tribus) asked Shannon what he had thought about when he had finally confirmed his famous measure. Shannon replied: "My greatest concern was what to call it. I thought of calling it 'information,' but the word was overly used, so I decided to call it 'uncertainty.' When I discussed it with John von Neumann, he had a better idea. Von Neumann told me, 'You should call it entropy, for two reasons. In the first place your uncertainty function has been used in statistical mechanics under that name, so it already has a name. In the second place, and more important, no one knows what entropy really is, so in a debate you will always have the advantage.'

> Myron Tribus and Edward C. McIrvine [312, p. 180]

---

[3] Lavis stresses that aspect in [210, Sect. 6.1].

[4] In Sect. 8.8 we will discuss the relation between this derivation of the second law and the fact that, under its "natural" time evolution, the Gibbs entropy is constant, as shown in Proposition 8.1.

7.7.1   Information Content of a Message

An elementary example of the type of problems to be solved in communication is to find the most efficient way to code an ordinary text (say, in English) into sequences of 0's and 1's, where "efficient" means using as little time (or costing as little money) as possible. We are not concerned here with secret encoding, which is a separate question: most messages sent over telephone or any other means of transmission are not secretly encoded.

A text is a sequence of letters and symbols; some letters occur more frequently than others: for example, in English, a is more frequent than z. It makes sense intuitively to code frequent letters by shorter sequences of 0's and 1' than the rare ones. The Morse code relies on this idea.

Let us try to systematize this intuition. We will assume that we know the frequency of all the symbols (say, in a given human language) labelled $i = 1, \ldots, K$ and we will model them as a sequence of independent random variables with probabilities p_i given by their frequencies (so, the probabilities here are interpreted as frequencies and not in any other sense). The "independence" assumption is of course a simplification: because of words like be, it, is, the, an, some pairs of letters are more frequent than the product of the probabilities of each of those letters, and more sophisticated methods can be used to deal with that, but here we want only to explain the simplest ideas in the coding of messages.

Suppose we have a sequence of such symbols of length N, $\mathbf{i} = (i_1, \ldots, i_N) \in \Omega_N$, $\Omega_N = \{1, \ldots, K\}^N$. Its probability is:

$$p(i_1, \ldots, i_N) = \prod_{\alpha=1}^{N} p_{i_\alpha} = \prod_{j=1}^{K} p_j^{n_j} \tag{7.7.1}$$

where n_j is the number of times the symbol j appears in the sequence $i_1, \ldots, i_N$.

How many times do we expect the symbol j to appear in that sequence? Since the frequency of the symbol j is p_j, it will typically appear approximately $p_j N$ times. So, for those typical sequences, we can rewrite (7.7.1) as:

$$p(i_1, \ldots, i_N) \approx \prod_{j=1}^{K} p_j^{p_j N}. \tag{7.7.2}$$

Since $p(i_1, \ldots, i_N)$ is a product of N factors less than 1, one expects this product to decrease exponentially with N: $p(i_1, \ldots, i_N) \approx \exp(-CN)$. But what is the value of C? For typical sequences, we get from (7.7.2), since $\log \prod_{j=1}^{K} p_j^{p_j N} = N \sum_{j=1}^{K} p_j \log p_j = -N S(\mathbf{p})$, with $\mathbf{p} = (p_j)_{j=1}^{K}$, and $S(\mathbf{p})$ defined in (7.2.1), that:

$$p(i_1, \ldots, i_N) \approx 2^{-N S(\mathbf{p})}, \tag{7.7.3}$$

The bound (7.7.3) can be made precise through the law of large numbers[5]:

Proposition 7.6 *Let* $G(N, \epsilon) = \{\mathbf{i} = (i_1, \ldots, i_N) \in \Omega_N \| \frac{1}{N} \log p(i_1, \ldots, i_N) - (-S(\mathbf{p}))| \leq \epsilon\}$ *be the set of "good" sequences (using the same terminology as in Sect. 2.3). Then, $\forall \epsilon > 0$, $\exists N_\epsilon$ such that $\forall N \geq N_\epsilon$,*

$$P(\mathbf{i} \notin G(N, \epsilon)) \leq \epsilon. \tag{7.7.4}$$

Proof To prove (7.7.4), use formula (7.7.1) to write $\log p(i_1, \ldots, i_N) = \sum_{\alpha=1}^{N} \log p_{i_\alpha}$ and apply to law of large numbers (2.3.8), (2.3.9) to that last sum. That law implies that $\frac{1}{N} \sum_{\alpha=1}^{N} \log p_{i_\alpha}$ converges to its average, as $N \to \infty$, and its average is $-S(\mathbf{p})$. To show that we can use the same ϵ in the definition of $G(N, \epsilon)$ and in (7.7.4) is left as an exercise. □

The meaning of this proposition, called the "Asymptotic equipartition" principle, is that most sequences have the same probabilities (in the sense that their rate of exponential decay in N is approximately the same),[6] and so the number of sequences in $G(N, \epsilon)$ is roughly the inverse of that rate:

$$|G(N, \epsilon)| \approx 2^{NS(\mathbf{p})}.$$

We can also say that $G(N, \epsilon)$ is a set of "typical" sequences (see (2.3.1)). The relevance of the set of typical sequences is that one can define an algorithm that assign a distinct sequence of 0' and 1's to each $\mathbf{i} \in G(N, \epsilon)$. Since almost all sequences $\mathbf{i}$ belong to $G(N, \epsilon)$ for large N, we obtain a coding with a negligible risk of errors (since $p(\mathbf{i} \notin G(N, \epsilon)) \leq \epsilon$). On the other hand, one cannot encode all the sequences of symbols $\mathbf{i} \in G(N, \epsilon)$ into less than $2^{N(S(\mathbf{p})-\epsilon)}$ sequences of 0's and 1's, for any $\epsilon > 0$.

If $S(\mathbf{p}) < \log K$, we can, for large N, encode much fewer messages than if we encoded all messages of length N composed of symbols in $\{1, \ldots, K\}$, whose number is $K^N = 2^{N \log K} >> 2^{NS(\mathbf{p})}$. For example, if $K = 2$, $p(1) = p$, $p(2) = 1 - p$, we have $S(\mathbf{p}) = -p \log p - (1 - p) \log(1 - p)$, whose shape is drawn in Fig. 6.1 and whose maximum is reached for $p = \frac{1}{2}$ and where $S(\mathbf{p}) = 1 = \log 2$. We see in that figure that, if $p \neq \frac{1}{2}$, $S(\mathbf{p}) < 1$.

---

[5] See Khinchin [192] for a detailed discussion of such results.

[6] This concerns the rate of decay in N; there is still a factor $\exp(\mathcal{O}(\sqrt{N}))$ between the probability of the most probable and of the least probable sequences in the set $G(N, \epsilon)$, see MacKay [234, p. 83–84].

7.7.2   Encoding Messages

In the previous subsection, we analyzed how to compress messages if we allow for a negligible number of errors. Let us now see what happens if we try to encode a set of messages without making any error.

Let us define a *code* for the set of symbols $i = 1, \ldots, K$ as a map:

$$C : \{1, \ldots, K\} \rightarrow \{0, 1\}^{\mathbb{N}}.$$

from that set to sequences of 0's and 1's.

To each message $i_1, \ldots, i_N$, with $i_\alpha \in \{1, \ldots, K\}, \forall \alpha \in \{1, \ldots, N\}$, we associate the string of 0's and 1's:

$$C(i_1, \ldots, i_N) = C(i_1)C(i_2) \ldots C(i_N) \equiv \prod_{\alpha=1}^{N} C(i_\alpha)$$

A code is *uniquely decodable* if $\forall i, j \in \{1, \ldots, K\}, i \neq j, C(i) \neq C(j)$. This is an obvious condition for a code to be at all useful.

But a code would not be really useful if it was not easily decodable. A way to ensure that latter property is to demand that no codeword is a prefix of another codeword, meaning that one does not have codewords c and d and a tail string t of the codeword d so that $ct = d$. For example, $c = 1$ is a prefix to the codeword $d = 101$ and $c = 10$ is also such a prefix. The codes satisfying that property (no codeword is a prefix of an other codeword) are called *prefix codes* or *prefix-free codes* or *self-punctuating codes*. Codes with four symbols such as $\{0, 10, 110, 111\}$ or $\{00, 01, 10, 11\}$ are prefix codes. Those codes are easily decodable, because one can simply decode a message $\prod_{\alpha=1}^{N} C(i_\alpha)$ from left to right without looking ahead to the next codeword, since the end of a codeword is immediately recognizable.

The expected length of a code is $L(\mathbf{p}, C) = \sum_{i=1}^{K} l_i p_i$, where l_i is the length of $C(i)$ and p_i the probability (or frequency) of symbol i.

Now let us ask which constraints are imposed on $L(\mathbf{p}, C)$ if we want our code to be a uniquely decodable code.

Before answering that question, we need to prove *Kraft's inequality*:

Proposition 7.7 *For any uniquely decodable code, the codeword lengths must satisfy:*

$$S = \sum_{i=1}^{K} 2^{-l_i} \leq 1. \tag{7.7.5}$$

Proof write $S^M = \sum_{i_1=1}^{K} \sum_{i_2=1}^{K} \cdots \sum_{i_M=1}^{K} 2^{-(l_{i_1} + l_{i_2} + \ldots l_{i_M})}$. The exponent $l_{i_1} + l_{i_2} + \ldots l_{i_M}$ is the length of the string of 0's and 1's encoding the sequence of symbols $i_1 i_2 \ldots i_M$. For every such sequences of length M there is one term in the above sum. Let A_l the number of sequence of M symbols whose encoding have length l. So we have

$$S^M = \sum_{l=Ml_{\min}}^{l=Ml_{\max}} 2^{-l} A_l,$$ (7.7.6)

where $l_{\min}$ and $l_{\max}$ are respectively the smallest and the largest length of a codeword.

Since there are 2^l distinct strings of 0's and 1's of length l and if C is uniquely decodable, we must have $C(i) \neq C(j)$ for $i \neq j$, this implies that $A_l \leq 2^l$. Inserting this in (7.7.6) yields:

$$S^M \leq \sum_{l=Ml_{\min}}^{l=Ml_{\max}} 2^{-l} 2^l \leq Ml_{\max}.$$

But this bound is linear in M, and that proves that S must be less than or equal to 1, otherwise S^M would grow exponentially with M. □

It turns out that a reverse of Kraft's inequality holds: for any set of codeword lengths satisfying (7.7.5), there exists a prefix code having those lengths,[7] see MacKay [234, p. 104].

Now we arrive at the main result of this section: the entropy of the set of symbols gives a lower bound on the expected length of any uniquely decodable code. We have the *Shannon's source coding theorem for symbol codes*:

Proposition 7.8 *For any uniquely decodable code C, we have:*

$$L(\mathbf{p}, C) \geq S(\mathbf{p})$$ (7.7.7)

Moreover, there exists a uniquely decodable code C such that:

$$L(\mathbf{p}, C) \leq S(\mathbf{p}) + 1$$ (7.7.8)

Proof Let $q_i = \frac{2^{-l_i}}{S}$ (where $S = \sum_{j=1}^{K} 2^{-l_j}$), so $0 \leq q_i$ and $\sum_{j=1}^{K} q_i = 1$. We have $l_i = \log \frac{1}{q_i} - \log S$. Gibbs' inequality ($\sum_{i=1}^{K} p_i \log \frac{1}{q_i} \geq \sum_{i=1}^{K} p_i \log \frac{1}{p_i}$, which is (7.3.8) with p_i and q_i exchanged) gives:

$$L(\mathbf{p}, C) = \sum_{i=1}^{K} p_i l_i = \sum_{i=1}^{K} p_i \log \frac{1}{q_i} - \log S \geq \sum_{i=1}^{K} p_i \log \frac{1}{p_i} - \log S \geq S(\mathbf{p}),$$

since, by Kraft's inequality, $-\log S \geq 0$. This proves (7.7.7).

To prove (7.7.8), choose a code so that l_i is the smallest integer greater or equal to $\log \frac{1}{p_i}$. We have $\log \frac{1}{p_i} \leq l_i \leq \log \frac{1}{p_i} + 1$

Kraft's inequality is satisfied for this choice of set of lengths: since $l_i \geq \log \frac{1}{p_i}$,

---

[7] To be precise, Kraft proved that if (7.7.5) holds, then there exists a prefix code with those given codeword lengths and McMillan proved that if a code is uniquely decodable, then (7.7.5) holds, so Proposition 7.7 is really due McMillan, see MacKay [234, p. 95].

$$\sum_{i=1}^{K} 2^{-l_i} \leq \sum_{i=1}^{K} 2^{\log p_i} = 1$$

and thus, by the reverse of Kraft's inequality, there exist a prefix code, hence a uniquely decodable code, with those lengths.

Since $l_i \leq \log \frac{1}{p_i} + 1$, we also have:

$$L(\mathbf{p}, C) = \sum_{i=1}^{K} l_i p_i \leq \sum_{i=1}^{K} (\log \frac{1}{p_i} + 1) p_i = \sum_{i=1}^{K} (- \log p_i + 1) p_i = S(\mathbf{p}) + 1,$$

which proves (7.7.8).   □

Let us end this section by giving a concrete example of coding, the *Huffman coding algorithm*:

1. Take the two least probable symbols in the set of symbols to be coded. Assign a 0 and a 1 to each of them.
2. Add the probabilities of these two symbols to combine them in a new symbol and repeat the operation.
3. The codewords to be assigned to each symbol are obtained by concatenating the binary digits in reverse order. In that way, the least probable symbols are assigned the longest codewords.

As an exercise the reader can check that if $K = 5$, with $p(1) = 0.25$, $p(2) = 0.25$, $p(3) = 0.2$, $p(4) = 0.15$, $p(5) = 0.15$ then we start by assigning 1 to 5 and 0 to 4 (or vice versa of course). Now we have a set with $K = 4$ and $p(1) = 0.3$, $p(2) = 0.25$, $p(3) = 0.25$, $p(4) = 0.2$, where $p(1) = 0.15 + 0.15$ results from the combination of 4 and 5. Repeating this, we get the code:

1. $C(1) = 00$
2. $C(2) = 10$
3. $C(3) = 11$
4. $C(4) = 010$
5. $C(5) = 011$

Although this coding is in some sense optimal, it is not always so in practice; for a discussion, see MacKay [234, Chap. 5].

7.8   Evaluation of the Information Theoretic Approach

7.8.1   *What Do Probabilities Mean Here?*

There are many attractive features of the information theoretic approach. One assumes only *one* principle governing the choice of probability distributions in equilibrium statistical mechanics and one derives from it all the usual ensembles, the correct formula for the thermodynamic entropy, as well as, in some sense, the second law (see Sect. 7.6). Moreover, that unique principle can be justified on the basis of an objective Bayesian reasoning. What more could one ask for?

First of all, one should eliminate the emphasis put on "information" in that approach. Reading the literature on theories of information or on the information theoretic approach to statistical mechanics or on quantum information or even on quantum mechanics, one might have the impression that the whole of science deals only with "information". But information, in the usual meaning of the word, is always information about something different from itself. Information means information about the world: if we know that it will rain tonight or that the stock market will crash or that the battle of Waterloo took place in 1815, this is "information" but it refers to things that are not themselves information.

Of course, we can quantify information, as we saw above. But what we saw is a purely syntactic notion: there is not much difference from the viewpoint of quantifying information between the sentences "it will rain tomorrow" and "your mother died", but, obviously, there is a lot of difference between those two sentences from a normal human viewpoint.

If we accept that scientific theories study the world and not information about it, the information theoretic approach should be seen as a tool in that study, not a goal in itself. But a tool for what? That approach seems solely concerned with selecting the appropriate probability measure in equilibrium situations. So, to see what this tool achieves, one needs to clarify which concept of probabilities is used, at least implicitly, in that approach. Is it a frequentist one, a subjective one or something else? Our impression is that the answer to that question depends very much on the author.

One way to think about the probability density $\rho(\mathbf{x})$ is as a continuous approximation of a swarm of points in phase space, each of which represents a microscopic state, and all those states share the same macroscopic properties. And one uses the measure with density $\rho(\mathbf{x})$ to describe the system, because we do not know in which microstate our system is, since the only information we have about the system is macroscopic.

Then we use this probability to compute averages or expectation values that give the right physical values for the pressure, the temperature etc.

But the issue is: why should we be interested in *averages*? What we observe are never averages but rather typical behaviors of individual systems.

Of course, because of the fundamental formula (6.6.21) those two quantity agree to a very high degree of approximation and thus the use of averages is justified.

But it is nevertheless misleading, from a fundamental point of view, to think that statistical mechanics deals with probability measures. It deals with the behavior of individual systems that are typical with respect to appropriate probability measures.

7.8.2   Jaynes' Approach

A justification of the information theoretic approach, which has been promoted by E. T. Jaynes and W.T. Grandy [159], and may be called the rational or objective Bayesian approach, consists in viewing the probability measures with maximum entropy, given the constraints provided by our information, as simply a form of rational inference that allows us to make predictions, like computing averages not included in the constraints (for example the pressure or the density or fluctuations of the latter) or predicting the increase of entropy between two equilibrium states in a closed system, as we saw in Sect. 7.6.

Jaynes [180, Chap. 14, p. 416] refers to his method as "predictive statistical mechanics" and said that:

"[It] is not a physical theory, but a form of statistical inference [...] instead of seeking the unattainable [it] asks a more modest question: 'Given the partial information that we do in fact have, what are the best predictions we can make of observable phenomena?'"

From this point of view, one could say that our information, although it is very limited (we have no idea about the real microstate of the system, we only know its corresponding macrostate) is *sufficient* to predict the usual quantities predicted in equilibrium statistical mechanics. Provided of course that we assign the right probability distribution to our system, namely the one satisfying the maximum entropy principle. If we chose any other probability distribution, satisfying the macroscopic constraints but not the maximum entropy principle, we would in general get the wrong predictions because that probability would, at least implicitly, include information that we do not have. Indeed the maximum entropy distribution is, among all distributions, the one that is the most indifferent or the less biased or the one that contains the less information, namely only the one included in the constraint.

A great advantage of this viewpoint, as emphasized by Jaynes, is when the predictions fail to agree with the observations. A spectacular example of such a failure was provided by the observation of the specific heats of solids at the end of the 19th century (the specific heat of a substance is the amount of heat needed to raise the temperature of one gram of that substance by one degree Celsius). The predictions for the latter, based on ordinary (i.e. classical at that time) statistical mechanics were wrong at low temperatures: while the observed specific heats were going to 0 as $T \to 0$, the theory predicted that they were constant in T. It is remarquable that, far from renouncing the usual rules of probabilistic computation, which were at the basis of statistical mechanics, physicists such as Einstein and later Debye were willing to modify the basic laws of physics. Instead of considering the energy as continuous, as it is classically, they were willing to make it discrete in a perfectly ad hoc way,[8]

---

[8] They followed the example of Planck who, in 1900, introduced the idea of discrete "quanta" of energy to account for a specific form of radiation, the black-body one.

in order to obtain a curve of the dependence of the specific heat of solids on the temperature that fitted the data.[9]

In turned out that those ad hoc models, as well as the Bohr model for the atom, which was also an ad hoc way to account for the radiation spectra of atoms, were later justified, in the late 1920s by quantum mechanics. So, what eventually changed in this whole episode were not the laws of statistical reasoning but the basic laws of physics!

If we think in terms of Cournot's principle (Sect. 2.5), what this example shows is that apparently improbable events do occur, such as the observed behavior of specific heats, since the classical statistical mechanics prediction is the result of a probabilistic reasoning. But this does not invalidate Cournot's principle, but is rather an indication that there is something wrong in our reasonings (here, the whole of classical physics!).

So, even though we maintain our criticism in the previous subsection of the information theoretic approach, it must be said that the point made by Jaynes here is interesting and worth keeping in mind.

7.9 Summary

In this chapter, we discussed an approach to statistical mechanics which, unlike the one in the rest of the book, does not start from microstates and macrostates but is focused on probability measures. This approach is also concerned more with predictions than explanations and is strongly linked to the theory of information.

We first introduced the Shannon entropy in Sect. 7.2, which measures how spread out a probability distribution is. This entropy is uniquely defined, given some natural conditions on what the "spreading" of a probability distribution means. In that section, we limit ourselves to probability distributions defined on discrete sets.

In Sect. 7.3, we introduced the basic "principle" of that approach, the one of maximum entropy: given some information about a system, assign to that system, among all the probability distributions that incorporate that information, the one having the largest entropy. This can be seen as a natural generalization of the indifference principle of Sect. 2.2.1.

Given that principle, one recovers the ensembles of Sect. 6.6 as being those that maximize the Shannon entropy, given some information about the system: either no information yielding the micro-canonical ensemble, or information about the average

---

[9] The 1907 model of Einstein predicted an exponential decrease of those specific heats C_V as $T \to 0$, $C_V \approx \exp(-\frac{c}{T})$, while the more sophisticated model of Debye, in 1911 predicted a decrease of the form T^3, which was in agreement with observations except at very low temperatures. See exercises 6.13, 6.14 for a sketch of Debye's argument (in those exercises we get that the energy is proportional to T^4 at low temperatures, which corresponds to a behavior proportional to T^3 for the specific heat which is proportional to the derivative of the energy with respect to the temperature); see Amit and Verbin [12] for a detailed discussion of specific heats of crystals at low temperatures..

energy yielding the canonical ensemble, or about that and the average number of particles, yielding the grand-canonical ensemble.

In Sect. 7.5, we extended the notion of Shannon entropy to continuous probability distributions.

In Sect. 7.6 we give a simple argument "deriving" the second law of thermodynamics from the maximum entropy principle; however, unlike the Boltzmann approach discussed in the next chapter, this derivation is concerned with what happens when one passes from one equilibrium state to another one, but not with the dynamics of what goes on in between those states.

However, the main usage of Shannon's entropy is in the theory of communication: it has many applications in the way one can efficiently encode a message and we digressed from the main subject of this book in order to give in Sect. 7.7 a brief illustration of those applications.

Finally, in Sect. 7.8 we evaluated this approach; we are basically critical of any approach that considers probabilities as being the fundamental object of study in any given branch of physics.

7.10   Exercises

Some of the exercises in this section are taken from MacKay [234], from Sator and Pavloff [284]

7.1. Compute the entropy of a uniform distribution on an interval $[a, b]$, according to definition (7.5.1) and determine when it is not positive.

7.2. Compute the Shannon entropy of the binomial and the multinomial distributions (11.1.11). The binomial distribution being given by (11.1.11) for $L = 2$.

7.3. Show that we can use the same ϵ in the definition of $G(N, \epsilon)$ in Proposition 7.6 and in (7.7.4).

7.4. Verify the coding given at the end of Sect. 7.7.

7.5. Consider two random variables X and Y taking values in $\{1, \ldots, N\}$ and let P_{XY} be the joint probability of those variables

   a. Express the marginal probabilities P_X and P_Y of the X and Y variables in terms of P_{XY}.

   b. Write the Gibbs entropies S_{XY} associated to P_{XY}, S_X associated to P_X and S_Y associated to P_Y.

   c. Prove that
$$S_{XY} \leq S_X + S_Y \tag{7.10.1}$$

   d. Give a necessary and sufficient condition for the equality in (7.10.1) to hold.

7.6. Consider four symbols a, b, c, d, with probabilities $p(a) = \frac{1}{2}, p(b) = \frac{1}{4}, p(c) = \frac{1}{8}, p(d) = \frac{1}{8}$. Compute Shannon's entropy of that distribution. Consider three different codes:

1. $C(a) = 0, C(b) = 10, C(c) = 110, C(d) = 111$.
2. $C(a) = 00, C(b) = 01, C(c) = 10, C(d) = 11$.
3. $C(a) = 0, C(b) = 1, C(c) = 00, C(d) = 11$.

For each of these codes, determine if is a prefix code, compute its expected length and compare it with the Shannon entropy of the distribution.

7.7. Construct a Huffman code for an alphabet $1, 2, 3, 4, 5, 6, 7$, with probabilities $p(1) = 0.01$, $p(2) = 0.24$, $p(3) = 0.05$, $p(4) = 0.20$, $p(5) = 0.47$, $p(6) = 0.01$, $p(7) = 0.02$.

Appendix

7.A   Proof that (7.2.6) Implies $A(N) = k \log N$

From (7.2.6), we obtain, by iteration, that $A(N^K) = K A(N)$, for $K, N \in \mathbb{N}$.

Now, for any two integers $S, T \geq 2$, we get that, for any large L, there is a K so that:

$$\frac{K}{L} \leq \frac{\log T}{\log S} < \frac{K+1}{L}, \tag{7.A.1}$$

or $S^K \leq T^L \leq S^{K+1}$.

Since $A(N)$ is increasing in N, we have $A(S^K) \leq A(T^L) \leq A(S^{K+1})$, or, since $A(N^K) = K A(N)$,

$$K A(S) \leq L A(T) \leq (K + 1) A(S),$$

which can be written as:

$$\frac{K}{L} \leq \frac{A(T)}{A(S)} \leq \frac{K+1}{L}. \tag{7.A.2}$$

Comparing (7.A.1) and (7.A.2), we see that:

$$\left| \frac{A(T)}{A(S)} - \frac{\log T}{\log S} \right| \leq \frac{1}{L},$$

which can be written as:

$$\left| \frac{A(T)}{\log T} - \frac{A(S)}{\log S} \right| \leq \frac{A(S)}{L \log T}.$$

For large L, $\frac{A(S)}{L \log T}$ is arbitrarily small, which proves that $\frac{A(T)}{\log T}$ is constant in T, and this implies the uniqueness of $A(N) = k \log N$.

However, this would *not hold* if we did not assume that $A(N)$ increases with N. Indeed, write $N = \prod_{i=1}^{M} N_i^{K_i}$, the prime number decomposition of N.

Then, from (7.2.6), we get:

$$A(N) = \sum_{i=1}^{M} K_i A(N_i). \tag{7.A.3}$$

But that means that one can assign *arbitrary values* to $A(N_i)$, for all prime numbers N_i, define $A(N)$ by (7.A.3), and (7.2.6) will be satisfied.

7.B   Jensen's Inequality

Let $F : \mathbb{R}^n \to \mathbb{R}$ be a convex function, i.e. that satisfies

$$F(\lambda \mathbf{x} + (1 - \lambda)\mathbf{x}) \leq \lambda F(\mathbf{x}) + (1 - \lambda)F(\mathbf{y}). \tag{7.B.1}$$

Then, for any measure on $\mathbb{R}^n$ with $\mu(\mathbb{R}^n) = 1$, we have *Jensen's inequality*:

$$F(\int_{\mathbb{R}^n} \mathbf{x} d\mu(\mathbf{x})) \leq \int_{\mathbb{R}^n} F(\mathbf{x}) d\mu(\mathbf{x}) \tag{7.B.2}$$

Note that μ could be a discrete measure, in which case the integral is reduced to a sum:

$$F(\sum_{i=1}^{N} \lambda_i \mathbf{x}_i) \leq \sum_{i=1}^{N} \lambda_i F(\mathbf{x}_i) \tag{7.B.3}$$

To prove (7.B.2), note that (7.B.3) follows from the definition (7.B.1) of convexity and then, approximate integrals by sums.

If F is concave ($-F$ is convex) (7.B.1) holds with the inequality sign reversed.

Chapter 8
Approach to Equilibrium

In this chapter we will explain how one passes from a non-equilibrium state to an equilibrium one. Since we consider isolated systems, the initial non-equilibrium state is created by a constraint imposed on that isolated system. The simplest example being a gas in a box which is constrained by a wall to be in a subset of the box. If one removes that wall at time $t = 0$, the gas will expand throughout the box and its density will converge towards a uniform one in that box. A similar phenomenon occurs if one puts two parts of a fluid at different temperatures in contact with each other: they will evolve towards a uniform temperature. Or if one puts two substances made of different types of molecules in contact with each other: they will tend to mix.

These evolutions are *irreversible* in the sense that we never see the gas going back to the subset of the box in which it was constrained to be at $t = 0$ or a body at a uniform temperature separating itself into a hot and a cold part. There is a large literature about the apparent "paradox" or "contradiction" between this irreversible behavior and the reversibility of the microscopic equations of motion expressed by (3.5.1), (3.5.2).

We want to explain here why there is no such paradox or contradiction. The basic ideas were already given by Boltzmann (see [34] for a detailed exposition of his work).

8.1 Boltzmann's Scheme

In 1872, Boltzmann claimed that (at low density) the state of a gas must evolve according to his equation [32], discussed in Sect. 8.6. Later, Boltzmann, answering various objections (to be discussed below), introduced a more probabilistic approach [33], that we will explain here in modern terms.

© Springer Nature Switzerland AG 2022
J. Bricmont, *Making Sense of Statistical Mechanics*, Undergraduate Lecture Notes in Physics, https://doi.org/10.1007/978-3-030-91794-4_8

8.1.1 Microstates and Macrostates

In Sect. 6.1, we introduced the notion of a *microstate* of a classical mechanical system on N particles:

$$\mathbf{x}(t) = (\vec{q}_1(t), \vec{q}_2(t), \ldots \vec{q}_N(t), \vec{p}_1(t), \vec{p}_2(t), \ldots, \vec{p}_N(t)) \in \mathbb{R}^{6N},$$

where $\vec{q}_i(t) \in \mathbb{R}^3$ and $\vec{p}_i(t) \in \mathbb{R}^3$ are the position and the momentum of the ith particle at time t. Its time evolution T^t was defined in Sect. 3.3.1.

We also introduced the notion of a *macrostate*, which is a map $M : \Omega \to R^L$ with L, the number of macroscopic variables, being much smaller than N: $L << N$. We gave several examples of macrostates, see (6.1.3), (6.1.5), (6.1.7). In (6.1.9), we showed how the evolution of the macrostate is induced by the one of the microstate. We also said that, if that induced evolution was autonomous, i.e. independent of the $\mathbf{x}(0)$ mapped onto $M(0)$, then the evolution of $M(t)$, which is called a *macroscopic law*, has been *reduced to* or *derived from* the microscopic one $\mathbf{x}(0) \to \mathbf{x}(t)$, in a straightforward way.

But we also said in Sect. 6.1 that such an autonomous evolution is, in general, impossible, without explaining why. Here is the reason: the evolution $\mathbf{x}(0) \to \mathbf{x}(t)$ is *reversible*, a notion defined in (3.5.1), (3.5.2): let I denotes the operation:

$$I(\mathbf{x}(t)) = (\mathbf{q}_1(t), \mathbf{q}_2(t), \ldots \mathbf{q}_N(t), -\mathbf{p}_1(t), -\mathbf{p}_2(t), \ldots, -\mathbf{p}_N(t)), \qquad (8.1.1)$$

then:
$$T^t I T^t \mathbf{x}(0) = I\mathbf{x}(0), \qquad (8.1.2)$$

or, in words, letting the system evolve according to the dynamical laws during a time t, then reversing the velocities (or the momenta), and, finally, letting the system evolve according those same laws for the same amount of time t, one gets the initial state with the velocities reversed.

But the evolution $M(0) \to M(t)$ is often *irreversible*, for example if M is the density and if one starts with a non-uniform density, the evolution of M tends to a uniform density and will not return to a non-uniform one. Consider for example the linear diffusion equation,[1] in one dimension (for simplicity):

$$\frac{\partial u(t, x)}{\partial t} = \frac{1}{2} \frac{\partial^2 u(t, x)}{\partial^2 x}. \qquad (8.1.3)$$

It is well-known that, given an integrable initial condition $u(0, x) \geq 0$, the solution of (8.1.3) in $\mathbb{R}$ tends to 0 as $t \to \infty$ and, in a box of side L, with $u(0, \pm L) = 0$ to a

---

[1] The more physical version of that equation is nonlinear, with the right hand side of (8.1.3) replaced by $\nabla D(x, u(t, x))\nabla u(t, x)$, where the diffusion constant D is replaced by a function $D(x, u(t, x))$ of x and of $u(t, x)$.

constant.[2] The function u here is a continuous approximation to the function $M_1(\mathbf{x})$ defined in (6.1.3). Since functions such as u belong to an infinite dimensional space, the restriction $L << N$ introduced in the definition $M : \Omega \to R^L$ of a macrostate does not hold if we replace M by u. But, as we said in Sect. 6.1, if u is a smooth function, we can approximate it by a few of its Fourier coefficients (or the coefficients of its expansion in another basis), take those coefficients as our macroscopic variables and recover the condition $L << N$.

Yet the reversibility argument shows that, since changing the sign of the velocities does not change the density, for each microstate $\mathbf{x}(t) = T^t \mathbf{x}(0)$ giving rise to a given value of $M(t) = M(\mathbf{x}(t))$, there may exist another microstate $I(\mathbf{x}(t))$ giving rise to the same value of $M(t) = M(I(\mathbf{x}(t)))$ but such that the future time evolution of $M(t)$ will be markedly different depending on whether it is induced by $\mathbf{x}(t)$ or by $I(\mathbf{x}(t))$. Hence, the evolution of the macrostate cannot be autonomous in the sense given above.

This seems to imply that one cannot derive a macroscopic law from a microscopic one and in particular that one cannot give a microscopic derivation of the second law of thermodynamics implying that the entropy monotonically increases. Yet, as we will explain now, this can be done, but not in a straightforward way.

8.1.2 Derivation of Macroscopic Laws from Microscopic Ones

The basis of the solution to the apparent difficulty mentioned in the previous subsection is that, as we discussed in Sect. 6.2, the map M is many to one in a way that depends on value taken by M. This was illustrated in Fig. 6.2.

To understand how $M(t)$ can evolve irreversibly even though its evolution is induced by a reversible microscopic evolution, consider Fig. 8.1, which illustrates what one expects to happen: the microstate $\mathbf{x}(t)$ evolves towards larger and larger regions of phase space and eventually ends up in the "thermal equilibrium" region. Therefore, the induced evolution of $M(t) = M(\mathbf{x}(t))$ should tend towards equilibrium. However, at the level of generality considered here, our expectation is simply based on the fact that some regions are (much) bigger than others, and so it would be natural for the microstate $\mathbf{x}(t)$ to evolves towards those bigger regions if no external constraint like a piston or an adiabatic wall prevents it from doing so.

There are several caveats here: one is that this scenario is what one expects or hopes for. We will give in Sect. 8.7 several examples (but rather artificial ones),

---

[2] The explicit solution in $\mathbb{R}$ is:

$$u(t, x) = \frac{1}{\sqrt{2\pi t}} \int_{\mathbb{R}} \exp(-\frac{(x - y)^2}{2t}) u(0, y) dy.$$

It is easy to see that, if $\int_{\mathbb{R}} |u(0, y)| dy < \infty$, then $\sup_{x \in \mathbb{R}} u(t, x) \to 0$ as $t \to \infty$, as $\mathcal{O}(\frac{1}{\sqrt{t}})$.

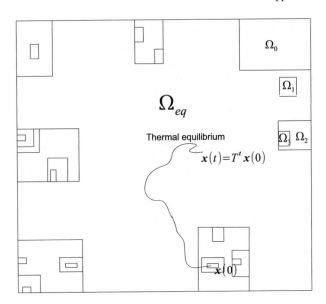

Fig. 8.1 The curve $\mathbf{x}(t) = T^t\mathbf{x}(0)$ describes a possible evolution of a microstate, which tends to enter regions of larger volume until it enters the region Ω_{eq} of thermal equilibrium

where this scenario can be demonstrated in detail. This will show that this scenario is certainly possible and even plausible, but it is certainly not demonstrated in any degree of generality in physically natural situations.

The more important caveat is that, even if this scenario is true, the desired evolution is definitely not true for all microstates $\mathbf{x}(t) = T^t\mathbf{x}(0)$ giving rise to a given value $M(t) = M(\mathbf{x}(t))$. That follows from the reversibility argument given in Sect. 8.1: for every microstate $\mathbf{x}(t)$ that induces the irreversible evolution of $M(t) = M(\mathbf{x}(t))$ there exists another microstate $I(\mathbf{x}(t))$ so that $M(t) = M(I(\mathbf{x}(t)))$ at time t, but $I(\mathbf{x}(t))$ induces a different evolution than the irreversible one for times later than t.

Let us illustrate this explanation of irreversibility in a concrete physical example. Consider a gas that is initially compressed by a piston in one-half of the box Λ, and that then expands into the whole box. Let M be the density of the gas. Initially, it is equal to 1 (say) in one half of the box and to 0 in the other half. After some time t, it is (approximately) equal to $\frac{1}{2}$ everywhere and remains so for the foreseeable future. The explanation of the irreversible evolution of M is that the overwhelming majority of the microscopic configurations corresponding to the gas in one-half of the box, will evolve deterministically so as to induce the observed evolution of M. There may of course be some exceptional configurations, for which all the particles stay in the left half. For example, their velocities could be perfectly parallel to the piston; then, assuming that there are no particle collisions, the particles will simply bounce back and forth between the walls of the box that are perpendicular to the piston. And a very slight variation of those initial conditions could keep the particles in that box for an arbitrarily long time (if the variation is small enough).

The claim made here is that those configurations are extraordinarily rare, and that we do not expect to see even one of them appearing when we repeat the experiment many times. So, the microscopic configurations that lead to the expected macroscopic behavior will be *typical* in the sense of (2.3.1).

Let us define the *good configurations* (up to a certain time T) as being those configurations that induce the macroscopic law up to time T, and the *bad configurations* (up to a certain time T) the other configurations; we will use indices G and B for the corresponding sets of configurations.[3]

8.1.3 Solution of the (Apparent) Reversibility Paradox

Take all the good microscopic configurations in one-half of the box, and let them evolve up to a time $t << T$, when the density is approximately uniform. Now, reverse all the velocities. We get a set of configurations that still determines a density approximately $\frac{1}{2}$ in the box, so the value of the density function M, defined by (6.1.2), is unchanged. However, those configurations are not good any more. Indeed, from now on, if the system remains isolated, the density just remains uniform according to the macroscopic laws. But for the configurations just described, the gas will move back to the part of the box from which it started, see (8.1.2), leading to a gross violation of the macroscopic law. What is the solution to this apparent paradox? Simply that those "reversed-velocities" configurations form a very tiny subset of all the microscopic configurations giving rise to a uniform density. And, of course, the original set of configurations, those coming from the initial half of the box, also form such a small subset. Most configurations corresponding to a uniform density do not go to one-half of the box, either in the future or in the past.

To express this idea in formulas, let Ω_t be the set of the configurations giving to the function M its value $M(t)$ at time t, that evolves according to a macroscopic law. In other words, Ω_t is the pre-image of $M(t)$ under the map M. Let $\Omega_{t,G}$ be the set of *good configurations*, at time t, namely those that lead to a behavior of M following the macroscopic laws, up to some time T.

One expects that, in general, $\Omega_{t,G}$ is a very large subset of Ω_t (meaning that $\frac{|\Omega_{t,B}|}{|\Omega_t|} = \frac{|\Omega_t \setminus \Omega_{t,G}|}{|\Omega_t|} << 1$, with $\Omega_{t,B}$ being the set of bad configurations), but is not identical to Ω_t.

In our example of the gas initially compressed in one-half of the box Λ, the set Ω_0 consists of all the configurations in one-half of the box at time zero, and $\Omega_{0,G}$ is the subset consisting of those configurations whose evolution lead to a uniform density at time t, which means that $T^t(\Omega_{0,G}) \subset \Omega_t$.

---

[3] We introduce the time upper bound T here, because, if we wait long enough, almost all configurations will be bad, since the Poincaré's recurrence theorem (see Sect. 4.2) implies that they will all come back arbitrarily close to their initial conditions. In the example of the gas in the box, this means that all the particles will come back simultaneously to the half-box in which they were initially, which is contrary to the behavior predicted by the macroscopic laws. See Sect. 8.2 for a discussion of this "paradox".

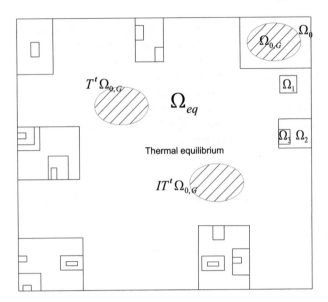

Fig. 8.2 Ω_0 are the configurations in one-half of the box at time zero; $\Omega_{0,G}$ are the good configurations in Ω_0 whose evolution lead to a uniform density at time t: $T^t(\Omega_{0,G}) \subset \Omega_t$; $I(T^t(\Omega_{0,G}))$ are the configurations of $T^t(\Omega_{0,G})$ with velocities reversed whose evolution after time t belongs to Ω_0: $T^t(I(T^t(\Omega_{0,G})) \subset \Omega_0$

Microscopic reversibility says that $T^t(I(T^t(\Omega_{0,G})) = I(\Omega_{0,G}) \subset I(\Omega_0) = \Omega_0$ (this is just (8.1.2) applied to the set $\Omega_{0,G}$). The last equality holds because the set of configurations in one-half of the box is invariant under the change of the sign of the velocities.

A paradox would occur if $T^t(I(\Omega_t)) \subset I(\Omega_0) = \Omega_0$. Indeed, this would mean that, if one reverses the velocities of all the configurations at time t corresponding to a uniform density, and let them evolve for a time t, one would get a set of configurations in one-half of the box (with velocities reversed). Since the operation I preserves the Lebesgue measure $|I(\Omega_t)| = |\Omega_t|$, this would imply that there are as many configurations corresponding to a uniform density (configurations in Ω_t) as there are configurations that will evolve back to one-half of the box in time t (those in $I(T^t(\Omega_{0,G}))$. Or, in other words, one would have $I(\Omega_t) \subset \Omega_{t,B} = \Omega_t \setminus \Omega_{t,G}$, which combined with $|I(\Omega_t)| = |\Omega_t|$ makes the bound $\frac{|\Omega_{t,B}|}{|\Omega_t|} = \frac{|\Omega_t \setminus \Omega_{t,G}|}{|\Omega_t|} << 1$ impossible.

But Ω_t is *not at all equal*, in general, to $T^t(\Omega_{0,G})$, so that $T^t(I(T^t(\Omega_{0,G})) = I(\Omega_{0,G}) \subset I(\Omega_0) = \Omega_0$ *does not imply* $T^t(I(\Omega_t)) \subset I(\Omega_0) = \Omega_0$. In our example, $T^t(\Omega_{0,G})$ is a tiny subset of Ω_t, because most configurations in Ω_t were not in half of the box at time zero.

This is illustrated in Fig. 8.2: Ω_0 is the set of configurations with all the particles in one-half of the box, and $\Omega_{0,G}$ the subset of those that evolve towards a uniform distribution after some time t. Thus $T^t(\Omega_{0,G})$ is a subset of Ω_t, which is the set of thermal equilibrium states. Reversing the velocities of every configurations in

$T^t(\Omega_{0,G})$ yields the set $I(T^t(\Omega_{0,G}))$, which is also a subset of Ω_t, but a "bad" subset, namely one that does not stay in equilibrium but moves back to the half box where the particles were to start with. So, the map M applied to those configurations will not evolve according to the usual macroscopic laws.

Of course it should be emphasized once more that the subsets in Fig. 8.2 are not drawn to scale: the sets Ω_0, $T^t(\Omega_{0,G})$ and $I(T^t(\Omega_{0,G}))$ are minuscule compared to the set of equilibrium configurations Ω_t.

In fact, one knows from Liouville's theorem (see Sect. 3.4) that the size of $\Omega_{0,G}$ and $T^t(\Omega_{0,G})$ are equal: $|\Omega_{0,G}| = |T^t(\Omega_{0,G})|$. Since the operation I also preserves the size of a set, we have: $|\Omega_{0,G}| = |T^t(\Omega_{0,G})| = |I(T^t(\Omega_{0,G}))|$, which is illustrated in Fig. 8.2.

Since $\Omega_{0,G} \subset \Omega_0$, we have $|\Omega_{0,G}| \leq |\Omega_0|$ and we already observed that the ratio $\frac{|\Omega_0|}{|\Omega_t|} \approx 2^{-N}$ (since each of the N particles can be in either half of the box in Ω_t, while those in Ω_0 can only be in one-half of the box). Thus,

$$\frac{|I(T^t(\Omega_{0,G}))|}{|\Omega_t|} = \frac{|T^t(\Omega_{0,G})|}{|\Omega_t|} = \frac{|\Omega_{0,G}|}{|\Omega_t|} \leq \frac{|\Omega_0|}{|\Omega_t|} \approx 2^{-N}, \qquad (8.1.4)$$

which is astronomically small for $N \approx 10^{23}$.

What one would like to show is that the good configurations are *typical* in the sense of (2.3.1). More precisely, one want to show that, for all times t not too large,

$$\frac{|\Omega_{t,G}|}{|\Omega_t|} \to 1$$

as $N \to \infty$.

This is not easy to show for realistic dynamical systems, but will be shown to be true in artificially simple ones in Sect. 8.7.

8.1.4 Irreversibility and Probabilistic Explanations

The above explanation of the irreversible behavior of $M(t)$ is again probabilistic: the vast majority of microstates $\mathbf{x}(0)$ corresponding to the macrostate $M(0)$ induce, through their deterministic evolution $\mathbf{x}(0) \to T^t(\mathbf{x}(0)) = \mathbf{x}(t)$, the expected time evolution of $M(0) \to M(t)$.

What else could one ask for? One could wish to show that the expected time evolution of $M(0) \to M(t)$ is induced by *all* microstates $\mathbf{x}(0)$ corresponding to the macrostate $M(0)$. But we showed that, because of (8.1.2), this is impossible.

So, our explanation is the best one can hope for. But is it satisfactory? It is, provided that one accepts the notion of probabilistic explanation given in Sect. 2.6. And if one does not accept it, it is not clear what notion of explanation one has in mind and how one could justify it.

Nevertheless, by speaking of a "vast majority of microstates", we did not say what we meant by the "vast majority". Since there are uncountably many such states, one needs a measure on Ω in order to make sense of that notion. The measure on subsets of Ω that we used here is the size of the set or its Liouville measure, i.e. the restriction of the Lebesgue measure to Ω (see Sect. 3.4.1). And the justification for that usage is the same as the one given for the equilibrium situation in Sect. 6.8.

8.1.5 Time Dependent Boltzmann Entropy and the Second Law

In Sect. 6.4 we defined the Boltzmann entropy as:

$$S_B(\mathbf{x}) = k \ln |\Omega(M(\mathbf{x}))| = k \ln |M^{-1}(M(0)))| \qquad (8.1.5)$$

where $M(\mathbf{x}) = M^{-1}(M(0)))$ is the value of the macrostate to which $\mathbf{x}$ belongs and $|M^{-1}(M(0)))|$ the Lebesgue measure of the set of microstates corresponding to the macrostate $M(\mathbf{x}) = M(0)$. It was emphasized that the Boltzmann entropy is a function of the microstate, depending on the choice of the macrostate.

We showed in Sects. 6.6 and 6.6.2 that the Boltzmann entropy coincides, *in equilibrium* and in the thermodynamic limit, with the Gibbs entropy and with the thermodynamic one.

The natural extension of (8.1.5) to a time dependent context is:

$$S_B(\mathbf{x}(t)) = k \ln |M^{-1}(M(\mathbf{x}(t))|, \qquad (8.1.6)$$

which can also be written as $S_B(t) = k \ln |\Omega_t|$ with $\Omega_t = M^{-1}(M(t))$, and $M(t) = M(\mathbf{x}(t))$.

As we just saw, one expects the microstate $\mathbf{x}(t)$ to evolve towards regions corresponding to larger and larger values of $|\Omega(M)|$ and therefore one expects the Boltzmann entropy to increase in time, for "typical" points $\mathbf{x}(t)$. This is a "derivation" of the second law, which is correct to the extent that our conjecture on the evolution of the microstate is correct.

By contrast, the Gibbs entropy is a function of *probability measures* and was defined in (6.6.7):

$$S_G(\nu) = -k \int_\Omega \ln \rho(\mathbf{x}) \rho(\mathbf{x}) d\mathbf{x}, \qquad (8.1.7)$$

(See (6.6.8) for the discrete case).

But the time evolution of that entropy seems to be radically different from the one of the Boltzmann entropy. Indeed, we have:

Proposition 8.1 *Define ν_t the Liouville evolution of an initial measure ν_0 at time t by: for A a Borel set in phase space,*

$$v_t(A) \equiv v_0(T^{-t}(A)) \tag{8.1.8}$$

Then, as a special case of Corollary 3.6,

$$S_G(v_t) = S_G(v_0). \tag{8.1.9}$$

Indeed, $S_G(v_t) = -k \int_{\mathbb{R}^{6N}} \rho_t(\mathbf{x}) \ln \rho_t(\mathbf{x}) d\mathbf{x}$ is, since $\rho_t(\mathbf{x}) = \rho_0(T^{-t}\mathbf{x})$, of the form $\int_{\mathbb{R}^{6N}} F(\rho_0(T^{-t}\mathbf{x})) d\mathbf{x}$ with $F(y) = -ky \ln y$ and is therefore constant in time.

This is *prima facie* a big difference between those two definitions of entropies and is sometimes thought (by those who prefer definition (8.1.7)) to be a problem in the derivation of the second law.

But, first, one might ask: why should probability measures on phase space evolve according to microscopic dynamical laws, as in (8.1.8)? If we think of the probability density $\rho(\mathbf{x})$, where $dv = \rho(\mathbf{x})d\mathbf{x}$ as a continuous density representing a swarm of points in phase space, all of which are "macroscopically" the same (meaning, in our language, that they all belong to the same set $M^{-1}(M)$ for a given value M of the macrostate), then it makes sense to let all these point evolve according to microscopic dynamical laws and that would justify (8.1.8).

But there are other ways to can think about probabilities: as a reflection of our knowledge/ignorance, as a tool to predict the future given some partial information about the present, or as a way to distinguish between typical and atypical sets of configurations. But if one makes any of these choices, then it is not at all clear that probabilities should evolve according to microscopic dynamical laws. We will return to this question in Sect. 8.8.1, where we will suggest another possible time evolution of measures on phase space.

8.2 Answers to the Classical Objections

8.2.1 Objections from Loschmidt and Zermelo

At the time of Boltzmann there were two kind of objections raised against his scheme: one due to the Austrian physico-chemist Josef Loschmidt (1821–1895) and another one due to the German mathematician Ernst Zermelo (1871–1953).[4]

The objection of Loschmidt[5] [232] which was based on the reversibility "paradox" was answered in detail in Sect. 8.1.3. The objection of Zermelo was based on Poincaré's recurrence theorem (see Sect. 4.2 and Remark 4.5).[6] This theorem implies that, if all the particles of a gas start in a half-box, they *must* return, all at the same time, to the half-box that they started from "if we wait long enough."

---

[4] Zermelo is also, with Igor Fraenkel, one of the founders of modern axiomatic set theory.

[5] According to Harman [167, p. 141], this objection was first noticed by Maxwell.

[6] Actually, Zermelo's objection was made previously by Poincaré himself [262].

Boltzmann's answer was simple and based on the end of the previous sentence: "you should live that long"![7]

Even elementary estimates of the time it takes for such Poincaré's recurrence to occur is easily seen to be much larger than the age of the universe. In footnote 4.3, we made a rough estimate of these times: assuming that the dynamics of the gas is ergodic (see Sect. 4.3), the time spent between two consecutive visits to the region in phase space where all the particles are in a half-box is of the order of $2^{10^{23}}$, which is much larger than the age of the universe (the units of time, seconds or hours, do not matter much for this conclusion to hold, as long as they are reasonable).

So there is no contradiction, from a physical point of view, between Boltzmann's explanations and Poincaré's theorem.

Loschmidt's objection has been repeated even relatively recently. According to the Nobel Prize in chemistry Ilya Prigogine, Poincaré did not recommend reading Boltzmann, because his conclusions were in contradiction with his premises ([273], p. 23). Indeed, Poincaré wrote [262]: "The problem is so complicated that it is impossible to treat it with complete rigor. We are therefore forced to make some simplifying hypotheses; are they legitimate, are they even reconcilable between them? I do not believe so. I don't want to discuss them here; but there is no need for a long examination to distrust a reasoning where the premises are at least apparently in contradiction with the conclusion, where one finds reversibility in the premises and irreversibility in the conclusion."

Moreover, discussing our example of a gas expanding in a container, Prigogine observes that "if irreversibility was only that, it would indeed be an illusion, because, if we wait even longer, then it may happen that the particles go back to the same half of the container. In this view, irreversibility would simply be due to the limits of our patience" ([273], p. 24). If "the limits of our patience" are much longer than the age of the universe, that way of speaking is rather misleading.

However, there is still a mathematical problem: if one tries to rigorously derive an irreversible macroscopic equation from the microscopic dynamics and suitable assumptions on initial conditions, the Poincaré recurrence time will put a limit on the length of the time interval over which these statements can be proven, and that would make such a derivation very difficult to do rigorously. That is one of the reasons why one discusses these derivations in suitable limits (e.g. when the number of particles goes to infinity) where the Poincaré recurrence time becomes infinite. But one should not confuse the fact that one takes a limit for mathematical convenience and the source of irreversibility. In the Kac model discussed in Sect. 8.7, one sees clearly that there are very different time scales: one over which convergence to equilibrium occurs, and a much larger one, where some analogue of Poincaré's recurrence takes place. But the first time scale is not an "illusion". In fact, it is on that time scale that all phenomena that we can possibly observe do take place.

---

[7] For the exchanges between Boltzmann and Zermelo, see [35, 36, 334, 335]. In [35], Boltzmann gives rough estimates on the return time.

8.2.2 An Objection from Poincaré

There was another objection, a rather strange one, due again to the great French mathematician Henri Poincaré. As we saw in the last subsection, he did share Loschmidt's and Zermelo's objections to Boltzmann, but in 1889 he came up with another objection [260].

He proved that there is no function of the positions and momenta of an isolated dynamical system that increases monotonically towards a maximum, hence there cannot be an entropy function satisfying Boltzmann's requirements for such a function.

The details of Poincaré's argument need not concern us, since his fundamental misconception is that he identifies this purported maximum of the entropy function with a *mechanical equilibrium* of the mechanical system, i.e. a point of phase space where the velocities vanish and the potential function reaches a stationary value (where $\nabla_{\vec{q}_i} V(\mathbf{q}) = 0$, $\forall i = 1, \ldots, N$ and therefore, by (3.2.6), the particles are at rest).

But it is enough to consider a gas "in equilibrium" in the sense of Boltzmann to realize that this has nothing to do with a mechanical equilibrium: far from being at rest, particles bounce back and forth between the walls of the box. Moreover, thermodynamic equilibrium corresponds to a set of configurations (see Fig. 6.2), not to a single one.

Of course it is true that Boltzmann's entropy, as a function of the microstate, does not always increase, partly because of Poincaré's recurrences, partly because of exceptional initial conditions. But one did not need Poincaré's argument to know that.

It is not clear what Poincaré had in mind when trying to prove this "impossibility result" but it illustrate the lack of understanding that sometimes exists between physicists and mathematicians, even the greatest ones.

8.2.3 Objections to Typicality Arguments

In a series of papers, Frigg and Frigg and Werndl, [132–135] and Uffink [316] have criticized the Boltzmannian approach outlined here by invoking "the measure zero problem.", as Lazarovici and Reichert call it [212].

Consider any given microscopic trajectory $T^t \mathbf{x}_0$ of a physical system and consider its complement $\Omega \setminus \{T^t \mathbf{x}_0 \mid t \in \mathbb{R}\}$ on the phase space Ω or on a given energy surface $S_E \setminus \{T^t \mathbf{x}_0 \mid t \in \mathbb{R}\}$. Obviously, with respect to any reasonable measure which is not concentrated on that trajectory, the set $\{T^t \mathbf{x}_0 \mid t \in \mathbb{R}\}$ is of measure 0 and thus, from our definition, its complement is typical. Yet, by definition, the trajectory does not enter that complement; hence, concludes the objection, the fact that a set of configurations is typical does not mean that a trajectory goes into it.

More generally, any argument that is supposed to apply to all trajectories except those in a set of small measure or even of measure zero can be "refuted" by pointing out that any given trajectory is such a set of measure zero.

But one should remember what we said about Cournot's principle: improbable events do not occur, provided we specify in advance how "probable" is defined. If one tosses a thousand coins, any specific sequence of results will have a very small probability (2^{-1000}) of occurrence, yet one such sequence necessarily occurs. But that does not refute the idea that "typically" no sequence of results will have 900 heads and 100 tails and, in practice, such results do not occur, unless the coins are biased or the machine doing the tosses is particularly fine-tuned (see [99]). And the same is true for trajectories: the fact that only one of them occurs for a given system at a given time, does not refute the idea that some trajectories are typical and the others are not and that, in practice, we only observe the typical ones.

Of course, the authors who raise these objections are right in wishing to have more detailed dynamical arguments proving that most trajectories converge to equilibrium. And it is easy to provide counterexamples where the dynamics is so special that convergence to equilibrium will not take place: free particles (with no collisions) will conserve their distribution of velocities, hence will not converge to a Maxwellian distribution (6.2.20) (but see Sect. 8.7.4 for the convergence to equilibrium of the distribution of particle's positions, even in that case). Or take uncoupled harmonic oscillators, all with the same frequency: they will oscillate in phase and not converge to any equilibrium distribution (but that is not true for anharmonic oscillators, even uncoupled ones, see [55]).

The problem is that, because of these (trivial) counterexamples, it is not easy to give general dynamical properties that will guarantee convergence to equilibrium and the Boltzmannian scheme offered here is the best one can do, together with Lanford's derivation of Boltzmann equation (see Sect. 8.6).

In particular, notions such as ergodicity (or mixing) do not allow us to derive convergence to equilibrium in a sense that would be relevant for statistical mechanics, as we will see in the next section.[8]

8.3 Ergodicity, Mixing, and Other Wrong Turns

In this section, we shall examine several "explanations" of irreversibility, different from the Boltzmannian one, and evaluate them.

---

[8] This is also true for the notion of ϵ-ergodicity, a notion weaker than ergodicity, proposed by Frigg and Werndl [134, 135] as a dynamical property that would, according to them, explain convergence to equilibrium (see Lazarovici and Reichert [212, 215] and Reichert [275] for a refutation of their argument). For a critique of other objections to typicality reasoning, see Dürr and Struyve [115].

8.3.1 Ergodicity

Among mathematicians and mathematical physicists the idea that ergodicity (defined in Sect. 4.3) accounts for irreversibility has been rather widespread.[9]. Let us explain why ergodicity is *neither necessary nor sufficient* to explain irreversibility.[10]

As we saw in Sect. 4.3, a dynamical system is *ergodic* if the average time spent by a trajectory in any region of the phase space is proportional to the volume of that region. To be more precise: average means in the limit of infinite time and this property has to hold for all trajectories, except (possibly) those lying in a subset of measure zero. This property implies that, if one considers the partition of phase space in Fig. 6.2, almost all trajectories will spend in each subset of that figure a time proportional to the size of that subset.[11]

Then, the argument goes, the measurement of any physical quantity will take some time. This time is long compared to the typical time scale of molecular processes. Hence, we can approximately regard it as infinite.[12] Therefore, the measured quantity, a time average, will approximately equal the average over phase space of the physical quantity under consideration (see (4.3.8)). But this latter average is exactly what one calls the equilibrium value of the physical quantity. So, if a dynamical system is

---

[9] For a history of the concept of ergodicity, and some very interesting modern developments, see Gallavotti [140, 141]. It seems that the (misleading) emphasis on the modern notion of ergodicity goes back to the Ehrenfests' paper [121], more than to Boltzmann, see Brush [62, p. 505]. A careful, but nevertheless exaggerated interest in ergodicity and mixing is found in the work of Khinchin [191] and Krylov [200]; it is also found e.g. in Chandler [76, p. 57], Hill [171, p. 16], Ma [233, Chap. 26], Thompson [308, p. 26], Dorfman [109, Chap. 4], and Dunford and Schwartz [111, p. 657], (but see below the quote of Schwartz 8.3.3 for a self-criticism of [111]).

[10] The role of ergodicity in foundations of statistical mechanics has been criticized for a long time by, e.g. Jaynes [180, p. 106], with arguments similar to those given here. This approach to foundations of statistical mechanics has also been criticized by Earman and Rédei [118], who give references to discussion of that issue in the philosophical literature.

[11] There was originally some ambiguity about the word 'ergodicity', in that it might have meant a stronger property than the one discussed here, namely that the trajectory would visit *every point* of the phase space. While it is possible to have continuous curves that do that (the Peano curves), it is not possible for solutions of the mechanical equations of motion (3.3.3), (3.3.4), that, by definition, do not intersect themselves, see Rosenthal and Plancherel [259, 277].

[12] For example, Khinchin wrote [191, p. 44–45] (for him, a phase function is a function defined on phase space): "An experiment or an observation which gives the measurement of a physical quantity is performed not instantaneously, but requires a certain interval of time which, no matter how small it appears to us, would, as a rule, be very large from the point of view of an observer who watches the evolution of our physical system. This system will be subjected during this interval of time to perturbations (such as mutual collisions of molecules) which may change essentially the values of the corresponding phase function. Thus we will have to compare experimental data not with separate values of phase functions, but with their averages taken over very large intervals of time. In other words, [...] with time averages of phase functions over a trajectory which represents the evolution of our physical system." In the French textbook of Vauclair [320], one reads a rather typical statement: "One considers that during the time δt of the measurement, the system has gone through all the possibly accessible states, and that it spent in each state a time proportional to its probability" ([320, p. 11]). And: "Only the systems having this property (mixing) tend to an equilibrium state, when they are initially in a state out of equilibrium" ([320, p. 197]).

ergodic, it converges towards equilibrium. This is in general how appeal to ergodicity is used to justify convergence to equilibrium.

Let us see the problems with this argument: a first, but relatively minor, problem is that it is very hard to give a mathematical proof that a realistic mechanical system is ergodic. But let us take such a proof for granted, for the sake of the discussion. Here is a more serious problem. Assume that the argument given above is true: how would it then be possible to observe or measure *any non-equilibrium* phenomenon? In the experiment with the box divided in two halves, we should not be able to see any intermediate stage, when the empty half gets filled, since the time for our measurements is supposed to be approximately infinite. So, where is the problem? We implicitly identified the time scale of molecular processes with what one might call the "ergodic time", i.e. the time taken by the system to visit all regions of phase space sufficiently often so that the replacement of time averages by averages over phase space is approximately true. But, whatever the exact meaning of the word "time scale of molecular processes" (for a few molecules) is, the ergodic time is certainly enormously longer. If we consider the partition in Fig. 6.2, some regions will correspond to all the particles being in some small subset of the box, other regions to all the particles being in another small subset of the box, etc. The time it takes for a given trajectory to visit each region, even once, will depend on the size of the regions. By taking finer and finer partitions, we can make that time as large as one wishes. So, if one were to take the argument outlined above literally, the "ergodic time" is infinite, and identifying it loosely to a relaxation time is misleading.

This argument also shows that ergodicity is not *necessary* because all we want to show is that typical trajectories go, over not too long time scales, towards the equilibrium region and stay there. We do not need to prove that, if we wait long enough (many times the age of the universe, as in Poincaré's recurrences) the trajectory will visit every set in Fig. 6.2 and, in particular that all the particles will simultaneously be in any given subset of the box. At this point of the discussion, defenders of the ergodicity approach often say that we do not need the time and phase space average to be (almost) equal for *all functions*, but only for the physically relevant ones (like the energy or particle densities). This is correct, but shows that we need a much weaker notion than ergodicity. All we need is that the microscopic configuration evolves in phase space towards those regions where the relevant macroscopic variables take their equilibrium values, which was illustrated in Fig. 6.2. The Kac model (see Sect. 8.7) perfectly illustrates this point: it is not ergodic in any sense, yet, on proper time scales, the macroscopic variables evolve towards equilibrium. One can also prove that most trajectories of systems with an equilibrium macrostate that occupies most of the phase space volume, will be, most of the time, in this part of the phase space, see Reichert [275]. Similar statements are given without proof by Khinchin [191, p. 50], and by Landau and Lifshitz [201, p. 5].

8.3.2 Mixing

Some people consider that ergodicity is indeed not the correct notion to justify irreversibility and that one needs a stronger property: mixing.

To explain the argument, let $d\mu = \frac{f(\mathbf{x})dv_m(\mathbf{x})}{\int_\Omega f(\mathbf{x})dv_m}$, where $f \geq 0$ is a v_m integrable function with v_m being the micro-canonical distribution. Then, if the system (S_E, T^t, dv_m) is mixing (see 4.4.4), for any bounded function g, we have:

$$\lim_{t \to \infty} < g(T^t(\mathbf{x}) >_\mu =< g(\mathbf{x}) >_{v_m} . \tag{8.3.1}$$

The idea is that μ can be considered as a non-equilibrium state: the function f could be, for example, the indicator function of some small subset in Fig. 6.2, and it would thus be concentrated on non-equilibrium configurations, like configurations where all the particles are in one-half of the box.

What (8.3.1) says is that, as $t \to \infty$, the trajectories starting in the support of μ tend to spread homogeneously all over the phase space; if, for example, f is the indicator function of a small subset Ω_0 of Ω in Fig. 8.1, then, if mixing holds, the set $T^t \Omega_0$ has the same size as Ω_0 (by Liouville's theorem) but its shape, for t large, will look like a very thin ribbon stretched over the entire phase space, which then leads to (8.3.1).

Since the right hand side of (8.3.1) is the same for all μ of the form $d\mu = \frac{f(\mathbf{x})dv_m(\mathbf{x})}{\int_\Omega f(\mathbf{x})dv_m}$, (8.3.1) seems to imply a strong form of "convergence to equilibrium" (hence its popularity as an explanation of that phenomenon among some people, mostly mathematicians) since, if we start with configurations concentrated on *any subset* of phase space, we get eventually the same asymptotic distribution of the trajectories, given by the micro-canonical distribution v_m.

But this mixing property is again neither necessary nor sufficient, and for the same reasons that made ergodicity neither necessary nor sufficient: is is not necessary because we do not need the convergence in (8.3.1) to hold for *every* function g. If g is the characteristic function of a small subset A of Ω, then (8.3.1) says that, for t large, the image under T^t of Ω_0 will have an intersection with A proportional to the size of A (which is equal to $< g(\mathbf{x}) >_{v_m}$ in this situation).

But we do not need to know that: we are only interested in convergence of the macroscopic quantities. And because (8.3.1) holds for every function f and g, we have no hope to control over which time scale the limit in (8.3.1) is approached. This shows that mixing is not sufficient either, since we want the macroscopic quantities to approach their limit over "reasonable" time scales, not over time scales much larger than the age of the universe, as will be the case in (8.3.1) for some functions f and g.

The problem with both approaches (ergodicity and mixing) is that they try to give a purely mechanical criterion for "irreversible behavior", which holds for microscopic systems as well as macroscopic ones.

Here is the basic dilemma: either we are willing to introduce a macro/micro distinction and to give a basic role to initial conditions in our explanation of irreversibility or we are not. If we make the first choice, then, as explained in Sect. 8.1, there is no deep problem with irreversibility, and subtle properties of the dynamics (like ergodic properties) play basically no role. On the other hand, there is no explanation of irreversibility that would hold for *almost all* initial conditions or apply to *all* functions on configuration space (therefore avoiding the micro/macro distinction). So, we have to make the first choice.

Another critique of the "ergodic" approach is that systems with one or few degrees of freedom may very well be ergodic, or mixing, as we saw in Sect. 4.6. But it makes no sense to speak about irreversibility for those systems; think, for example, about the motion of a billard ball: the reversed motion will look as natural as the original one, while reversed macroscopic motions, like two substances separating themselves, look impossible. So, this is another sense in which the notion of ergodicity or mixing is not sufficient.

8.3.3   The Brussels-Austin School

There is a school of thought centered around Ilya Prigogine (who taught at Brussels and Austin), that tries to explain irreversibility using mixing or other "chaotic" properties of dynamical systems but by making a radical move: the idea that trajectories should be abandoned, and replaced by probabilities. What does that mean? Prigogine writes: "Our leitmotiv is that the formulation of the dynamics for chaotic systems must be done at the probabilistic level" ([273], p. 60). For chaotic systems, "*trajectories are eliminated from the probabilistic description* …The statistical description is *irreducible*" ([273], p. 59). Or: "We must therefore eliminate the notion of trajectory from our microscopic description. This actually corresponds to a realistic description: no measurement, no computation lead strictly to a point, to the consideration of a *unique* trajectory. We shall always face a *set* of trajectories" ([273], p. 60). Moreover: "the notion of chaos leads us to rethink the notion of 'law of nature'" ([273], p. 15). The existence of chaotic dynamical systems supposedly marks a radical departure from a fundamentally deterministic world-view, makes the notion of trajectory obsolete, and offers a new understanding of irreversibility.

Let us first see how reasonable it is to "eliminate the notion of trajectory" for chaotic systems by considering, for example, billard balls on a sufficiently smooth table, so that we can neglect friction (for some time), and assume that there are suitable obstacles and boundaries so that the system is sensitively dependent on initial conditions (see Sect. 4.5). Although we are unable to predict the trajectory of the ball, we know that it *has a trajectory*, by simply looking at it. And if we started by describing the initial position of the ball by an absolutely continuous probability distribution,[13] we also know that the time evolution under the usual dynamics of that probability distribution will lead, after some time, to a probability distribution spread

---

[13] If one considers probabilities given by delta measures defined in (2.A.2), it is equivalent to considering trajectories.

out over the whole billard table, precisely because the dynamics is mixing. Equation (8.3.1) gives an exact formulation of this idea for mixing systems: the probability distribution on the right hand side of (8.3.1) is uniform over phase space and the average values of any function evaluated over trajectories at time t, and averaged with any absolutely continuous distribution over initial conditions, will converge as $t \to \infty$ to that uniform average.

So, unless one invents a sort of "reduction of the probability distribution" when someone looks at the position of the billard ball, analogous to the "reduction of the quantum state" in quantum mechanics, which has no classical justification whatsoever, one has to admit that the "irreducible" statistical description is not a complete description at all, at least for macroscopic objects whose time evolution is governed by a chaotic dynamics.

The probability distribution spread out over the whole billard table describes adequately our knowledge (or our ignorance) of the system, obtained on the basis of our initial information. But it would be difficult to commit the Mind Projection Fallacy (see Sect. 4.8) more radically than to confuse the objective position of the ball and our best bet for it. In fact, chaotic systems illustrate this difference: if all nearby initial conditions followed nearby trajectories, the distinction between probabilities and trajectories would not matter too much. But chaotic system show exactly how unreasonable is the assignment of "irreducible" probabilities, since the latter quickly spread out over the phase space in which the system evolves.

Of course, nobody in the Brussels-Austin school will deny that the ball is always somewhere. But this example raises the following the question: what does it mean then to "eliminate trajectories"? Either the dynamics is expressed directly at the level of probability distributions, and we run into the difficulties mentioned in the previous paragraphs, or the dynamics is *fundamentally* expressed in terms of trajectories, probabilities are a very useful tool, whose properties are *derived* mathematically from those of the trajectories, and nothing radically new has been done.

One cannot stress strongly enough the difference between the role played by probabilities here and in the classical Boltzmannian solution. In the latter, we use probabilities as in the coin-throwing experiment. We have some macroscopic constraint on a system (the coin is fair; the particles are in the left half of the box), corresponding to a variety of microscopic configurations. We predict that the behavior of certain macroscopic variables (the average number of heads; the average density) will be the one induced by the vast majority of microscopic configurations, compatible with the initial constraints. That's all. But it works only because a large number of variables are involved, *in each single physical system*. However, each such system is described by a point in phase space (likewise, the result of many coin tosses is a particular sequence of heads and tails). In the "intrinsic irreversibility" approach, a probability distribution is assigned to *each single physical system*, as an "irreducible" description.

The temptation to consider a probabilistic description of physical systems as irreducible confuses again determinism and predictability: just because we humans cannot predict where a billard ball will be in a few seconds does not mean that it will not be *somewhere*, and the same is true for the microstate of macroscopic systems.

To end this critique of the ergodic/mixing approach to irreversibility, let us quote Jacob T. Schwartz, a ex-pure mathematician who became rather self-critical with respect to the usefulness of rigorous mathematical reasoning in the natural sciences[14]:

> The intellectual attractiveness of a mathematical argument, as well as the considerable mental labor involved in following it, makes mathematics a powerful tool of intellectual prestidigitation – a glittering deception in which some are entrapped, and some, alas, entrappers. Thus, for instance, the delicious ingenuity of the Birkhoff ergodic theorem has created the general impression that it must play a central role in the foundations of statistical mechanics. (This dictum is promulgated, with a characteristically straight face, in Dunford-Schwartz, *Linear Operators*, Vol. 1, Chap. 7.) Let us examine this case carefully, and see. Mechanics tells us that the configuration of an isolated system is specified by choice of a point p in its phase surface, and that after t seconds a system initially in the configuration represented by p moves into the configuration represented by $M_t p$. The Birkhoff theorem tells us that if f is any numerical function of the configuration p (and if the mechanical system is metrically transitive), the time average tends (as $t \to \infty$) to a certain constant; at any rate for all initial configurations p not lying in a set e in the phase surface whose measure $\mu(e)$ is zero; μ here is the (natural) Lebesgue measure in the phase surface. Thus, the familiar argument continues, we should not expect to observe a configuration in which the long-time average of such a function f is not close to its equilibrium value. Here I may conveniently use a bit of mathematical prestidigitation of the very sort to which I object, thus paradoxically making an argument serve the purpose of its own denunciation. Let $v(e)$ denote the probability of observing a configuration in the set e; the application of the Birkhoff theorem just made is then justified only if $\mu(e) = 0$ implies that $v(e) = 0$. If this is the case, a known result of measure theory tells us that $v(e)$ is extremely small wherever $\mu(e)$ is extremely small. Now the functions f of principal interest in statistical mechanics are those which, like the local pressure and density of a gas, come into equilibrium, i.e., those functions for which $f(M_t p)$ is constant for long periods of time and for almost all initial configurations p. As is evident by direct computation in simple cases, and as the Birkhoff theorem itself tells us in these cases in which it is applicable, this means that $f(p)$ is close to its equilibrium value except for a set e of configurations of very small measure μ. Thus, not the Birkhoff theorem but the simple and generally unstated hypothesis "$\mu(e) = 0$ implies $v(e) = 0$" necessary to make the Birkhoff theorem relevant in any sense at all tells us why we are apt to find $f(p)$ having its equilibrium value. The Birkhoff theorem in fact does us the service of establishing its own inability to be more than a questionably relevant superstructure upon this hypothesis.
>
> Jacob T. Schwartz [287]

Let us finish this section by mentioning that *there is* a notion of mixing that is relevant for convergence to equilibrium, but is not the one considered here. It will be explained in a concrete model in Sect. 8.7.3.

---

[14] To explain the relation of the notations in that quote with those of this book: the Birkhoff ergodic theorem is theorem 4.7. The book he refers to is Dunford and Schwartz [111]; what is calls p is our $\mathbf{x}$; M_t is our T^t; the "known result in measure theory" is that, in his notation, if $\mu(e) = 0$ implies that $v(e) = 0$, then the measure v is absolutely continuous with respect to μ and so one can write (2.A.19), with v instead of ν, which implies that $v(e)$ is extremely small wherever $\mu(e)$ is extremely small.

8.3.4 Real Systems Are Never Isolated

It is sometimes alleged that, for some reason (the Poincaré recurrences, for example) a truly isolated system will never reach equilibrium. But it does not matter, since true isolation never occurs and external ("random") disturbances will always drive the system towards equilibrium.[15] This is true but irrelevant.[16]

In order to understand this problem of non-isolation, we have to see how to deal with idealizations in physics. Boltzmann compares this with Galilean invariance (see [38, p. 170]). Because of non-isolation, Galilean (or Lorentz) invariance can never be applied strictly speaking (except to the entire universe, which is not very useful). Yet, there are many phenomena whose explanation involve Galilean (or Lorentz) invariance. We simply do as if the invariance was exact and, then, we argue that the fact that it is only approximate does not spoil the argument. The same idealization holds when one uses the law of conservation of energy.

One can use a similar reasoning in statistical mechanics. If we can explain what we want to explain (e.g. irreversibility) by making the assumption that the system is perfectly isolated, then we do not have to introduce the lack of isolation in our explanations. We have only to make sure that this lack of isolation does not conflict with our explanation. And how could it? The lack of isolation should, in general, speed up the convergence towards equilibrium. Also, if we want to explain why a steamboat cannot use the kinetic energy of the water surrounding it to move, we apply irreversibility arguments to the system boat+water, even though the whole system is not really isolated.

Another way to see that lack of isolation is true but irrelevant is to imagine a system being more and more isolated. Is irreversibility going to disappear at some point? That is, will different fluids not mix themselves, or will they spontaneously unmix? It is difficult to think of any example where this could be argued.

---

[15] One can even invoke a theorem to that effect: the ergodic theorem for Markov chains (see e.g. Seneta [288, Chap. 4]), or for Markov processes, where the dynamics is probabilistic instead of deterministic and of which a special case will be used in Sect. 8.7.1. But this is again highly misleading. This theorem says that probability distributions will converge to an "equilibrium" distribution (for suitable chains). This is similar, and related, to what happens with mixing dynamical systems. But it does not explain what happens to a single system, unless we are willing to distinguish between microscopic and macroscopic variables, in which case the ergodic theorem is not necessary.

[16] Borel [40] tried to answer the reversibility objection, using the lack of isolation and the instability of the trajectories. As we saw in Sect. 8.1, this objection is not relevant, once one introduces the micro/macro distinction. And Fred Hoyle wrote: "The thermodynamic arrow of time does not come from the physical system itself...it comes from the connection of the system with the outside world" [174], quoted in [203]. See also Cohen and Stewart ([84], p. 260) for similar ideas.

8.4   Maxwell's Demon

In his book *Theory of Heat* [237], James Clerck Maxwell wrote[17]:

> But if we conceive of a being whose faculties are so sharpened that he can follow every molecule in its course, such a being, whose attributes are as essentially finite as our own, would be able to do what is impossible to us. For we have seen that molecules in a vessel full of air at uniform temperature are moving with velocities by no means uniform, though the mean velocity of any great number of them, arbitrarily selected, is almost exactly uniform. Now let us suppose that such a vessel is divided into two portions, A and B, by a division in which there is a small hole, and that a being, who can see the individual molecules, opens and closes this hole, so as to allow only the swifter molecules to pass from A to B, and only the slower molecules to pass from B to A. He will thus, without expenditure of work, raise the temperature of B and lower that of A, in contradiction to the second law of thermodynamics.

James Clerck Maxwell [237, pp. 308–309]

He did not call this being a demon, he called it a valve or a finite being (the expression 'demon' was introduced later by William Thomson (Lord Kelvin) [309]), but it became known and discussed under that name ever since.

There is an enormous literature on that demon, mostly trying to show or explain why, in actual fact, it does not violate the second law.

The first question is of course, what did Maxwell have in mind? What did he try to prove? Maxwell emphasizes that his purpose in invoking this thought experiment was to explain that the second law of thermodynamics has only a statistical certainty. Indeed, we can only control macroscopic variables, but if we were an homunculus of very small size able to observe single molecules and act as his being would, then violations of the second law would occur.

It should be stressed that it is not only "us" humans who cannot perform such manipulations but also any known (isolated) natural process. So the short answer to the apparent paradox created by the demon is simply that it does not exist in the physical world.

But it is not entirely clear, in Maxwell's statement, whether he had in mind a physical or non-physical being: how does that being, for example, "see" the individual molecules? Of course, if we allow non physical beings to intervene in physics, then anything becomes possible, like miracles, and we would just have stopped doing physics. So, let us consider, from now on, only "demons" subjected to ordinary physical laws.

One possibility is to have a small door in front of the hole that oscillates parallel to the membrane separating the regions A and B whose motion is deterministic but chaotic.[18] Let us assume that, for long periods of time, one can neglect the friction linked to that motion (just as Maxwell assumed that his demon opens and closes the hole without doing any work).

---

[17] This idea appeared already in a 1867 letter of Maxwell to his friend Tait [239, p. 331–332].

[18] The rest of the argument would work also if the motion of the door was periodic, see Albert [9, p. 39].

And assume also that the chaotic motion of the door just happens to be so that, each time a fast molecule comes in front of the hole in region A, the door is open and, when a slow molecule comes in front of the hole in region B, the door is also open. But if a slow molecule comes in front of the hole in region A, the door is closed and when a fast molecule comes in front of the hole in region B, the door is also closed. Then the effect of this door would be exactly the same as the one of the demon.

The reader may think that this "accidental" coordination between the chaotic motion of the door and the motion of the gas molecules is extremely improbable. Yes, but not impossible. One can imagine that there is a very tiny subset of the phase space of the combined system gas + door in which the motions just described take place. But this would not refute the second law, at least not as understood by Boltzmann: it would just be another example of exceptional, but very rare, initial conditions leading to an anti-thermodynamic behavior.

Thus, so far, the "lessons" one can draw from the demon are those that we know already: one must introduce a micro/macro distinction and some exceptional initial conditions may reverse the "irreversible" motion.

However, around the beginning of the 20th century, various schemes have been suggested in order to create purely physical Maxwell's demons. The idea being to exploit "random" fluctuations that occur naturally in the gas and try to force them to act in a well-defined direction. The simplest example of random fluctuations is given by the Brownian motion: a relatively large molecule being constantly bombarded by collisions with smaller molecules moves in a random way, because of the fluctuations in those collisions.

But could one connect these random motions to a mechanical device that would force the motions to go in a given direction? If that was possible, one would obtain a violation of the second law, since then the thermal agitation of the smaller molecules would be converted entirely into a mechanical motion. In 1912, Smoluchowski [300] gave a series of arguments showing that such devices would necessarily be subjected to thermal fluctuations that would make their functioning impossible.

Probably the most famous such device is "the ratchet and the pawl" analyzed by Richard Feynman in [129, Sect. 46]. We will consider a simpler example due to Smoluchowski [300], Smoluchowski's trapdoor (see Norton [246]).

Consider again a vessel divided into two parts separated by a membrane in which there is a hole and, in front of the hole, there is a door that can close the hole but that can only move in one direction, see Fig. 8.3.

Now, if particles hit the door coming from one side of the vessel, they will be able to push the door and go on the other side, while this is impossible for the particles coming from that other side. Eventually, this will lead to a difference of density of the gases in the two sides of the vessel, which would imply that one can go from an equilibrium situation (both sides with equal density) to a non-equilibrium one.

The reason why this won't work in the long run (and it would have to work for a rather long time in order to create a non-equilibrium situation, not just a small fluctuation) is that, for this to work, the door has to be very light and easy to move. But then, the door will come into thermal equilibrium with the gas and will have its

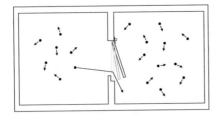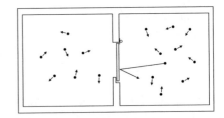

Fig. 8.3 Smoluchowski's trapdoor: particles coming from the left can push the door open and go on the other side while those coming from the right cannot. A similar picture can be found in Norton [243]

own fluctuations: it will bounce back and forth towards the membrane and that will allow molecules to move in both directions.

There are many other devices that have been suggested that would try to orient fluctuations in a given direction but they all suffer from similar problems. In short, physical Maxwell's demons do not exist.

There is even an abstract argument against the possibility of Maxwell's demons: if one considers the closed system made of the gas plus the would-be demon, then the total system should be a mechanical (Hamiltonian) system that moves the entire region of equilibrium into a smaller volume of the total phase space, because that is what going from an equilibrium state to a non equilibrium one means. It should move all the points in the equilibrium region towards the smaller volume at the same time, because that is what the demon is supposed to do: create a non equilibrium situation, starting from *any* equilibrium configuration. But such a mechanical motion is impossible because of Liouville's theorem: volumes in phase space are invariant under Hamiltonian evolution so that one cannot "push" a large set of configurations into a smaller one (see Norton [247] for details of the argument).

Of course, one can create a non equilibrium situation, starting from an equilibrium one, for example, by introducing a piston that pushes the gas in one half of the box. But then, the system is not isolated.

There is a rather large literature on Maxwell's demon that follows a totally different line: one that does not concern itself with a physically realizable demon but considers it from the point of view of information theory (an approach that was discussed in Chap. 7). The argument is that, for the demon to "see" the fast and the slow molecules, it must store some information, but, since its memory is necessarily finite, it must also erase that information after letting a molecule going from one side to the other. And, so the argument goes, that erasure corresponds to an increase of entropy that compensates the decrease of entropy created by the operation on the gas. This is called Landauer's principle [202] which has been developed by Charles Bennett [25]. A previous information theoretic argument trying to rule out Maxwell's demon is due to Leo Szilard [307], and also Léon Brillouin [59]. That argument was based on the amount of entropy linked to the information that the demon must gain (rather than erase) in order to operate.

We will not discuss that approach because it is not clear whether the demon here is physical or not. If it is physical, we already know that it cannot work and if it is not physical, it should not concern us.[19]

For a critique of those approaches see the papers by Norton [243–248], and by Earman and Norton [119].

8.5 The Origins of the Low Entropy States

Even if Boltzmann's scheme is accepted, there remains a deep problem in "explaining" irreversibility.[20]

To understand it, consider the gas in the box when it is expanding in the whole box but has not yet reached equilibrium. Let us ask what is the most probable past state of the gas if all we know is that partly expanded state and if we assume that the system has been isolated for a very long time.

By the reversibility of the microscopic equation of motion, we must do the same reasoning to infer the past state of the gas as we did to infer its future state. And we must then conclude that its past state was a homogeneous distribution in the box. How could that be true? Coming back to the usual time direction, it would mean that the gas underwent an enormous fluctuation out of its equilibrium state into the inhomogeneous state that is only partly expanded in the box.

Of course, such fluctuations are extremely rare and we would not expect them to occur on time scales less than the age of the universe. But if we had to explain why the gas started concentrated in one half of the box, which is what happens in our scenario, we would have to assume an even bigger fluctuation out of equilibrium.

Of course, all these reasonings assume that the box is isolated, which we know is not true. The initial state with all the particles in one half of the box was created by an experimentalist who pushed the particles there with a piston. But the experimentalist herself is obviously a out-of-equilibrium system (otherwise she would be dead). That system is maintained out of equilibrium (i.e. in a relatively low entropy state) by food that ultimately depends on the photosynthesis of plants (animals either eat plants or

---

[19] Since Smoluchowski argued that a physical demon cannot function, an objection based on a demon not subjected to physical laws came up in a question put to Smoluchowski by Kaufmann at the conclusion of Smoluchowski's lecture at the 84th Meeting of Natural Scientists in Münster in 1912. It would, Kaufmann suggested, be: "a conclusion that one possibly could regard as proof, in the sense of the neovitalistic conception, that the physico-chemical laws alone are not sufficient for the explanation of biological and psychic occurrences" [300, p. 1018]. Smoluchowski, taken somewhat aback by the question, replied: "What was said in the lecture certainly pertains only to automatic devices, and there is certainly no doubt that an intelligent being, for whom physical phenomena are transparent, could bring about processes that contradict the second law. Indeed Maxwell has already proven this with his demon" Quoted in Norton [248]. Norton adds: "He [Smoluchowski] then recovered and proceeded to suggest that perhaps even an intelligent being is constrained by normal physics, so that some neglected physical process would still protect the second law".

[20] This is a problem that is explicitly rejected as a problem by adherents of the "information theoretic" approach such as Grandy [159, Chap. 11].

eat other animals that eat plants). But the latter also depends on a source of low entropy which comes from the fact that the sun sends us high frequency photons that the earth reemits as low frequency photons. Although we will not do the calculation here, high frequency photons have a lower entropy than low frequency ones. So, the sun does not send us energy per se (it is reemitted, otherwise we would eventually burn) but it sends us energy at low entropy that is reemitted in a form of higher entropy.

But the hen and the egg problem does not stop here: how come that the sun sends low entropy photons? Through its thermonuclear reactions, that also increase its entropy. But then, the sun must have started in a low entropy state. So, we go back to the formation of the solar system and, ultimately, to the beginning of the universe, the Big Bang. The logic of our reasoning leads us to assume that the universe started in a low entropy state or more precisely in a microstate typical relative to a macrostate of low entropy, like a typical microstate of a gas initially constrained to be in a very small part of a box.

This assumption has been called by David Albert "the past hypothesis" (see [9, Chap. 4]): the microstate of the universe at the time of the Big Bang was typical (in the sense of Sect. 2.3.1) of a macrostate of very low entropy. In [255, p. 343] Roger Penrose estimates that the volume occupied by that initial macrostate must be a fraction $10^{-10^{123}}$ of the total volume of the phase space of the universe. He also shows a picture of 'God' aiming at the "absurdly tiny volume of the phase space of possible universes."

Feynman stated this necessary assumption very clearly when he wrote [128, p. 116]: "Therefore I think it necessary to add to the physical laws the hypothesis that in the past the universe was more ordered, in the technical sense, than it is today - I think this is the additional statement that is needed to make sense, and to make an understanding of the irreversibility."

But low entropy also means "improbable." So, although we can account for irreversibility and the arrow of time by showing that this evolution, starting from a non-equilibrium state, is extremely *probable* we must assume, in order to "explain" the existence of this non-equilibrium state in the first place an extremely *improbable* initial condition of the universe.

This is obviously not very satisfactory, to say the least, but this problem has nothing to do with the usual objections or alternatives to the Boltzmannian approach discussed previously.

We will not go into the cosmological hypotheses made in order to account for this initial low entropy state of the universe, see Penrose [255, Chap. 7] or Tumulka [314, Sect. 9.3].

There is another approach to that problem, relying on spontaneous fluctuations, but that Feynman sharply criticized. If a system is isolated for a long enough time, fluctuations of all sizes will occur. Now consider the universe as a whole. If it is eternal, a rather common assumption before the Big Bang, and if you assume that it is most of the time in equilibrium, then fluctuations of all sizes will necessarily occur sometimes.

Why not assume that our current universe is on its way back to equilibrium starting from some gigantic fluctuation away from it that occurred in the past?

That sounds like a plausible hypothesis until one thinks seriously about it. Indeed, smaller fluctuations are more likely and thus more frequent than large ones. As Feynman says, if you have a gas in equilibrium in a vessel and see a small subset of the vessel where it is out of equilibrium, you will naturally assume that the rest of the gas is still in equilibrium and that the fluctuation occurred only in the small subset of the vessel.

Coming back to the universe, instead of assuming that it is the result of a gigantic fluctuation, why not assume that there was smaller fluctuation in the past, giving rise only to the solar system, the rest of the universe being still in equilibrium (no stars, no galaxies). That would be enough to account for the existence of life and of humans. But then, how to account for our perception of distant stars and galaxies? Well, note that all these perceptions are not direct: we do not go out of the solar system to see what is going on. All those perceptions depend on signals that exist within the solar system, and in fact that occur on Earth. If everything in the solar system is exactly as it is, but results from a "small" fluctuation giving rise to it alone and not from one creating the entire universe, it will also include all those signals; simply the latter will be part of the "small" fluctuation and will not come from the universe outside the solar system. Of course, those signals will mislead us about what is going on in the universe, but assuming that those signals arise from a fluctuation is more likely than assuming that they come from distant stars and galaxies that themselves arose from fluctuations, since smaller fluctuations are more probable than big ones.

But now we are on a dangerous path. Indeed, one can ask, what is the smallest fluctuation, which is also the most probable and thus the most frequent one that is necessary to account for our experiences? The answer is a so-called Boltzmann brain, namely just my brain with nothing else in the universe, not even my body. By "my brain" one means that brain with all its current sensations, including my memories. After all, all this must be encoded in my neurons and their connections.

Of course, such a brain, without a body and without blood influx would not survive more than a split second, but at any given instant, one can imagine that one *is* nothing more than a Boltzmann brain. Physics seems to justify a "materialist" version of solipsism: solipsism says that only my sensations exist and everything else (including my body and my brain) is an illusion. In the Boltzmann brain scenario, sensations are not detached from the body or at least not from the brain, but nevertheless, everything that we believe about the outside world is an illusion.

Why should we not believe that we are Boltzmann's brains? More or less for the same reason that we should not believe in solipsism. If we do not think that our memories can be trusted, then why believe our scientific theories in the first place? All the arguments in favor of our scientific theories are ultimately based on observations, but if the latter are all illusory so is the whole of science.

Besides, Boltzmann brains die almost instantly, so that, if you believe that you had a past (as you should), you are not a Boltzmann brain.

Even is we assume larger fluctuations than just a brain, say a collection of humans or only the Earth or only the solar system, but all arising from a fluctuation, then we would not expect the correlations that we observe: why would all books about, for example, Napoleon say that he became emperor in 1804 and was defeated at Waterloo on June 18, 1815, if they all arose out of fluctuations? And why would we observe what we do about the distant universe if in reality it is all in thermal equilibrium and other stars or galaxies do not exist?[21] Of course a fluctuation of the size of a human brain is very unlikely to occur during the approximately 13 billion years since the Big Bang but remember that here we work within the pre-Big Bang hypothesis of an eternal universe.

There is yet another possible scenario, introduced by Carroll and Chen [70], which does not require a past hypothesis, and where there are only few Boltzmann brains. One assumes time to be infinite in the past, but also that entropy is unbounded, hence the universe never reaches equilibrium; one also assumes that the energy surface of the universe has infinite Liouville measure, hence that Poincaré's recurrences never occur. But even in that new scenario, we have to assume that the Universe had a lower entropy in the past than now in order to account for the observed irreversible behavior, see [314, p. 105], [153, 214], and references there for further discussion.

8.6 The Boltzmann Equation

The Boltzmann equation governs the evolution of Boltzmann's f function.[22] We will not go into the mathematical derivation of that equation, that can be found in many places (see e.g. Chap. 8 of Tumulka [314] or Chap. 2 of Cercignani, Illner and Pulvirenti [73]), but we will focus on its conceptual structure, in particular the meaning of the f function (and we will follow here Sects. 3.4–3.6 of Goldstein, Lebowitz, Tumulka, and Zanghì [155] and Chap. 8 of Tumulka [314].).

The f function is simply a limit of the macroscopic quantity (6.1.7) introduced in Sect. 6.1. Boltzmann considered N molecules of radius a, modeled as billard balls, in a container Λ (so that the phase space of the system is $(\Lambda \times \mathbb{R}^3)^N$) in the limit $N \to \infty$, $a \to 0$ and the sizes of the cells of the partition of phase space $|\Delta_i|$ going to 0, in such a way that the number of particles per cell goes to infinity:

$$N \to \infty \qquad a \to 0, \qquad 4Na^2 \to \lambda$$

---

[21] There is still a problem with Boltzmann brains: if we assume that the universe has no end in time and is eventually in equilibrium, then Boltzmann brains will necessarily occur and will be far more numerous than ordinary humans (if the length of time when the universe is in equilibrium, hence without humans, is infinite, there will be infinitely many Boltzmann brains). Even though the individual lifetime of a Boltzmann brain is short, the combined lifetime of all those Boltzmann brains is longer than the one of ordinary humans Then, one should rationally believe that one is a Boltzmann brain, but 'living" in a distant future, and with illusions about its own past. For a discussion of this problem, where one has to take into account the possibility that the universe has an end and not only a beginning, or that the universe is expanding, or else that quantum mechanics may play a role, see Tumulka [314, Sect. 9.3], [315] for further discussion and references.

[22] The reader may want read first Sect. 8.7.2, on the Kac ring model, which is conceptually similar to what is done here, but simpler.

This is called the *Boltzmann-Grad limit* [158]. Note that the total volume occupied by the molecules, $\frac{4\pi}{3}a^3 N$ goes to zero in that limit, so this is a low density limit. In that limit, the macroscopic quantity (6.1.7) converges towards a continuous function $f(\mathbf{q}, \mathbf{v})$, $\mathbf{q} \in \Lambda$, $\mathbf{v} \in \mathbb{R}^3$.

The time evolution of $f(\mathbf{q}, \mathbf{v}, t)$ is, as argued by Boltzmann:

$$\left(\frac{\partial}{\partial t} + \mathbf{v} \cdot \nabla_{\mathbf{q}} \right) f(\mathbf{q}, \mathbf{v}, t) = Q(\mathbf{q}, \mathbf{v}, t) \tag{8.6.1}$$

where $Q(\mathbf{q}, \mathbf{v}, t)$ is the "collision term":

$$Q(\mathbf{q}, \mathbf{v}, t)$$
$$= \lambda \int_{\mathbb{R}^3} d^3\mathbf{v}_* \int_{S^2} d^2\omega 1_{\omega \cdot (\mathbf{v} - \mathbf{v}_*) > 0} (\omega \cdot (\mathbf{v} - \mathbf{v}_*) \tag{8.6.2}$$
$$\times [f(\mathbf{q}, \mathbf{v}', t) f(\mathbf{q}, \mathbf{v}'_*, t) - f(\mathbf{q}, \mathbf{v}, t) f(\mathbf{q}, \mathbf{v}_*, t)],$$

where S^2 is the unit sphere in $\mathbb{R}^3$ and, from the conservation of momenta and energies during elastic collisions (see e.g. Cercignani [74, Appendix 4.1] or Cercignani, Illner and Pulvirenti [73, Chap. 2]) we have:

$$\mathbf{v}' = \mathbf{v} - [(\mathbf{v} - \mathbf{v}_*) \cdot \omega]\omega$$

and

$$\mathbf{v}'_* = \mathbf{v}_* + [(\mathbf{v} - \mathbf{v}_*) \cdot \omega]\omega,$$

that are the outgoing velocities of a collision between two balls with incoming velocities $\mathbf{v}$ and $\mathbf{v}_*$ and where ω is a unit vector oriented along the centers of the two balls (thus the integral over ω in (8.6.2) is over the unit sphere S^2 in $\mathbb{R}^3$).

We have to add to (8.6.1) boundary conditions along the boundary of Λ, on which the balls are reflected elastically.

Here we assume that the collisions occur when both particles are at the same point $\mathbf{q}$, but remember that we also assumed that the Boltzmann-Grad limit has been taken, in which the radii of the particles $a \to 0$.

To understand (8.6.1), consider first what happens in the absence of collisions: then, particles of velocity $\mathbf{v}$ have a linear motion and $f(\mathbf{q}, \mathbf{v}, t) = f_0(\mathbf{q} - \mathbf{v}t, \mathbf{v})$, from which one obtains by differentiation that the left hand side of (8.6.1) equals 0.

The Boltzmann equation is obtained as an approximation to the same equation as (8.6.1), but with another kernel Q':

$$Q'(\mathbf{q}, \mathbf{v}, t)$$
$$= \lambda \int_{\mathbb{R}^3} d^3\mathbf{v}_* \int d^2\omega 1_{\omega \cdot (\mathbf{v} - \mathbf{v}_*) > 0} (\omega \cdot (\mathbf{v} - \mathbf{v}_*) \tag{8.6.3}$$
$$\times [f_2(\mathbf{q}, \mathbf{v}', \mathbf{q}, \mathbf{v}'_*, t) - f_2(\mathbf{q}, \mathbf{v}, \mathbf{q}, \mathbf{v}_*, t)]$$

where $f_2(\mathbf{q}, \mathbf{v}, \mathbf{q}, \mathbf{v}_*, t)$, $f_2(\mathbf{q}, \mathbf{v}', \mathbf{q}, \mathbf{v}'_*, t)$ are defined in the same way as $f(\mathbf{q}, \mathbf{v}, t)$ but as the empirical density of pairs of particles with position $\mathbf{q}$ and velocities $\mathbf{v}, \mathbf{v}_*$ or $\mathbf{v}', \mathbf{v}'_*$.

One passes from (8.6.3) to (8.6.1) by the crucial *hypothesis of molecular chaos*:

$$f_2(\mathbf{q}, \mathbf{v}', \mathbf{q}, \mathbf{v}'_*, t) = f(\mathbf{q}, \mathbf{v}', t) f(\mathbf{q}, \mathbf{v}'_*, t)$$
$$f_2(\mathbf{q}, \mathbf{v}, \mathbf{q}, \mathbf{v}_*, t) = f(\mathbf{q}, \mathbf{v}, t) f(\mathbf{q}, \mathbf{v}_*, t) \tag{8.6.4}$$

Note that the word "chaos" here has nothing to that notion in the modern theory of dynamical systems discussed in Chap. 4, which came much after Boltzmann.

There are two main mathematical properties of Boltzmann's equation:

1. The only stationary solutions of (8.6.1) are functions $f(\mathbf{q}, \mathbf{v})$ that are independent of $\mathbf{q}$ and Gaussian in $\mathbf{v}$ (it is called a global Maxwellian):

$$f(\mathbf{q}, \mathbf{v}) = \left(\frac{m\beta}{2\pi}\right)^{\frac{3}{2}} \exp\left(-\frac{\beta m |\mathbf{v}|^2}{2}\right),$$

   Those are the f's in thermal equilibrium. Taking any other function f as initial condition corresponds to a non equilibrium situation.

2. Let H be Boltzmann's H-function:

$$H(t) = k \int_{\Lambda \times \mathbb{R}^3} d^3\mathbf{q} d^3\mathbf{v} f(\mathbf{q}, \mathbf{v}, t) \ln f(\mathbf{q}, \mathbf{v}, t).$$

   Then, if $f(\mathbf{q}, \mathbf{v}, t)$ is a solution of Boltzmann's equation:

$$\frac{dH(t)}{dt} \leq 0, \tag{8.6.5}$$

   with equality if and only if $f(\mathbf{q}, \mathbf{v}, t)$ is equal to a local Maxwellian i.e. of the form:

$$f(\mathbf{q}, \mathbf{v}, t) = n(\mathbf{q}, t) \left(\frac{m\beta(\mathbf{q}, t)}{2\pi}\right)^{\frac{3}{2}} \exp\left(\frac{-\beta(\mathbf{q}, t) m (\mathbf{v} - \mathbf{u}(\mathbf{q}, t))^2}{2}\right), \tag{8.6.6}$$

   which is a Gaussian distribution, but with a locally varying density $n(\mathbf{q}, t) = \int_{\mathbb{R}^3} f(\mathbf{q}, \mathbf{v}) d\mathbf{v}$, locally varying average velocity $\mathbf{u}(\mathbf{q}, t) = \frac{\int_{\mathbb{R}^3} \mathbf{v} f(\mathbf{q}, \mathbf{v}, t) d\mathbf{v}}{n(\mathbf{q}, t)}$ and locally varying temperature $kT(\mathbf{q}, t) = \frac{2}{3}\left[\frac{e(\mathbf{q}, t) - \frac{1}{2} m n(\mathbf{q}, t) \mathbf{u}(\mathbf{q}, t)^2}{n(\mathbf{q}, t)}\right]$, with $e(\mathbf{q}, t) = \frac{1}{2n(\mathbf{q}, t)} \int_{\mathbb{R}^3} m \mathbf{v}^2 f(\mathbf{q}, \mathbf{v}, t) d\mathbf{v}$.

Since H has a sign opposite to the one of S, the decrease of H is equivalent to the increase of the entropy. Thus (8.6.5) is a derivation of the second law for the macrostate defined by (6.1.7).

When we spoke of mathematical properties 1 and 2 above, we were somewhat formal because we did not discuss the existence of solutions to Boltzmann's equation. This is a difficult subject whose results are not entirely satisfactory. It is known that differential equations have, under very general conditions, unique solutions at least for a short time interval (see e.g. Kolmogorov and Fomin [197, Sect. 2.4]). Boltzmann's equation is an integro-differential equation and under some mild assumptions on the initial condition $f_0(\mathbf{q}, \mathbf{v})$ (continuity and some Gaussian bound) one can prove that the solution exists and is unique for a short time interval, see Lanford [208], Cercignani, Illner and Pulverenti [73] or Tumulka [314, Sect. 8.6].

Although the result holds only for small times, there is no reason to believe that it would not hold for all times, but that is a typical restriction on proofs of existence of solutions for differential or integro-differential equations. Global existence in time is an open problem, but there are some special cases where it is known: if $f_0(\mathbf{q}, \mathbf{v})$ is independent of $\mathbf{q}$ or almost independent of $\mathbf{q}$ or is sufficiently close to a Maxwellian, see Cercignani, Illner and Pulverenti [73] and Tumulka [314, Sect. 8.6].

A deeper issue is whether Boltzmann's equation can be rigorously derived from the microscopic dynamics of billiard balls, under suitable assumptions on initial conditions. A truly remarkable result in that direction was proven by Oscar Lanford in 1975–1976 [206, 208]. We follow Goldstein, Lebowitz, Tumulka, and Zanghì [155] in paraphrasing that result[23]:

We need to introduce a "coarse graining" of the one-particle phase space $\Lambda \times \mathbb{R}^3$ into cells $\{\Delta(\vec{u}_i)\}$ (small cubes of equal volume). Introduce the macrostate (similar to $M_3(\mathbf{x})$ defined by (6.1.7)):

$$\tilde{M}_3(\mathbf{x}) = \left[\frac{|\{j \in 1, \ldots, N \mid (q_j, v_j) \in \Delta(\vec{u}_i)\}|}{N \Delta f |\Delta(\vec{u}_i)|} \right] \Delta f \tag{8.6.7}$$

where $[\cdot]$ is the integer part.

Lanford's "Theorem"[24]

Let λ in (8.6.2) equal $4Na^2$. Then, for any nice initial density $f_0(\mathbf{q}, \mathbf{v})$, N large, a very small, and for a coarse graining into cells $\{\Delta(\vec{u}_i)\}$ that are small but not too small (they must contain many particles) most initial points $(\mathbf{q}_0, \mathbf{v}_0)$ with approximate empirical distribution f_0 (relative to the coarse graining (8.6.7)) evolve in such a way that the empirical density of $(\mathbf{q}(t), \mathbf{v}(t))$ is close to the solution $f(\mathbf{q}, \mathbf{v}, t)$ of Boltzmann's equation with initial datum $f_0(\mathbf{q}, \mathbf{v})$, for small enough times.

The restriction to small times is again a technical restriction related to the limitations in the proof of the existence of solutions to Boltzmann's equation.

Lanford's theorem shows that Boltzmann's ideas can be expressed and proven mathematically (at least with the restriction on small times) because it does show that most microstates corresponding to a given macrostate evolve in such a way that

---

[23] See also Tumulka [314, Sect. 8.7], Cercignani, Illner and Pulverenti [73] or Spohn [304, Chap. 4] for precise statements and Volchan [323] for a discussion similar to the one given here.

[24] We use quotation marks here because, although Lanford's theorem is a true theorem, our paraphrasing of it here is not precise enough to be one.

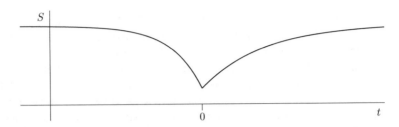

Fig. 8.4 For most phase points, entropy increases in both time directions, albeit not necessarily at the same rate. Figure taken from Tumulka [314, p. 95]

the induced evolution of the macrostate (8.6.7) is independent of that microstate and follows the relevant macroscopic evolution (here, Boltzmann's equation).

What happens to the Boltzmann's equation if we consider the reversibility of the microscopic equations of motion? Suppose that we reverse all the velocities: $\mathbf{v}_i \to -\mathbf{v}_i, \forall i = 1, \ldots, N$ at some time $t = t_0$. Then, we have, for the macrostate, the transformation: $f(\mathbf{q}, \mathbf{v}) \to f^r(\mathbf{q}, \mathbf{v}) = f(\mathbf{q}, -\mathbf{v})$. Since this exchanges the velocities of the outgoing particles (after a collision) and the incoming velocities, it amounts to exchanging the two terms quadratic in f in (8.6.2), or, equivalently, to change λ into $-\lambda$ in (8.6.1).

Thus, if $f(\mathbf{q}, \mathbf{v}, t)$ is a solution of Boltzmann's equation with a constant λ, $f^r(\mathbf{q}, \mathbf{v}, -t)$ is a solution of Boltzmann's equation with a constant $-\lambda$. This means that $H(t)$ decreases (and entropy increases) in the negative time direction, see Fig. 8.4. But, by Lanford's theorem, this implies that most phase space points whose macrostate is $f^r(\mathbf{q}, \mathbf{v})$ at time $t = t_0$ will have their entropy increasing towards the past, meaning that it decreases in the usual time direction.

But that of course, is not what we observe; the solution of this "paradox" is the same as the one we gave before about the reversibility paradox: we need to assume that the past had a lower entropy than the present.

Another "paradox" is that, if we start with an initial phase point x whose macrostate is $f(\mathbf{q}, \mathbf{v})$, evolve it for some time $t: \mathbf{x} \to T^t\mathbf{x}$, then reverse its velocities, $T^t\mathbf{x} \to (T^t\mathbf{x})^r$ and evolve it again for the same amount of time t, we get the original $\mathbf{x}$ with velocities reversed: $T^t(T^t\mathbf{x})^r = \mathbf{x}^r$, so that during the second evolution $t \to 2t$, entropy decreases. However, as we just saw, the macrostate corresponding to $(T^t\mathbf{x})^r$, namely $f^r(\mathbf{q}, \mathbf{v}, t)$ sees its entropy increasing in the past, which means that the evolution of the microstate (whose entropy decreases) will not follow the one of its corresponding macrostate.

This "paradox" is resolved by observing that $(T^t\mathbf{x})^r$ is *not a typical point* for $f^r(\mathbf{q}, \mathbf{v}, t)$, because, in $(T^t\mathbf{x})^r$, the velocities of the incoming particles (which were the outgoing ones in $T^t\mathbf{x}$) are highly correlated. See Sect. 8.7.2.2 for a discussion of this fact in the simpler Kac ring model.

Finally, there is the crucial question of the limits of solutions to the Boltzmann equation. Since we do not have a proof of the existence of global (in time) solutions, we obviously do not have a rigorous answer to that question, but here is what one expects, according to Goldstein and Lebowitz, since local Maxwellians (8.6.6) are not stationary: "it is expected and partially proven [69, 72, 97], that starting with an initial $f_0(\mathbf{q}, \mathbf{v})$, which can be far from a local Maxwellian, $f(\mathbf{q}, \mathbf{v}, t)$ will rapidly approach an f which is close to a local Maxwellian and will stay close to it while the local variables in (8.6.6) change on a slower time scale. As the gradients become smaller this evolution will be hydrodynamic, i.e. $n(\mathbf{q}, t)$, $\mathbf{u}(\mathbf{q}, t)$, $e(\mathbf{q}, t)$ will evolve according to the compressible Navier-Stokes equations, which will then bring the gas to equilibrium" [150] (quote adapted to our notation). In particular, Desvillettes and Villani showed that, if solutions to the Boltzmann equation exist globally in time and are sufficiently nice, then they converge to equilibrium [97, 321].

8.6.1 The Relation Between the Boltzmann Equation and the Full Evolution of Measures

Let us now briefly discuss another notion of f: that of a marginal distribution.[25]

Let $\rho(\mathbf{q}_1, \mathbf{q}_2, \ldots \mathbf{q}_N, \mathbf{v}_1, \mathbf{v}_2, \ldots \mathbf{v}_N)$ be some probability density on the total phase space of the N particles, $(\Lambda \times \mathbb{R}^3)^N$. The time evolution of that probability density is given by Liouville's equation (3.4.8):

$$\frac{\partial \rho_t(\mathbf{q}(t), \mathbf{p}(t))}{\partial t} = \sum_{i=1}^{N} \frac{\partial \rho(t, \mathbf{q}(t), \mathbf{p}(t))}{\partial \vec{p}_i} \cdot \nabla_{\vec{q}_i} H - \frac{\partial \rho(t, \mathbf{q}(t), \mathbf{p}(t))}{\partial \vec{q}_i} \cdot \nabla_{\vec{p}_i} H$$

(8.6.8)

Then one can define the one particle density $\tilde{f}(\mathbf{q}_1, \mathbf{v}_1)$:

$$\tilde{f}(\mathbf{q}_1, \mathbf{v}_1)$$
$$= \int_{(\Lambda \times \mathbb{R}^3)^{N-1}} d^3\mathbf{q}_2 d^3\mathbf{q}_3 \ldots d^3\mathbf{q}_N d^3\mathbf{v}_2 d^3\mathbf{v}_3 \ldots d^3\mathbf{v}_N \rho(\mathbf{q}_1, \mathbf{q}_2, \ldots \mathbf{q}_N, \mathbf{v}_1, \mathbf{v}_2, \ldots \mathbf{v}_N),$$

and the two particles density:

$$\tilde{f}_2(\mathbf{q}_1, \mathbf{q}_2, \mathbf{v}_1, \mathbf{v}_2)$$
$$= \int_{(\Lambda \times \mathbb{R}^3)^{N-2}} d^3\mathbf{q}_3 d^3\mathbf{q}_4 \ldots d^3\mathbf{q}_N d^3\mathbf{v}_3 d^3\mathbf{v}_4 \ldots d^3\mathbf{v}_N \rho(\mathbf{q}_1, \mathbf{q}_2, \ldots \mathbf{q}_N, \mathbf{v}_1, \mathbf{v}_2, \ldots \mathbf{v}_N),$$

and so on, for all the l particles densities $\tilde{f}_l(\mathbf{q}_1, \mathbf{q}_2, \ldots \mathbf{q}_l, \mathbf{v}_1, \mathbf{v}_2, \ldots \mathbf{v}_l)$, for all $l < N$.

---

[25] The distinction between these two notions is emphasized by Goldstein, Lebowitz, Tumulka and Zanghì [150, 155].

One can write down *exact* equations, derived from the microscopic dynamics (8.6.8), relating the time evolutions of the various $\tilde{f}_l$'s. This is called the BBGKY hierarchy (after Bogolyubov, Born, Green, Kirkwood and Yvon), see e.g. Cercignani [74, Appendix 4.3], Cercignani, Illner and Pulvirenti [73, Chap. 2] or Spohn [304, Chap. 4] for its derivation. The problem is that the equation giving the time evolution of $\tilde{f}_l$ involves $\tilde{f}_{l+1}$ so that this hierarchy is not really easier to solve than the full Liouville equation (8.6.8). One possibility is to truncate the hierarchy for some l by modeling $\tilde{f}_{l+1}$ in terms of functions $\tilde{f}_m$, with $m \leq l$.

If one does that for $l = 1$ (thus modeling $\tilde{f}_2$ in terms of $\tilde{f}_1$, which means that one uses (8.6.4) with f_2, f_1 replaced by $\tilde{f}_2, \tilde{f}_1$) one arrives again at Boltzmann's equation. Thus, we get the same equation for f_1 and $\tilde{f}_1$, so that few people distinguish between the two (we introduced the tilde notation to stress that distinction, but this is highly unusual). But conceptually, these quantities are distinct: one is the empirical density for typical phase points and the other is the marginal probability density, relative to some measure with density $\rho(\mathbf{q}_1, \mathbf{q}_2, \dots \mathbf{q}_N, \mathbf{v}_1, \mathbf{v}_2, \dots \mathbf{v}_N)$.

Equation (8.6.8) is often taken to be the starting point of the whole of non equilibrium statistical mechanics, its "rigorous" basis. But this raises the usual problem: what is the meaning of a probability distribution on the full phase space $(\Lambda \times \mathbb{R}^3)^N$?

If probabilities reflect our knowledge/ignorance, then it is not clear why they should evolve according to physical dynamical laws. Our knowledge/ignorance evolves in time, but that depends on the information we gain or possibly loose about the system and that information does not depend only on the system's dynamics.

If, on the other hand, one views $\rho(\mathbf{q}_1, \mathbf{q}_2, \dots, \mathbf{q}_N, \mathbf{v}_1, \mathbf{v}_2, \dots, \mathbf{v}_N)$ as a continuous density defined by an infinity of copies of "identical systems" but with different initial conditions, then it makes sense to evolve it according to physical dynamical laws.

But what is the physical meaning of these copies? And, more importantly, what does it mean to compute "averages" over that density of functions defined on the full phase space?

In [155], Goldstein, Lebowitz, Tumulka, and Zanghì contrast the viewpoints of Boltzmann and of the mathematician Marc Kac, who had a great interest in statistical physics, (this viewpoint was also the one of C. Cercignani, R. Illner, and M. Pulvirenti [73] and Tolman [310]):

> For example, Kac ([184], first page) wrote: $f(r, v)drdv$ is the average number of molecules in $drdv$, whereas Boltzmann (1898, paragraph 3 p. 36) wrote: let $f(\xi, \eta, \zeta, t)d\xi, d\eta, d\zeta$ [. . .] be the number of m-molecules whose velocity components in the three coordinate directions lie between the limits ξ and $\xi + d\xi$, η and $\eta + d\eta$, ζ and $\zeta + d\zeta$.

> S. Goldstein, J.L. Lebowitz, R. Tumulka, N. Zanghì [155]

In [155], Goldstein, Lebowitz, Tumulka, and Zanghì continue by explaining why mathematicians might prefer studying the marginal distribution:

> So why would anybody want the marginal distribution? Kac aimed at a rigorous derivation of the Boltzmann equation in whatever context long before Lanford's theorem, and saw better chances for a rigorous proof if he assumed collisions to occur at random times at exactly the rate given by Boltzmann's hypothesis of molecular chaos. This setup replaces the time

evolution in phase space (or rather, since Kac dropped the positions, in 3N-dimensional velocity space) by a stochastic process, in fact a Markov jump process. (By the way, as a consequence, any density ρ on $3N$-space tends to get wider over time, and its Gibbs entropy increases, contrary to the Hamiltonian evolution).[26] So the mathematician's aim of finding statements that are easier to prove leads in a different direction than the aim of discussing the mechanism of entropy increase in nature.

S. Goldstein, J.L. Lebowitz, R. Tumulka, N. Zanghì [155]

And, finally there is the issue of mathematical rigor vs the sloppy notation of physicists[27]:

Another thought that may lead authors to the marginal distribution is that $f(\mathbf{q}, \mathbf{v}) d^3\mathbf{q} d^3\mathbf{v}$ certainly cannot be an integer but must be an infinitesimal, so it cannot be the number of particles in $d^3\mathbf{q} d^3\mathbf{v}$ but must be the average number of particles. Of course, this thought neglects the idea that as long as N is finite, also the cells C_i should be kept of finite size and not too small, and the correct statement is that f_i vol C_i is the number of particles in C_i (or, depending on the normalization of f, N^{-1} times the number of particles); when followers of Boltzmann express the volume of C_i as $d^3\mathbf{q} d^3\mathbf{v}$ they merely express themselves loosely.

S. Goldstein, J.L. Lebowitz, R. Tumulka, N. Zanghì [155]

Of course, since we have the fundamental formula (6.6.21) which means that typical values $\approx$ average values, one should have $f_1 \approx \tilde{f}_1$ and it is not surprising that both functions satisfy the same equation and are in fact identified most of the time in the literature.

Remark 8.2 A lot of confusion is due to the identification between the "general" Boltzmann entropy defined in (8.1.6), and the approximation to it given by (minus) the H-function (as emphasized by Lebowitz in [217]). Another frequent confusion about Boltzmann's equation is to mix two conceptually different ingredients entering in its derivation: one is an assumption about initial conditions and the other is to make a particular approximation i.e. one considers the Boltzmann-Grad limit, see Spohn [304, Chap. 4], in which the equation becomes exact (in the Kac model of Sect. 8.7.2 below, this limit reduces simply to letting N go to infinity for fixed t). To account for irreversible behavior, one has always, as we saw, to assume something about initial conditions, and the justification of that assumption is statistical. But that part does not require, in principle, any approximation. To write down a concrete (and reasonably simple) equation, as Boltzmann did, one uses the approximation given by the Boltzmann-Grad limit. Failure to distinguish these two steps leads one to believe that there is some deep problem with irreversibility outside the range of validity of that approximation.

To make a vague analogy, in equilibrium statistical mechanics, one has the concept of phase transition (see Chap. 9). Mean field theory gives an approximate description of phase transitions. But the concept of phase transition is much wider than the range of validity of that approximation.

---

[26] We will see that phenomenon of increase of the Gibbs entropy in the Ehrenfest model, which is a stochastic model, in Sect. 8.7.1 below. (Note of J.B.)

[27] The cells C_i in the quote here are those that we denote $\Delta(\vec{u}_i)$.

Remark 8.3 In [178], Jaynes gives examples of initial states for which Boltzmann's H function does not decrease. But he assumes a strong potential; in fact he assumes a large initial kinetic energy that is then converted into potential energy. But Goldstein and Lebowitz reply in [150, Sect. 5] that those initial microstates are not typical given the macrostate defined by the initial f function f_0, because the large initial kinetic energy is larger than what it would be for an equilibrium system with the given initial total energy, so that Jaynes' example does not contradict the Boltzmannian scheme.

Remark 8.4 In [150] Goldstein and Lebowitz show that the one-particle marginal distribution $\tilde{f}_1$ does not necessarily evolves like Boltzmann's f function. Consider noninteracting particles moving among a periodic array of scatterers; since there is no interaction, $\tilde{f}_1$ evolves according to the one-particle Liouville evolution (3.3.8), (3.3.9) and $\int \tilde{f}_1 \ln \tilde{f}_1$ remains constant in time, see Proposition 8.1 unlike $\int f \ln f$.

8.7 Simple Models

In this section, we shall consider several artificially simple models, two well-known ones and two less well-known ones, in which we will be able to illustrate the ideas of this chapter.

8.7.1 Ehrenfest's Urns

We have two urns, with $2N$ balls located in two urns, A and B, the balls being numbered $i = 1, \ldots, 2N$. At each time $t \in \mathbb{N}$, one chooses at random a number i: if the ball with number i is in urn A one moves it to urn B and vice versa. The fact that this choice is "random" makes this dynamics stochastic, something that we have not discussed in this book. However, we will not need much background on stochastic dynamics to analyze this model and, at the end of this section, we will explain how to replace this stochastic model by a deterministic one.

Intuitively, one expects the number of balls in each urn to converge towards "equilibrium" namely an equal number of balls in each urn, with small fluctuations around those numbers: indeed, if one urn contains more balls than the other, then there is a higher probability that the ball picked at random and moved to the other urn is in that urn rather than in the one with fewer balls.

Let us write that in equations. Here the macrostate is the number of balls in urn A, which we denote by $N + n$, with $-N \leq n \leq N$ (and thus the number of balls in urn B is $N - n$), and a microstate is the set of indices $i_1, \ldots, i_{N+n}$ of the balls in A.

Let $P(n, t)$ be the probability of having $N + n$ balls in urn A at time t. We have the following recursion relation:

$$P(n, t) = P(n - 1, t - 1)\frac{N - (n - 1)}{2N} + P(n + 1, t - 1)\frac{N + n + 1}{2N} \quad (8.7.1)$$

since we can have $N + n$ balls in urn A at time t if there were $N - n - 1$ balls in urn A at time $t - 1$ (which has probability $P(n - 1, t - 1)$), and thus $N - (n - 1)$ balls in urn B at time $t - 1$, and we picked a ball from B and moved it to A (which has probability $\frac{N - (n-1)}{2N}$, since the numbers of the balls are picked at random) or, if there were $N + n + 1$ balls in urn A at time $t - 1$ (which has probability $P(n + 1, t - 1)$) and we picked a ball from A and moved it to B (which has probability $\frac{N+n+1}{2N}$). We have the boundary condition $P(N + 1, t) = P(-N - 1, t) = 0, \forall t$.

Equation (8.7.1) can be written in matrix form:

$$\mathbf{P}(t) = \mathbf{P}(t - 1)\mathcal{P}, \quad (8.7.2)$$

where the vector $\mathbf{P}(t) = (P(-N, t), \ldots, P(N, t))$ and the $(2N + 1) \times (2N + 1)$ matrix $\mathcal{P}$ has entries $\mathcal{P}_{n,n\pm1} = \frac{N\mp n}{2N}$ (or $\mathcal{P}_{n\pm1,n} = \frac{N\pm n+1}{2N}$) and $\mathcal{P}_{i,j} = 0$ for $j \neq i \pm 1$. Note that this implies $\mathcal{P}_{N,N+1} = \mathcal{P}_{-N,-N-1} = 0$, so that the boundary condition $P(N + 1, t) = P(-N - 1, t) = 0$ is preserved in time.

One can check that the matrix $\mathcal{P}$ is *stochastic*, which means that all its entries are positive and that, $\forall n$,

$$\sum_{m=-N}^{N} \mathcal{P}_{n,m} = \sum_{m=n\pm1} \mathcal{P}_{n,m} = \frac{N - n}{2N} + \frac{N + n}{2N} = 1. \quad (8.7.3)$$

This implies the conservation of probability under the evolution (8.7.2): $\sum_{n=-N}^{N} P(n, t) = \sum_{n=-N}^{N} \sum_{m=n\pm1} P(n, t - 1)\mathcal{P}_{n,m}$, which equals 1 if $\sum_{n=-N}^{N} P(n, 0) = 1$.

From (8.7.2), we get $\mathbf{P}(t) = \mathbf{P}(0)\mathcal{P}^t$.

The following binomial distribution is stationary,

$$P^*(n) = \frac{(2N)!}{(N - n)!(N + n)!}\left(\frac{1}{2}\right)^{2N} \equiv G_N(n)\left(\frac{1}{2}\right)^{2N}, \quad (8.7.4)$$

which means that $\mathbf{P}^*\mathcal{P} = \mathbf{P}^*$:

$$P^*(n) = P^*(n - 1)\frac{N - (n - 1)}{2N} + P^*(n + 1)\frac{N + n + 1}{2N}, \quad (8.7.5)$$

an identity easy to check explicitly.

Note that this distribution is the one we would obtain if the balls were distributed "at random" between the two urns.

This probability distribution plays the role of the Liouville measure in physical systems and its stationarity is the analogue of Liouville's theorem (theorem 3.4).

This model is reversible in an appropriate (probabilistic) sense:

$$P(n(t-1) = n|n(t) = m) = P(n(t+1) = n|n(t) = m). \qquad (8.7.6)$$

This can be shown by explicit computation, where the probabilities are defined in equilibrium, i.e. with $P(n(t) = n) = P^*(n)$. We have the same denominator on both sides of (8.7.6), so (8.7.6) reduces to showing that $P(n(t-1) = n, n(t) = m)$ is symmetric in n and m or symmetric in $t-1$ and t. We leave that verification to the reader (exercise 8.2).

Now, observe that

$$\mathcal{P}^{2N}_{n,m} > 0, \qquad (8.7.7)$$

$\forall n, m$ with $-N \le n, m \le N$. Indeed, this just means that there is a positive probability to go from $N + n$ balls in A to $N + m$ balls in A is less than $2N$ steps. One can achieve that by moving a ball from A to B or vice-versa at each step; the most extreme case being the transition from all balls in A ($n = N$) to all balls in B ($m = -N$), which takes $2N$ steps.

Given (8.7.7), we can use a purely algebraic result, the Perron-Frobenius theorem (see e.g. Seneta [288, Chap. 1]): let M be an $l \times l$ matrix satisfying $\sum_{j=1}^{l} M_{i,j} = 1$ (see (8.7.3)) and such that all entries of M^k are strictly positive for some k. Then, there exists a unique vector $\mathbf{v} \in \mathbb{R}^l$, such that $\mathbf{v}M = \mathbf{v}$; moreover $\mathbf{v}$ has positive entries, $\sum_{i=1}^{l} v_i = 1$ and, $\forall i, j$,

$$\lim_{t \to \infty} M^t_{i,j} = v_j,$$

with the rate of convergence being exponential in t.

Because of (8.7.3) and (8.7.7), we can apply this result to $M = \mathcal{P}$, with $l = 2N + 1$ and $k = 2N$; this implies that, $\forall \mathbf{P}(0)$,

$$\lim_{t \to \infty} \mathbf{P}(0)\mathcal{P}^t = \mathbf{P}^*, \qquad (8.7.8)$$

with $\mathbf{P}^*$ given by (8.7.4), since $\mathcal{P}$ has a unique eigenvector of eigenvalue 1, which therefore must be $\mathbf{P}^*$, and the convergence is exponential in t.

One could ask whether there is an analogue of Poincaré's recurrences in this model. One can prove that, with probability one, the system will come back infinitely often to its initial value of $N + n$. This is intuitively obvious: if an event has a probability $p > 0$ in a given experiment and we repeat that experiment infinitely often, with each experiment being independent of the others, then that event will necessarily occur, as explained in footnote 21 in Chap. 2.

Now, in any sequence of $2N$ exchanges of the balls, by (8.7.7), there will be some probability to return to the initial value of $N + n$. Now, consider such a sequence of $2N$ exchanges of the balls as a single experiment and apply the above argument to an infinite repetition of that experiment, with such sequences being independent of each other when they are consecutive.

But more can be said. In fact, the system is ergodic, in the appropriate sense here,[28] namely the following follows trivially from (8.7.8): $\forall\, \mathbf{P}(0),\, \forall n$, with $-N \leq n \leq N$,

$$\lim_{T\to\infty} \frac{1}{T} \sum_{t=0}^{T-1} (\mathbf{P}(0)\mathcal{P}^t)(n) = \mathbf{P}^*(n) = \frac{(2N)!}{(N-n)!(N+n)!} \left(\frac{1}{2}\right)^{2N}, \qquad (8.7.9)$$

which means that the average time spent by the system in state n is given by $\mathbf{P}^*(n)$, irrespective of the initial distribution $\mathbf{P}(0)$. If $\mathbf{P}(0)$ was concentrated on state n, it would mean that not only does the system return infinitely often to that state, but that it spends an average time $\mathbf{P}^*(n)$ in that state.

This gives us also an estimate of the (expected) return time to an initial state, which is the inverse of the average time spent in that state. We leave it to the reader to check that this time is at most 2^{2N}. Using the estimates of Sect. 6.2, one can check that the average time spent by the system in state n is $\exp(-\mathcal{O}(N\epsilon^2))$ for $|n| > \epsilon$, so that return time will be $\exp(\mathcal{O}(N\epsilon^2))$.

Assuming that the time unit, when a jump of a ball occurs is the picosecond (10^{-12} s), the time 2^{2N} would equal 30 billions of billions of years for $2N = 120$, many times more than the age of the universe. We let the reader estimate what that time would be for N of the order of Avogadro's number. This is again enough to explain why we never see a Poincaré recurrence.

It is interesting to compare the Boltzmann and Gibbs entropies in this model. The Boltzmann entropy is a function of a microstate (a specification of which ball is in which urn) that depends only on the macroscopic variable n and is given (up to the factor k) by the number of microstate giving rise to the value n of the macroscopic variable. Thus:

$$S_B(n) = k \ln \frac{(2N)!}{(N-n)!(N+n)!}$$
$$\approx -k(N-n)\ln(N-n) - k(N+n)\ln(N+n) + C \qquad (8.7.10)$$

where $C = k2N \ln(2N)$ is a constant in n and the approximation uses Stirling's formula (6.A.3) for large N.

On the other hand, the Gibbs entropy is a function of a probability measure:

$$S_G(\mathbf{P}) = -k \sum_{n=-N}^{N} P(n) \ln \frac{P(n)}{G_N(n)} = -k \sum_{n=-N}^{N} P(n) \ln P(n) + k \sum_{n=-N}^{N} P(n) \ln G_N(n)$$
$$(8.7.11)$$

where $\frac{P(n)}{G_N(n)}$, with $G_N(n)$ defined in (8.7.4), plays the role of $\rho(\mathbf{x}) = \frac{d\nu}{d\mathbf{x}}$ in (6.6.7).

The equilibrium value of the Boltzmann entropy is the maximum value of (8.7.10) which occurs at $n = 0$ and that value is $k2N \ln 2$. The set of microstates where the value of n is exactly equal to 0 is not dominant, but, by the central limit theorem

---

[28] Compare this with the definition of ergodic dynamical systems in Sect. 4.3.

(2.1), the set of values where $|n| \le \epsilon(N)$, with $\frac{\epsilon(N)}{\sqrt{N}} \to \infty$ as $N \to \infty$ does include the vast majority of microstates.

The Gibbs entropy for the equilibrium measure $\mathbf{P}^*$ equals $-k \ln 2^{-2N} = k2N \ln 2$, see (8.7.4), and coincides with the Boltzmann entropy (for N large, i.e. within the approximation given by Stirling's formula).

But the time evolutions of both entropies are quite different: the Boltzmann entropy follows the evolution of the macrostate, so it tends towards the equilibrium value but not monotonically since the macrostate fluctuates and, even after having reached a value close to the equilibrium one, i.e. with $|n| \le \epsilon(N)$ as above, there are still rare but large deviations from the equilibrium value, since the system is ergodic in the sense of (8.7.9). But as noticed above, if the time needed for a large deviation from equilibrium is of the order $\exp(\mathcal{O}(N\epsilon^2))$, they will not occur over observable times.

On the other hand, one can prove that the Gibbs entropy, defined in (8.7.11), does increase monotonically towards its equilibrium value, see Klein [194]. This follows from the fact that the function $x \ln x$ is convex and that, by (8.7.1) $\frac{P(n,t+1)}{G_N(n)}$ is a linear combination of the form $\sum_m c_{m,n} \frac{P(m,t)}{G_N(m)}$ where the sum is over $m = n \pm 1$, with $c_{m,n} \ge 0$, $\sum_{m=n\pm1} c_{m,n} = 1$, and $c_{m,n} G_N(n) = c_{n,m} G_N(m)$. (check that as an exercise). By convexity of $x \ln x$ and Jensen's inequality, we get:

$$\frac{P(n,t+1)}{G_N(n)} \ln \frac{P(n,t+1)}{G_N(n)} \le \sum_{m=n\pm1} c_{m,n} \frac{P(m,t)}{G_N(m)} \ln \frac{P(m,t)}{G_N(m)}.$$

Multiply both sides by $G_N(n)$ and sum over $n = -N, \ldots, N$. We get:

$$\sum_{n=-N}^{N} P(n,t+1) \ln \frac{P(n,t+1)}{G_N(n)} \le \sum_{n=-N}^{N} \sum_{m=n\pm1} c_{m,n} G_N(n) \frac{P(m,t)}{G_N(m)} \ln \frac{P(m,t)}{G_N(m)}.$$

Using $c_{m,n} G_N(n) = c_{n,m} G_N(m)$ and $\sum_{n=m\pm1} c_{n,m} = 1$, we get:

$$\sum_{n=-N}^{N} P(n,t+1) \ln \frac{P(n,t+1)}{G_N(n)} \le \sum_{m=-N}^{N} G_N(m) \frac{P(m,t)}{G_N(m)} \ln \frac{P(m,t)}{G_N(m)}$$

$$= \sum_{n=-N}^{N} P(n,t) \ln \frac{P(n,t)}{G_N(n)},$$

where in the last identity, we sum over n instead of m. The inequality is strict unless $\frac{P(n,t)}{G_N(n)} = 1$, which proves that (8.7.11) increases in time (because of the $-$ sign in (8.7.11)).[29]

---

[29] Even though there is an analogue of Liouville's theorem in this model (8.7.5), there no analogue of one of its corollaries, namely the constancy in time of the Gibbs entropy.

Since everything here is discrete, the Gibbs entropy must reach its equilibrium value after a finite time and remain constant afterwards.

Note that here, only the distribution (8.7.4) is invariant and the Gibbs entropy is not invariant under the stochastic time evolution, unlike what happens for deterministic mechanical systems (see Proposition 8.1).

From this, it may seem that the Gibbs entropy has a more satisfactory behavior than the Boltzmann one, but the meaning of entropy here is unclear for the same reason that the meaning of probability distributions is unclear: does the latter reflect our knowledge? But then, what is its physical meaning? Does it reflect the average behavior of many "identical systems? But then what can one say about individual systems? The Boltzmann entropy answers that question for typical individual systems.

Concerning typicality, in [18], Baldovin, Caprini and Vulpiani show that the probability P_N that a deviation from the average value of order N^b, for some $b < 1$ occurs even once during an interval of time of order N goes to 0 as $N \to \infty$:

$$P_N(|n(t) - <n(t)|n(0)>| \le N^b \quad \forall t, 0 < t < cN) \to 1, \qquad (8.7.12)$$

as $N \to \infty$, where $<n(t)|n(0)>$ is the average value of $n(t)$ given a initial value $n(0)$. In their estimates, one can take $b = 0.8$.

The reader might ask whether (8.7.12) is compatible with the fact that eventually, by (8.7.9), every value of $n(t)$ is reached by any trajectory of the system, including those far away from equilibrium. The answer is that, in (8.7.12) one consider only time-scales of order N^b while large deviations from equilibrium occur on times scales of order $\exp(\mathcal{O}(N))$.

Baldovin, Caprini and Vulpiani in [18] also derive the following formula for $<n(t)|n(0)>$:

$$<n(t)|n(0)> = \frac{N}{2} + \left(1 - \frac{2}{N}\right)^t (n(0) - N), \qquad (8.7.13)$$

which means that the average number of ball in A converges exponentially to the equilibrium value $\frac{N}{2}$, but on a time scale $\mathcal{O}(N)$.

To summarize, the model does illustrate perfectly Boltzmann's ideas: it is reversible and yet approaches an equilibrium value, with rare deviations from it.

Finally, if the reader is worried that this model is unphysical because it is not deterministic, here is a variant of it that is purely deterministic: let $I_j = [\frac{j+N}{2N}, \frac{j+N+1}{2N}[$, $j = -N, \ldots, N - 1$, and consider the map $T : [0, 1[\to [0, 1[: Tx = 2Nx \mod 1$. Then, what we said in Sect. 4.6.1 about itineraries implies that moving from one urn to the other the ball of index n_t at time t where the sequence $(n_t)_{t=0}^\infty$ determined by $T^t x \in I_{n_t}$, is just as "random", for almost every $x \in [0, 1[$, as the Ehrenfest model.[30] The problem with that model is not that it is probabilistic but that it is artificial, specially when it is formulated as a deterministic model. Nevertheless it incorporates

---

[30] Even in this deterministic version of the model, the Gibbs entropy is not constant in time, because the model is not Hamiltonian.

Fig. 8.5 At each site i there is a particle that has either a plus or minus sign. During an elementary time interval each particle moves clockwise to the nearest site. If the particle crosses an interval marked with a cross (as the one between the sites i and $i + 1$), it changes sign but if it crosses an interval without a cross (as the one between the sites $i + 1$ and $i + 2$) it does not change its sign

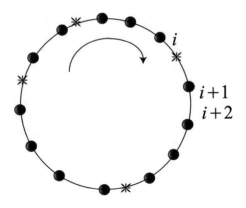

many features of what one would like to show for more realistic models: reversibility (in a probabilistic sense, see (8.7.6)), Poincaré's recurrence and even ergodicity (in the sense of (8.7.9)) and convergence to equilibrium (in the sense of (8.7.8)).

We will now turn to a naturally deterministic model, which also mimics better the behavior of real systems.

8.7.2 Kac Ring Model

This is a simple model, due to Mark Kac ([185] p. 99, see also Thompson ([308] p. 23) and Gottwald and Oliver [157]), which nicely illustrates Boltzmann's solution to the problem of irreversibility, and shows how to avoid various misunderstandings and paradoxes. We will use a slightly modified version of the model.

One considers N equidistant points on a circle; M of the intervals between the points are marked with a cross and form a set called S. The complementary set (of $N - M$ intervals) will be called $\bar{S}$. We will define the density of crosses as

$$\rho_c = \frac{M}{N}.$$

It will be convenient later to assume that

$$\rho_c < \frac{1}{2}. \tag{8.7.14}$$

Each of the N points there is a particle that can have either a plus sign or a minus sign (in the original Kac model, one spoke of white and black balls). During an elementary time interval each particle moves clockwise to the nearest site, obeying the following rule: if the particle crosses an interval in S, it changes sign upon completing the move but, if it crosses an interval in $\bar{S}$, it performs the move without changing sign.

Suppose that we start with all particles having a plus sign; the question is: what happens after a large number of moves. After (8.7.20) we shall also consider other initial conditions.

To formalize the model, introduce for each $i = 1, \ldots, N$, the variable[31]

$$
\alpha_i = \begin{cases} +1 \text{ if the interval in front of } i \in \bar{S} \\ -1 \text{ if the interval in front of } i \in S \end{cases} \tag{8.7.15}
$$

and we let $\eta_i(t) = \pm 1$ be the sign of the particle at site i and time t.

Then, we get the "equations of motion:"

$$
\eta_{i+1}(t+1) = \eta_i(t)\alpha_i. \tag{8.7.16}
$$

Let us first explain the analogy with mechanical laws. The particles are described by their positions and their (discrete) "velocity," namely their sign. One of the simplifying features of the model is that the "velocity" does not affect the motion (of course, that makes the model artificial to put it mildly!). The only reason one calls it a "velocity" is that it changes when the particle collides with a fixed "scatterer," i.e. an interval in S. Scattering with fixed objects tends to be easier to analyse than collisions between particles. The "equations of motion" (8.7.16) are given by the clockwise motion, plus the changing of signs.

These equations are obviously deterministic and reversible: if, after a time t, we change the orientation of the motion from clockwise to counterclockwise, we return after t steps to the original state.[32]

Moreover, the motion is strictly periodic: after $2N$ steps each interval has been crossed twice by each particle, hence they all come back to their original sign. This is analogous to the Poincaré cycles in mechanics, except that, here, the length of the cycle is the same for all configurations (there is no reason for this feature to hold in general mechanical systems).

Let $N_+(t)$ and $N_-(t)$ denote the total number of "plus" particles or "minus" particles at time t (i.e., after t moves, t being an integer). $N_+(t)$ and $N_-(t)$ are the macroscopic variables in this problem (since $N_+(t) + N_-(t) = N$ there is only one independent macroscopic variable).

It is easy to compute the number of (microscopic) configurations whose number of "plus" particles is $N_+(t)$. It is given by the binomial formula:

$$
\frac{N!}{N_+(t)!(N - N_+(t))!} \tag{8.7.17}
$$

---

[31] See for example Fig. 8.5 where $\alpha_i = -1$ and $\alpha_{i+1} = +1$.

[32] There is a small abuse here, because it seems that we change the laws of motion by changing the orientation (from clockwise to counterclockwise). But one can attach another discrete "velocity" parameter to the particles, having the same value for all of them, and indicating the orientation, clockwise or counterclockwise, of their motion. Then, the motion is truly reversible, and we have simply to assume that the analogue here of the operation I of (8.1.1) changes also that extra velocity parameter.

Fig. 8.6 An example of
distribution of crosses where
every configuration is
periodic of period 4

and this number reaches its maximum value for $N_+ = \frac{N}{2} = N_-$.

We can, as in Fig. 6.2, introduce a partition of the phase space according to the different approximate values of N_+, N_-. And what (8.7.17) shows is that different elements of that partition have very different number of elements, the vast majority corresponding to those near $N_+ = \frac{N}{2} = N_-$, and thus, by definition, this will be the set of configurations in "equilibrium."

It is easy to find special configurations which obviously do not tend to equilibrium: start with all particles being "plus" and let every other interval belong to S (with $M = \frac{N}{2}$). Then, after two steps, all particles are minus, after four steps they are all plus again, etc. The motion is periodic with period 4, see Fig. 8.6 for a simple example.

Turning to the solution, one can start by analyzing the approach to equilibrium in this model in a way similar to Boltzmann's equation.

8.7.2.1   Analogue of Boltzmann's Equation

Let $N_+(S; t)$, $N_-(S; t)$ be the number of "plus" particles or of "minus" particles which are going to cross an interval in S at time t.

We have the immediate conservation relations:

$$N_+(t+1) = N_+(t) - N_+(S; t) + N_-(S; t)$$
$$N_-(t+1) = N_-(t) - N_-(S; t) + N_+(S; t) \qquad (8.7.18)$$

If we want to solve (8.7.18), we have to make some assumption about $N_+(S; t)$, $N_-(S; t)$. Otherwise, one has to write down equations for $N_+(S; t)$, $N_-(S; t)$ that will involve new variables and lead to a potentially infinite regress.

So, following Boltzmann, we introduce the assumption analogous to the "Stosszahlansatz" or "hypothesis of molecular chaos":

$$N_+(S; t) = \rho_c N_+(t)$$
$$N_-(S; t) = \rho_c N_-(t),$$ (8.7.19)

with $\rho_c = \frac{M}{N}$ the density of crosses.

The intuitive justification for this assumption is that each particle is "uncorrelated" with the event "the interval ahead of the particle belongs to S", so we write $N_+(S; t)$ as equal to $N_+(t)$, the total number of "plus" particles, times the density ρ_c of intervals in S. This assumption looks completely reasonable. However, upon reflection, it may lead to some puzzlement: what does "uncorrelated" exactly mean? Why do we introduce a statistical assumption in a mechanical model? Fortunately here, these questions can be answered precisely and we shall answer them later by solving the model exactly. But let us return to the Boltzmannian story.

One obtains from (8.7.19):

$$N_+(t+1) - N_-(t+1) = (1 - 2\rho_c)(N_+(t) - N_-(t))$$

Thus

$$N^{-1}[N_+t) - N_-(t)] = (1 - 2\rho_c)^t N^{-1}[N_+(0) - N_-(0)],$$ (8.7.20)

which equals $(1 - 2\rho_c)^t$ if we start with $N_+(0) = N$, $N_-(0) = 0$.

Using (8.7.14) and (8.7.20), we obtain a *monotonic* approach to equal number of particles "plus" and particles "minus", i.e. to equilibrium. Note that we get a monotonic approach for *all* initial conditions $(N_+(0) - N_-(0))$ of the particles.

We can see here in what sense Boltzmann's solution is an approximation. The assumption (8.7.19) cannot hold for all times and for all configurations, because it would contradict the reversibility and the periodicity of the motion. Indeed we know that, if $t = 2N$, then $N_+(t) = N_+(0)$ again. And, if we were to apply (8.7.19) to the reversed motion described above, it would be wrong because then the sign of each particle will be highly correlated with the position of the crosses since those positions determined the sign of each particle. And, if we reverse the motion at some time t, we have $N_+(2t) = N_+(0)$. Both results contradict (8.7.20).

However, we will now show that the fact that (8.7.19) is an approximation does not invalidate Boltzmann's ideas about irreversibility.

8.7.2.2 Microscopic Analysis of the Model

Let us reexamine the model at the microscopic level, first mechanically, and then statistically. The solution of the equations of motion (8.7.16) is:

$$\eta_i(t) = \eta_{i-t}(0)\alpha_{i-1}\alpha_{i-2}\cdots\alpha_{i-t}$$ (8.7.21)

(where the subtractions in the indices are done modulo N). So we have an explicit solution of the equations of motion at the microscopic level.

We can express the macroscopic variables in terms of that solution:

$$N_+(t) - N_-(t) = \sum_{i=1}^{N} \eta_i(t) = \sum_{i=1}^{N} \eta_{i-t}(0)\alpha_{i-1}\alpha_{i-2}\cdots\alpha_{i-t}, \qquad (8.7.22)$$

and we want to compute $\frac{N_+(t)-N_-(t)}{N}$ for large N, for various choices of initial conditions $\{\eta_i(0)\}_{i=1}^{N}$ and various sets S (determining the α_i's). It is here that "statistical" assumptions enter. Namely, we fix an arbitrary initial condition $\{\eta_i(0)\}_{i=1}^{N}$ and consider all possible sets S with $M = \rho_c N$ fixed (one can of course think of the choice of S as being part of the choice of initial conditions). Then, for each set S, one computes the "curve" $\frac{N_+(t)-N_-(t)}{N}$ as a function of time.

The result of the computation, done in [185], is that, for any given t and for N large, the overwhelming majority of these curves will approach $(1 - 2\rho_c)^t$, i.e. what is predicted by (8.7.20). (to fix ideas, Kac suggests to think of N as being of the order 10^{23} and t of order 10^6). The fraction of all curves that will deviate significantly from $(1 - 2\rho_c)^t$, for fixed t, goes to zero as N^{-1}, when $N \to \infty$.

More precisely, we have:

Proposition 8.5 $\forall \epsilon > 0$ and any initial distribution of the signs of the particles $\{\eta_i(0)\}$

$$P\left(\left|\frac{N_+(t) - N_-(t)}{N} - (1 - 2\rho_c)^t \frac{N_+(0) - N_-(0)}{N}\right| \geq \epsilon\right) \leq \frac{2t}{N\epsilon^2}, \qquad (8.7.23)$$

where the probability distribution P assigns probability ρ_c to $\alpha_i = -1$ and $1 - \rho_c$ to $\alpha_i = +1$, independently for each α_i.

Proof To simplify the proof we will limit ourselves to the situation where $\eta_i(0) = 1, \forall i = 1, \ldots, N$. The general case is left as an exercise. We use the bound (2.B.2) with $\frac{1}{N}\sum_{i=1}^{N} f_i(x_i)$ replaced by

$$\frac{N_+(t) - N_-(t)}{N} - (1 - 2\rho_c)^t = \frac{1}{N}\sum_{i=1}^{N}\left(\eta_i(t) - (1 - 2\rho_c)^t\right)$$

$$= \frac{1}{N}\sum_{i=1}^{N}\left(\eta_{i-t}(0)\alpha_{i-1}\alpha_{i-2}\cdots\alpha_{i-t} - (1 - 2\rho_c)^t\right)$$

Using the bound (2.B.2) and setting $\eta_{i-t}(0) = 1, \forall i$,

$$P\left(\left|\frac{N_+(t) - N_-(t)}{N} - (1 - 2\rho_c)^t\right| \geq \epsilon\right)$$

$$\leq \frac{\mathbb{E}\left(\left(\frac{1}{N}\sum_{i=1}^{N}\alpha_{i-1}\alpha_{i-2}\cdots\alpha_{i-t} - (1 - 2\rho_c)^t\right)^2\right)}{\epsilon^2}. \qquad (8.7.24)$$

We have:

$$
E\left(\left(\frac{1}{N}\sum_{i=1}^{N}\alpha_{i-1}\alpha_{i-2}\cdots\alpha_{i-t}-(1-2\rho_c)^t\right)^2\right)
$$

$$
=\frac{1}{N^2}\mathbb{E}\left(\left(\sum_{i=1}^{N}\alpha_{i-1}\alpha_{i-2}\cdots\alpha_{i-t}\right)^2\right)-(1-2\rho_c)^{2t}
$$

(8.7.25)

because $\mathbb{E}(\alpha_{i-1}\alpha_{i-2}\cdots\alpha_{i-t})=(1-2\rho_c)^t$, since the variables α_i are independent and $\mathbb{E}(\alpha_i)=1-2\rho_c$. We write:

$$
\mathbb{E}\left(\left(\sum_{i=1}^{N}\alpha_{i-1}\alpha_{i-2}\cdots\alpha_{i-t}\right)^2\right)
$$

$$
=\sum_{i,j=1}^{N}\mathbb{E}(\alpha_{i-1}\alpha_{i-2}\cdots\alpha_{i-t}\alpha_{j-1}\alpha_{j-2}\cdots\alpha_{j-t}),
$$

so that the right hand side of (8.7.25) can be written as

$$
\frac{1}{N^2}\sum_{i,j=1}^{N}\left(\mathbb{E}(\alpha_{i-1}\alpha_{i-2}\cdots\alpha_{i-t}\alpha_{j-1}\alpha_{j-2}\cdots\alpha_{j-t})-(1-2\rho_c)^{2t}\right) \qquad (8.7.26)
$$

For the terms with $|i-j|>2t$, we have $\mathbb{E}(\alpha_{i-1}\alpha_{i-2}\cdots\alpha_{i-t}\alpha_{j-1}\alpha_{j-2}\cdots\alpha_{j-t})=(1-2\rho_c)^{2t}$, because then the variables $\alpha_{i-1}\alpha_{i-2}\cdots\alpha_{i-t}$ and $\alpha_{j-1}\alpha_{j-2}\cdots\alpha_{j-t}$ are independent, each with averages equal to $(1-2\rho_c)^t$.

So, all terms in (8.7.26) vanish except the $2tN$ terms where $|i-j|\le 2t$. Those terms are bounded in absolute value by 1, since $0\le\mathbb{E}(\alpha_{i-1}\alpha_{i-2}\cdots\alpha_{i-t}\alpha_{j-1}\alpha_{j-2}\cdots\alpha_{j-t})\le 1$ and $0\le(1-2\rho_c)^{2t}\le 1$. This proves that the right hand side of (8.7.25) is bounded by $\frac{2t}{N}$; inserting this in (8.7.24) proves (8.7.23).[33] □

Let us express what happens in this model in terms of the sets depicted in Fig. 8.2. The "phase space" Ω consists of all configurations of signs and scatterers (with $2M<N$), $\{\alpha_i,\eta_i\}_{i=1}^{N}$.

The "thermal equilibrium" set in Fig. 8.2 corresponds to the set of configurations of particles such that $\frac{N_+(t)-N_-(t)}{N}$ is approximately equal to 0 and to the set of configurations of scatterers such that $\frac{N_+(t)-N_-(t)}{N}$ remains close to 0 in the future, at least for $t\ll N$.

If one illustrates this model through Fig. 8.2, the set Ω_0 consists of all configurations of scatterers and of all particles being "plus": $\eta_i=+1,\ \forall i=1,\ldots,N$.

---

[33] For stronger estimates on the probability of $|\frac{N_+(t)-N_-(t)}{N}-(1-2\rho_c)^t|\ge\epsilon$, but with a longer proof, see De Bièvre and Parris [92].

The subset $\Omega_{0,G} \subset \Omega_0$ of good configurations consists of those configurations of scatterers such that $\frac{N_+(t) - N_-(t)}{N}$ tends to 0 and of all the particles having a plus sign.

Then $T^t(\Omega_{0,G}) \subset \Omega_t$ is a set of configurations with $\frac{N_+(t) - N_-(t)}{N}$ approximately equal to 0 but a set of scatterers that is special in the following sense: the configurations in $I(T^t(\Omega_{0,G}))$, where I changes the orientation of the motion from clockwise to counterclockwise, will evolve in a time t to a configuration with all the particles being plus.

So, although, as far as the signs of the particles are concerned, there is nothing special about the configurations in $T^t(\Omega_{0,G})$ ($\frac{N_+(t) - N_-(t)}{N}$ is close to 0), there is a subtle correlation between the configurations of the particles and the scatterers in $T^t(\Omega_{0,G})$, as shown by what happens if one applies the orientation-reversal operation I to those configurations. This is simply a "memory effect" due to the fact that the configurations in $T^t(\Omega_{0,G})$ were initially in $\Omega_0 \supset \Omega_{0,G}$ (this is similar to the memory effect of the particles of the gas in Sect. 8.1 that were initially in one half of the box).

But the configurations in $T^t(\Omega_{0,G})$ form a very small subset of Ω_t; indeed, $|T^t(\Omega_{0,G})| = |\Omega_{0,G}| \leq |\Omega_0|$, and $\frac{|\Omega_0|}{|\Omega|} = 2^{-N}$ (because there are two possible signs in Ω for each of the N sites, but only one sign, plus, in Ω_0). But, reasoning as in the proof of Proposition 8.5, one can show that $|\Omega_{eq}| \approx |\Omega|$ for N large, so that, $\frac{|\Omega_0|}{|\Omega_{eq}|} \approx 2^{-N}$, again for N large and thus $\frac{|\Omega_0|}{|\Omega_{eq}|}$ is extremely small in that limit.

Note that here, we define typical behavior by counting the number of configurations, see (8.7.17), which is the same as putting a probability equal to $\frac{1}{2}$ to each particle sign. This uniform probability is again the one following from the indifference principle.

Note also that one could invent an analogue of Maxwell's demon in Kac model: insert a "demon" somewhere on the circle that adds or removes a cross when a $+$ or a $-$ particle comes so as to make all particles $+$ after some time even if one starts with same number of $+$ and $-$ particles. But, as for the usual Maxwell's demon, it is difficult to see how this demon could act as it is supposed to do by purely physical means, because it has to "see" the sign of the incoming particle.

The variables $N_+(t)$, $N_-(t)$ play the role of macroscopic variables. We can associate to them a Boltzmann entropy:

$$S_B = k \ln \frac{N!}{N_+! N_-!}$$

i.e. the logarithm of the number of (microscopic) configurations whose number of plusses is $N_+(t)$. Since (8.7.17) reaches its maximum value for $N_+ = \frac{N}{2} = N_-$, we see that (8.7.23) predicts a monotone increase of S_B with time. We can also introduce a partition of the "phase space" according to the different values of N_+, N_- as in Fig. 6.2. And what the above formula shows is that different elements of the partition have very different number of elements, the vast majority corresponding to "equilibrium", i.e. to those near $N_+ = \frac{N}{2} = N_-$.

One should not overemphasize the importance of the Kac model. It has many simplifying features (for example, there is no conservation of momentum; the scatterers here are "fixed"). However, it has *all* the properties that have been invoked to show that mechanical systems cannot behave irreversibly, and therefore it is a perfect counterexample that allows us to refute all those arguments (and to understand exactly what is wrong with them): it is isolated (the particles plus the scatterers), deterministic, reversible and periodic, which is a stronger property than the existence of Poincaré cycles and that implies that the system is not ergodic.

This result, obtained in the Kac model, is exactly what one would like to show for general mechanical systems, in order to establish irreversibility. It is obvious why this is very hard. In general, one does not have an explicit solution (for an N-body system!) such as (8.7.16), (8.7.21), in terms of which the macroscopic variables can be expressed, as in (8.7.22).

If we prepare a Kac model many times and if the only variables that we can control are N_+ and M, then we expect to see the irreversible behavior obtained above, simply because this is what happens *deterministically* for the vast majority of microscopic initial conditions corresponding to the macroscopic variables that we are able to control.

8.7.3 Uncoupled Baker's Maps

Gibbs illustrated the idea of irreversibility with the picture of a droplet of ink in a glass of water (see Fig. 8.7): the droplet will diffuse in the glass and lead to a mixture of ink and water and remain so in the future. This involves a notion of mixing which is very different from the mixing in phase space discussed in Sect. 8.3.2. In order to explain the difference between these two notions, we will consider here a very simple purely dynamical model in which this difference is clear.[34] It will also show in what sense dynamical notions may enter into the derivation of irreversibility.

Let $\Omega = D^N = \prod_{i=1}^{N} D_i$ consists of N copies of the unit square $D = [0, 1]^2 \subset \mathbb{R}^2$, and, on each of them, let a copy of the baker map $T_{b,i}$ defined by (4.1.11) act. Let $\mathbf{T} = \prod_{i=1}^{N} T_{b,i}$. Ω is the "Big" phase space, usually called Γ space. D is the "small" phase space, called the μ space.[35]

To define the macrostate, we introduce a partition of the unit square D into small rectangles of equal size. Since the map is equivalent to a shift on symbol sequences of 0's and 1's $\Omega = \{\mathbf{a} = (a_k)_{k \in \mathbb{Z}}, a_k = 0, 1\}$ (see (4.6.3)), consider the latter, and define the partitions by considering the $L \equiv 2^{2n+1}$ sequences $\mathbf{a} = (a_k)_{k=-n}^{k=n}$. Each such finite sequence defines a small rectangle $C_{\mathbf{a}} \subset D$ labelled by $\mathbf{a}$, with $|C_{\mathbf{a}}| = 2^{-(2n+1)}$, where $|\cdot|$ is the Lebesgue measure in D.

For each $\mathbf{a}$, let $n_{\mathbf{a}}(\mathbf{x}) = \frac{N_{\mathbf{a}}(\mathbf{x})}{N}$, where $N_{\mathbf{a}}(\mathbf{x})$ is the number of particles, in a given configuration $\mathbf{x} = (x_i)_{i=1}^{N} \in \Omega$, situated in the rectangle $C_{\mathbf{a}}$ and N is the total number of particles. The set of $n_{\mathbf{a}}(\mathbf{x})$'s are our macroscopic variables.

---

[34] See Cerino, Cecconi, Cencini and Vulpiani [75, Sect. 4] for a somewhat realistic model of an ink drop analyzed from a Boltzmannian persective.

[35] For a Boltzmann-type equation for this model, see Dorfman [109, Sect. 7.2].

Fig. 8.7 Diffusion of a droplet of ink in a glass of water. Source: Zvonimir Loncaric, CC BY-SA 4.0 https://creativecommons.org/licenses/by-sa/4.0, via Wikimedia Commons

The map $M : \Omega \to \mathbb{R}^L$, with $L = 2^{2n+1}$, associates to each configuration, or microstate, $\mathbf{x} = (x_i)_{i=1}^N \in \Omega$ the macrostate $\mathbf{n} = (n_{\mathbf{a}}(\mathbf{x}))$. We can write:

$$n_{\mathbf{a}}(\mathbf{x}) = \frac{1}{N} \sum_{i=1}^N \mathbb{1}(x_i \in C_{\mathbf{a}}) \tag{8.7.27}$$

The equilibrium configurations are the typical configurations with respect to the micro-canonical measure, which is product of Lebesgue measures $\mu = \prod_{i=1}^N \mu_i$ with μ_i a copy of the Lebesgue measure on D or, equivalently, of the product measure on the set of symbol sequences $\tilde{\Omega}_2 = \{\mathbf{a} = (a_k)_{k \in \mathbb{Z}}, a_k = 0, 1\}$, see example 4 in Sect. 4.1, with equal probability for $a_k = 0, 1$.

By the law of large numbers, we have

$$\lim_{N \to \infty} n_{\mathbf{a}}(\mathbf{x}) = \mathbb{E}_{\mu_1}(\mathbb{1}(x_1 \in C_{\mathbf{a}})) = |C_{\mathbf{a}}| = 2^{-(2n+1)} \tag{8.7.28}$$

almost everywhere with respect to μ.

Let us now define the initial nonequilibrium state that we will study: we introduce a uniform measure $\nu_0 = \prod_{i=1}^N \nu_{i,0}$, where each $\nu_{i,0}$ is a copy of a measure ν_0 on D, which is the Lebesgue measure restricted to a given rectangle $C_{\mathbf{b}}$, with $|C_{\mathbf{b}}| = 2^{-(2l+1)}$, where l is not necessarily equal to n:

$$d\nu_0(x) = \frac{\mathbb{1}(x \in C_{\mathbf{b}})}{2^{-(2l+1)}} d\mathbf{x}. \tag{8.7.29}$$

Equivalently, ν_0 is distributed on symbol sequences, so that the symbols between $-l$ and l are fixed equal to $\mathbf{b} = (b_k)_{k=-l}^{k=l}$ and is a product measure on symbol sequences $(a_k)_{k \in \mathbb{Z}, |k| > l}$, with equal probability for $a_k = 0, 1$.

We can write, as in (8.7.27):

$$n_{\mathbf{a}}(\mathbf{x}(t)) = \frac{1}{N} \sum_{i=1}^{N} \mathbb{1}(T_{b,i}^t(x_i) \in C_{\mathbf{a}}),$$

where T_b is the baker's map.

Using again the law of large numbers and (8.7.29), one obtains:

$$\lim_{N \to \infty} n_{\mathbf{a}}(\mathbf{x}(t)) = \mathbb{E}_{\nu_0}(\mathbb{1}(T_{b,i}^t(x_1) \in C_{\mathbf{a}})) = 2^{2l+1}|C_{\mathbf{b}} \cap T_b^{-t} C_{\mathbf{a}}| = 2^{2l+1}|T_b^t C_{\mathbf{b}} \cap C_{\mathbf{a}}|,$$

(8.7.30)

almost everywhere with respect to ν_0, using the fact that T_b leaves the Lebesgue measure invariant.

This means that the time evolution of the macrostate is autonomous with probability close to 1 for N large.

Moreover, one can check, using the conjugation between the baker's map and the shift map (4.6.3) that:

$$\lim_{t \to \infty} 2^{2l+1}|C_{\mathbf{b}} \cap T_b^{-t} C_{\mathbf{a}}| = \lim_{t \to \infty} 2^{2l+1}|T_b^t C_{\mathbf{b}} \cap C_{\mathbf{a}}| = |C_{\mathbf{a}}|, \qquad (8.7.31)$$

This is left as an exercise[36] and means that $\lim_{t \to \infty} \lim_{N \to \infty} n_{\mathbf{a}}(\mathbf{x}(t)) = |C_{\mathbf{a}}|$ almost everywhere with respect to the initial non-equilibrium state ν_0, i.e. that the macrostate converges to the equilibrium one, see (8.7.28), for large N, as $t \to \infty$.

Equation (8.7.31) follows from the fact that the baker's map is mixing (see Sect. 4.4), where the T_b-invariant measure here is the Lebesgue measure. But, and that is a crucial difference, *mixing here is in D, not in* Ω, or, in the physicist terminology, in the μ space not in the Γ space.

This illustrates Gibbs' picture of mixing with ink poured into a glass of water, which then spreads throughout the glass and renders the whole liquid colored. That form of mixing is similar to what happens here: mixing in real space (here the glass) *not* in phase space.

Since the baker map conserves the Lebesgue measure, we have $|T_b^t C_{\mathbf{b}}| = |C_{\mathbf{b}}|$, $\forall t$, so the size of $T_b^t C_{\mathbf{b}}$ does not change, but its *shape* changes a lot! It starts being a small rectangle $C_{\mathbf{b}}$, and becomes a thin ribbon that extends though each rectangle $C_{\mathbf{a}}$, in such a way that (8.7.31) holds.

And the sense of "mixing" introduced here, that does depend on the properties of the dynamics (the baker map is mixing with respect to the Lebesgue measure on D), is relevant for the approach to equilibrium. We may not need in general such a strong property to hold but some sort of mixing in real space or in μ-space must hold; for example in the simple example of diffusion, a set of molecules well localized

---

[36] It can also be seen as a special case of the mixing property of the baker's map, see (4.4.3).

initially in a small region will spread throughout the whole system, but again in real space, not in phase space.

The time scale over which mixing takes place in μ-space (D) is very short compared to the time scale over which mixing in Γ space ($\Omega = D^N$) occurs, if it does.

The reason why so many mathematicians and mathematically inclined physicists have essentially confused this sort of physical mixing with the unphysical mixing in phase space is probably due to the "intellectual attractiveness of a mathematical argument" to quote again Jacob T. Schwartz [287], since that latter mixing depends only on the dynamics and not, as here, on the choice of the macrostate.

8.7.4 Ideal Gas in a Box

De Bièvre and Parris [92] considered a very simple mechanical model that also illustrates Boltzmann's ideas about approach to equilibrium, namely an ideal gas of N particles on a d-dimensional torus $\mathbb{T}^d = (S^1)^d$. The phase space is $(\mathbb{T}^d \times \mathbb{R}^d)^N$ and particles move freely. Since we are on a torus, there are not even refections on the boundary.

They consider the macroscopic variables given by (6.1.2), (6.1.3) and an initial probability $d\mu = \rho(x, p)dxdp$ on $\mathbb{T}^d \times \mathbb{R}^d$ with some smoothness property (in p), which can represent an initial distribution of particles confined in a small sub-box of $\mathbb{T}^d$. One can even assume that all the particles start at the same point ($\rho(x, p)$ is delta measure in x) provided that their velocity distribution is smooth enough.

De Bièvre and Parris prove that typical points with respect to the product measure on $(\mathbb{T}^d \times \mathbb{R}^d)^N$ given by a copy of μ on each factor $\mathbb{T}^d \times \mathbb{R}^d$ evolve in such a way that each sub-box is occupied by a fraction of the total number of particles equal to its size, after a time of order 1 (so that the particles initially in the spatial support of $\rho(x, p)$ have enough time to become dispersed in $\mathbb{T}^d$) and remain in such a state for a time exponential in N (after which deviations from equilibrium may occur). What is meant here by typical is that the probability of atypical initial configurations is exponentially small in N (see De Bièvre and Parris [92, Sect. 4] for precise estimates). This result again illustrates Boltzmann's ideas.

Of course this is what one would expect intuitively: even if the particles are initially constrained by $\rho(x, p)$ to be located in a small subset of $\mathbb{T}^d$ or even at a point in $\mathbb{T}^d$, they will have different velocities[37] and therefore will spread themselves more or less homogeneously over $\mathbb{T}^d$.

On the other hand, since the velocities of the particles do not change (because the particles do not interact with each other), there will be no convergence towards a Maxwellian distribution for the velocities. For that to happen, we need collisions, or at least interactions between the particles.

---

[37] That is why we need some smoothness of $\rho(x, p)$ in p: if, to take an extreme case, $\rho(x, p)$ was a delta measure in p, this spreading would not occur.

8.7.5 The Abiabatic Piston

We will end this list of examples with one which both very simple to explain but very hard to analyze. Consider a volume V divided into two parts of volumes V_1, V_2; each part contains N molecules of identical ideal gases in equilibrium at temperatures T_1, T_2. On both sides, the ideal gas law holds: $P_i V_i = N k T_i$, $i = 1, 2$.

Initially the two parts are separated by a fixed piston, which is supposed to be perfectly adiabatic, i.e. absolutely no exchange of energy is possible through the piston. Then, the piston is released and the question is: what will be the final values of P_i, V_i, T_i, $i = 1, 2$?

Here is what is expected and partly proven: over a relatively short time scale, a mechanical equilibrium will be established, meaning that $P_1 = P_2 = P$ (this was argued by Callen [66, p. 53–54] on the basis of thermodynamics). Then, we have $PV_i = NkT_i$ $i = 1, 2$, but, since the piston is perfectly adiabatic, there is no energy transfer between the two parts of the system and thus no equalization of the temperatures. So, we can still have $V_1 \neq V_2$ and $T_1 \neq T_2$ after the establishment of mechanical equilibrium. Yet, true thermal equilibrium (the state that maximizes the entropy of the system) requires $T_1 = T_2$ and thus $V_1 = V_2$. One expects that the fluctuations on the piston coming from both sides of the system (and that are asymetric since $T_1 \neq T_2$) will eventually drive it towards thermal equilibrium, but that is still an open problem (see Caglioti, Chernov, Lebowitz, Sinai [65, 80–82] for partial proofs and references to previous work).

8.8 Boltzmann Versus Gibbs Entropies

In Sect. 8.1.5 we emphasized the difference between the Boltzmann and the Gibbs entropies outside of equilibrium and mentioned the apparent problem of the constancy in time of the Gibbs entropy, see Proposition 8.1.

In order to "solve" that problem, many ideas have been proposed, for example to coarse grain that entropy over Ω (see e.g. Bais and Doyne Farmer [17]): partition Ω into "cells" of equal size $(C_i)_{i=1}^N$ and define the coarse-grained probability density $\bar{\rho}(\mathbf{x})$ as:

$$\bar{\rho}(\mathbf{x}) = \frac{\int_{C_i} \rho(\mathbf{y}) d\mathbf{y}}{|C_i|} \equiv <\rho>_i, \forall \mathbf{x} \in C_i.$$

so that the function $\bar{\rho}(\mathbf{x})$ is constant in each cell C_i. Define then the corresponding coarse-grained Gibbs entropy:

$$\tilde{S}_G(\rho) = S_G(\bar{\rho}) = -k \sum_{i=1}^N |C_i| <\rho>_i \ln <\rho>_i . \tag{8.8.1}$$

By the convexity of the function $x \ln x$ and Jensen's inequality, we have $< \rho >_i \ln <$ $\rho >_i \leq < \rho \ln \rho >_i$.   Thus, $S_G(\bar{\rho}) \geq -k \sum_{i=1}^{N} |C_i| < \rho \ln \rho >_i = -k \int_{\Omega} \rho(\mathbf{x}) \ln \rho(\mathbf{x}) d\mathbf{x}$ which proves that:

$$\tilde{S}_G(\rho) \geq S_G(\rho)$$

Now, if we let $\rho = \rho(t)$ evolve according to Liouville evolution (3.3.8), (3.3.9), the coarse-grained Gibbs entropy may increase in time (unlike the usual Gibbs entropy) but its physical meaning is totally unclear, because the partition does not correspond to any physically relevant notion. It is sometimes alleged that we introduce this partition because of the limitation of our measurements or observations, but should the definition of a concept such as entropy depend on those limitations? Moreover, as stressed by Cerino, Cecconi, Cencini and Vulpiani in [75], that coarse-grained entropy remains constant during a time depending on the size of the cells, which shows the artificial nature of that coarse-graining. By contrast, the coarse graining in Sect. 6.1 is physical because it is determined by the macrostate.

However, we saw in Sect. 8.5 that the Gibbs entropy *does increase* when a system goes from one equilibrium state to another. But there, we did not introduce any time evolution between those two equilibrium states.

So, there seems to be a paradox here: how can the Gibbs entropy increase between two equilibrium states while being constant in time?

8.8.1 Another Time Evolution of Measures

After Proposition 8.1, we asked whether the time evolution of measures is necessarily, or even naturally, given by the Liouville one (8.1.8), namely by the one induced by the microscopic dynamical laws. Here we will suggest another way to think about the temporal evolution of probability measures on phase space and therefore also of the Gibbs entropy.

Let us go back to the set $\Omega(M(\mathbf{x}(t)))$ that enters the definition (8.1.6) of the time dependent Boltzmann entropy: $S_B(\mathbf{x}(t)) = k \ln |M^{-1}(M(\mathbf{x}(t)))|$. If we had defined the time evolution of the Boltzmann entropy in the same way as we defined the (Liouville) time evolution of ν_t in (8.1.8), namely by: $\tilde{S}_B = k \ln \tilde{\Omega}(t)$, with

$$\tilde{\Omega}(t) = T^t \Omega(M(\mathbf{x}(0))),$$

we would have, by Liouville's theorem 3.4, $|\tilde{\Omega}(t)| = |\Omega(M(\mathbf{x}(0)))|$. So that, if we had defined the time evolution of the Boltzmann entropy using $\tilde{\Omega}(t)$ instead of what we did in (8.1.6), the Boltzmann entropy would have remained constant and for the same reason that the Gibbs entropy remains constant.

Thus, the difference between the time evolution of Gibbs and Boltzmann entropies is not mostly due to the difference between formulas (8.1.6) and (8.1.7) but is rather due to the fact that, in one situation, one uses the Liouville evolution and, in the other, an evolution "driven" by the evolution of the macrostate.

In fact, one can use formula (8.1.7) to define a Gibbs entropy that trivially increases with time. If our macrostate corresponds to a subset $\Omega_t = M^{-1}(M(t))$ of phase space, where $M(t)$ is the value of the macrostate at time t, and, if we define a time-dependent microcanonical measure $\nu_{m,t}$ as:

$$\frac{d\nu_{m,t}(\mathbf{x})}{d\mathbf{x}} = \frac{\mathbb{1}(\mathbf{x} \in \Omega_t)}{|\Omega_t|}, \qquad (8.8.2)$$

then, we get that the Gibbs entropy of $\nu_{m,t}$ is equal to the Boltzmann entropy[38]:

$$S_G(\nu_{m,t}) = -k \int_\Omega \frac{\mathbb{1}(\mathbf{x} \in \Omega_t)}{|\Omega_t|} \ln(\frac{\mathbb{1}(\mathbf{x} \in \Omega_t)}{|\Omega_t|}) d\mathbf{x} = k \ln |\Omega_t| \qquad (8.8.3)$$

From that point of view, the "mystery" of the constancy of the Gibbs entropy disappears naturally: as time evolves, the set $\Omega_t = M^{-1}(M(t))$ corresponding to a value $M(t)$ of the macrostate changes, its size increases and the Gibbs entropy, with a time evolution defined as here, increases also, since it equals the Boltzmann entropy.

It should be emphasized that, with this time evolution, the Liouville measure is no longer time invariant. One could say that it is covariant or equivariant: at each instant of time, we have a measure concentrated on the set of microstates in the inverse image of the value of the macrostate at that time, and that set evolves according to the evolution of the macrostate, not with the Liouville evolution.

One could also introduce a "canonical" version of the microcanonical measure defined by (8.8.2):

$$d\nu_{c,t}(\mathbf{x}) = \frac{\exp(-\lambda_t M(\mathbf{x}))d\mathbf{x}}{Z(\lambda_t)}$$

with $Z(\lambda_t) = \int_\Omega \exp(-\lambda_t M(\mathbf{x}))d\mathbf{x}$ and λ_t determined by $< M >_{\nu_{c,t}} = M(t)$; $\lambda_t F(\mathbf{x})$ is replaced by $\boldsymbol{\lambda}_t \cdot M(\mathbf{x})$ if M takes values in $\mathbb{R}^L$, for $L > 1$, with $\boldsymbol{\lambda}_t \in \mathbb{R}^L$.

The Gibbs entropy of this measure is, see (6.6.9), (6.6.10)

$$S_G(\nu_{c,t}) = k\lambda_t M(t) + k \ln Z(\lambda_t) \qquad (8.8.4)$$

Following the reasoning that led to (6.5.5), (6.5.6), one shows that this "canonical" Gibbs entropy coincides with the Boltzmann "microcanonical" one (8.8.2) in the $N \to \infty$ limit[39]:

---

[38] To evaluate the logarithm of the indicator function, use the fact that $0 \ln 0 = 0$ and $1 \ln 1 = 0$.

[39] At least for suitable functions $M(\mathbf{x})$; here we do not try make a rigorous statement. But see the example in the next section.

$$\lim_{N\to\infty}\frac{1}{N}S_G(\nu_{c,t}) = \lim_{N\to\infty}\frac{1}{N}S_G(\nu_{m,t}) = \lim_{N\to\infty}k\frac{1}{N}\ln|\Omega_t| = k\lim_{N\to\infty}\frac{1}{N}\ln S_B(\mathbf{x}(t))$$

$$= k\lim_{N\to\infty}\frac{1}{N}\ln|M^{-1}(M(\mathbf{x}(t)))|, \tag{8.8.5}$$

for almost all $\mathbf{x}(t)$ relative to the measure $\nu_{m,t}$.

One can also show, by following the arguments of Sect. 6.6.3 that the microcanonical $\nu_{m,t}$ and the canonical ensembles $\nu_{c,t}$ are equivalent.

We already discussed in Sect. 7.8 the limits of the measure theoretic approach to statistical mechanics, from a conceptual point of view (as opposed to a practical one), which also implies that the Boltzmann entropy is more fundamental that the Gibbs' one, unless they coincide, as they do when the Gibbs entropy (and its time evolution) is defined as above.

8.8.2   The Example of the Uncoupled Baker's Maps of Sect. 8.7.3

We will illustrate the different entropies introduced above in the example of the uncoupled baker's maps introduced in Sect. 8.7.3 and analyzed there.

Consider the different time dependent ensembles associated to the system of uncoupled baker's maps: first, the microcaconical ensemble:

$$\frac{d\nu_{m,t}(\mathbf{x})}{d\mathbf{x}} = \frac{\mathbb{1}(\mathbf{x}\in\Omega_t)}{|M^{-1}(\{n_{\mathbf{a}}(t)\}|} = \frac{\prod_{\mathbf{a}}\mathbb{1}(n_{\mathbf{a}}(\mathbf{x})=n_{\mathbf{a}}(t))}{|M^{-1}(\{n_{\mathbf{a}}(t)\}|},$$

with $\Omega_t = M^{-1}(\{n_{\mathbf{a}}(t)\})$, $n_{\mathbf{a}}(t) = \mathbb{E}_{\nu_0}(\mathbb{1}(T_b^t(x_1)\in C_{\mathbf{a}})) = 2^{2l+1}|T_b^t C_{\mathbf{b}}\cap C_{\mathbf{a}}|$, and where the product runs over all the values of $\mathbf{a} = 1,\ldots,L$.

We can define also the time dependent canonical ensemble:

$$\frac{d\nu_{c,t}(\mathbf{x})}{d\mathbf{x}} = \frac{\exp(-\sum_{\mathbf{a}=1}^{L}\lambda_{\mathbf{a}}(t)N_{\mathbf{a}}(\mathbf{x}))}{Z(t)} \tag{8.8.6}$$

where

$$Z(t) = \int_{\Omega}\exp\left(-\sum_{a=1}^{L}\lambda_{\mathbf{a}}(t)N_{\mathbf{a}}(\mathbf{x})\right)d\mathbf{x}$$

with $N_{\mathbf{a}} = n_{\mathbf{a}}N = \sum_{i=1}^{N}\mathbb{1}(x_i\in C_{\mathbf{a}})$, and where $\lambda_{\mathbf{a}}(t)$ is chosen so that $\mathbb{E}_{\nu_{c,t}}(n_{\mathbf{a}}(\mathbf{x})) = n_{\mathbf{a}}(t), \forall\mathbf{a} = 1,\ldots,L$.

Again, by arguing as in Sect. 6.6.3, we have equivalence of those ensembles in the sense of Proposition 6.6. We can introduce:

1. The Boltzmann entropy defined in (8.1.6):

$$S_B(\mathbf{x}(t)) = k \ln |M^{-1}(\{n_{\mathbf{a}}(\mathbf{x}(t))\}|, \qquad (8.8.7)$$

where set of $n_{\mathbf{a}}(\mathbf{x}(t))$'s define the macroscopic variables. So, the evolution of $S_B(\mathbf{x}(t))$ is determined by the one of the macrostate, which, for N large, is given by (8.7.30):

$$n_{\mathbf{a}}(\mathbf{x}(t)) \approx 2^{2l+1}|T'C_{\mathbf{b}} \cap C_{\mathbf{a}}| = n_{\mathbf{a}}(t),$$

which holds almost everywhere with respect to ν_0, when $N \to \infty$.

The usual calculations (see Sect. 6.2) give $|M^{-1}(\{n_{\mathbf{a}}\}| = \frac{N!}{\prod_{\mathbf{a}=1}^{L} N_{\mathbf{a}}!}$, and

$$\ln |M^{-1}(\{n_{\mathbf{a}}\}| \approx -N \sum_{\alpha=1}^{L} n_\alpha \ln n_\alpha,$$

with $L = 2^{2n+1}$. So, we get, since $n_{\mathbf{a}}(\mathbf{x}(t)) \approx n_{\mathbf{a}}(t)$:

$$S_B(\mathbf{x}(t)) \approx -kN \sum_{\mathbf{a}=1}^{L} n_{\mathbf{a}}(t) \ln n_{\mathbf{a}}(t).$$

2. The usual Gibbs entropy evolving according to Liouville's equation (see definition (6.5)):

$$S_{\text{Gibbs}}(\nu_t) = -k \int_\Omega d\nu_t(\mathbf{x}) \ln \left(\frac{d\nu_t(\mathbf{x})}{d\mathbf{x}} \right) = -kN \int_D d\nu_t(x) \ln \left(\frac{d\nu_t}{dx} \right),$$

where $\nu_t = T'\nu_0 = \prod_{i=1}^{N} \nu_{i,t}$, with each $\nu_{i,t}$ a copy of $\nu_t = T'\nu_0$, with ν_0 given by (8.7.29) (since ν is a product measure), i.e. the initial measures evolved through the Liouville evolution. This entropy is constant in time by Proposition 8.1 but is uninteresting physically and the measures ν_t do not coincide with $\nu_{c,t}$.
3. A modified Gibbs entropy, which is the Gibbs entropy of the canonical measure $\nu_{c,t}$. From (8.8.6), we get

$$S_{\text{Gibbs}}(\nu_{c,t}) = k\left(N \sum_{\mathbf{a}=1}^{L} \lambda_{\mathbf{a}}(t) \mathbb{E}_{\nu_{c,t}}(n_{\mathbf{a}})(\mathbf{x}) + \ln Z(t) \right)$$

$$= k\left(N \sum_{\mathbf{a}=1}^{L} \lambda_{\mathbf{a}}(t) n_{\mathbf{a}}(t) + \ln Z(t) \right).$$

This modified Gibbs entropy coincides, in the limit $N \to \infty$, with the Boltzmann entropy, as in (8.8.5), ν_0 almost everywhere, and therefore this Gibbs entropy increases in time.

Remark 8.6 One might think that it would be "more rigorous" to let the diameter of the rectangles $C_{\mathbf{a}}$ go to zero, i.e. in terms of symbol sequences, let $n \to \infty$. But this is not so, because the value of $n_a(\mathbf{x}(t))$ is *not* independent of the microstate $\mathbf{x}(t)$ if the diameter of C_a is too small, and therefore, the set $\{n_a(t)\}$ would not constitue valid macroscopic variables in that case. Physically we need $L \ll N$, i.e. L finite, since N is always finite.

Remark 8.7 The modified Gibbs entropy *does not* necessarily coincide with the coarse grained (in the Γ space Ω) naïve Gibbs entropy defined in (8.8.1), which may increase in time but which has no clear physical justification.

Remark 8.8 The evolution of the measure $\nu_{c,t}$ given by (8.8.6) *does not* coincide with the Liouville evolution $T^t \nu_0 = \prod_{i=1}^{N} \nu_{i,t}$. So, the problem is not with the use of the Gibbs formula for the entropy, but with the use of the Liouville evolution of measures on the Γ space. Measures on that space have no physical meaning and therefore there is no deep reason why they should evolve according to physical laws.

Remark 8.9 $S_{\text{Gibbs}}(\nu_{c,t})$ is solely a function of the *level of description*, i.e. of the choice of macroscopic variables $\{n_a\}_{a=1}^{L}$.

8.9 Conclusion: Are Entropy and Irreversibility Subjective?

One often hears the following sort of thoughts: irreversibility is basically the increase of entropy; entropy is linked to probability; probabilities are subjective (states of mind, or expectations); hence the whole statistical mechanical approach has an irreducibly subjective aspect. Or, irreversibility is based on a choice of macroscopic variables; these variables depend on which variables human beings are able to measure or to control. Hence, the whole approach is anthropomorphic, and thus, in some sense, subjective. Or, irreversibility is due to our ignorance; but ignorance is related to our cognitive capacities and therefore is subjective.

Let us give some examples of such ideas. Prigogine wrote: "In the classical picture, irreversibility was due to our approximations, to our ignorance" ([273], p. 37).[40]

Here are other examples; Heisenberg wrote: "Gibbs was the first to introduce a physical concept which can only be applied to an object when our knowledge of the object is incomplete. If for instance the motion and the position of each molecule in a gas were known, then it would be pointless to continue speaking of the temperature of the gas"([168], p. 38). And Max Born said: "Irreversibility is therefore a consequence of the explicit introduction of ignorance into the fundamental laws" ([43], p. 72). In his popular book, *The Quark and The Jaguar*, Gell-Mann wrote: "Entropy and

---

[40] Prigogine, as we saw in Sect. 8.3.3, thought that thanks to the existence of unstable dynamical systems, "the notion of probability that Boltzmann had introduced in order to express the arrow of time does not correspond to our ignorance and acquires an objective meaning" ([273, p. 42]). See also Prigogine and Stengers ([270] p. 284) for similar remarks. We already criticized in Sect. 8.3.3 this "solution".

information are closely related. In fact, entropy can be regarded as a measure of ignorance"([144], p. 219).

Let us consider these ideas one by one:

1. Macroscopic events existed before human beings appeared and they happen whether we know them or not. That has nothing to do with our knowledge or our ignorance What is true is that we ignore the initial conditions of the microstates of our system or the details of their time evolution, but what we argued in this chapter is that this ignorance does not prevent us from understanding why the system tends towards equilibrium, because this is result of the deterministic evolution of the overwhelming majority of microstates. In other words, ignorance could prevent us from understanding macroscopic phenomena, but, in this case, it does not.

2. If we adhere to a subjective approach to probabilities, there is nothing subjective in our use of probabilities in order to explain phenomena when the latter are typical, as we tried to explain in Sect. 2.5. Of course, what counts as an explanation for us, humans, is necessarily human (other beings might have other criteria) and what is a scientific explanation is even more specific culturally or historically. But that is true of the whole of science.

3. Finally, it is true that what the macrostate is depends on its definition; that is almost a tautology, but it is true that there is a choice of the variables defining the macrostate. In the simple example of the Kac ring model, one could divide the ring into two halves and take the number of plus particles in each half as our macroscopic variables. By following the analysis done in Sect. 8.7.2 one could show that this macrostate converges towards its equilibrium value and the corresponding entropy increases, just as they did when we had only the total number of plus particles as macrostate.

One could continue like that and divide the circle into quarters, one-eights etc. and the results would be similar as long as N is large (see exercise 8.6). Of course, if $N = 2^L$ for some $L \in \mathbb{N}$ and if we repeat this subdivision L times, we would end up with as many "macroscopic variables" as there are sites and the above analysis would obviously fail. There would be nothing macroscopic about those "macroscopic" variables. The entropy associated to each such variable would be constant and equal to 0, since there is only one configuration corresponding to a given number of plus particles on a given site (number which must be equal to 0 or 1).

This is the basis of the (rather common) statement made, for example, by Gell-Mann: "Indeed, it is mathematically correct that the entropy of a system described in perfect detail would not increase; it would remain constant"([144], p. 225). It is mathematically correct but physically incorrect to speak of entropies for such a fine-grained description.

In order to illustrate the importance of the choice of the macrostate, Jaynes invents in [182] the following scenario: in the next century (meaning the present one), one discovers two kinds of argon, A_1 and A_2 as well as a two substances (which he calls Whifnium and Whafnium) from which one can make a piston permeable to A_1 but not to A_2 and another piston permeable to A_2 but not to A_1. Then, someone who knows all this will be able to play a trick to a person ignoring the difference between the two kinds of argon, by using the fact that, thanks to the pistons, one can mix the two kinds of argon reversibly and therefore convert into work heat taken from a heat bath with which the system is in contact.

Since the transformation is done reversibly, the total entropy of the system plus the heat bath does not change; but for the observer ignorant of the two kinds of argon as well as of the Whifnium and Whafnium, an amount of heat has been extracted from the heat bath and converted entirely into work, in violation of the second law.

Jaynes likes to express this in terms of our information and then relates entropy to information, but that is not the only way to look at that imaginary experiment. It could be viewed as a perfectly objective process, independent of anybody's knowledge, and Jaynes' idea of using this process to play a trick to a person ignorant of the difference between the two kinds of argon, only illustrate the fact that many tricks could be played by a competent physicist to people ignorant of the laws of physics.

Jaynes also gives in [177] several physical examples of the multiplicity of possible choices of macroscopic variables: for a certain type of crystal salt, one could take the temperature, volume and pressure. But one could add the strain tensor and the electric polarization to the set of variables or a certain number of coefficients of the expansion of the strain tensor in an orthonormal basis.

Jaynes, following Wigner, calls these entropies "anthropomorphic" [177, p. 398]. A better word might be "contextual", i.e. they depend on the physical situation being considered and on its level of description.

But it is true that entropy is not as "objective" in some sense as the energy: the total energy of a system is a well-defined mechanical quantity, while the entropy is also well-defined, but only once we specify the set of macroscopic variables under consideration.

Jaynes rightly says that he does not know what is the entropy of a cat [177, p. 398]. The same thing could be said for a painting, an eye or a brain. The problem is that there is no well-defined set of macroscopic variables that is specified by the expression "a cat".

Finally, when speaking about our knowledge or our ignorance one should pay attention to two different uses of the word "knowledge". If "some demon" (as Popper would say [265, p. 106], see Sect. 2.4) were to provide us with a detailed knowledge of the microscopic state of the gas in one half of a box, nothing would change to the future evolution of that gas. But we may imagine situations where one can *control* more variables, as in the example mentioned here. If one wants to use the expression "to know" to mean "to control", then our knowledge matters, but this is a misleading use of that expression. What the example of Maxwell's demon shows is that knowledge here would imply control, if a purely physical Maxwell's demon existed.

In his detailed analysis of the alleged subjectivity of entropy (obviously related to the subjectivity discussed here) [96] Denbigh arrived at conclusions similar to ours, namely that the only "subjective" aspect of entropy (if one insists to use that term) is the choice of the macroscopic variables (see also Remark 8.9).

This 'trivial' (but important to keep in mind) aspect of the subjectivity of entropy is expressed by Lebowitz and Maes in a way that is similar but less misleading than the one of Jaynes; they imagine a dialogue between a scientist (S) and an angel (A) who has the answers to all the questions that the scientist may ask:

S: Then what would be an example of having different entropies in your sense?

A: There are many but here is a trivial one. Suppose that I have here red and green particles and that initially, all red particles were to the right of my box and all green particles were to the left. Same pressures, same temperatures. Now I let them go and they mix. I get something brown. Did the entropy increase?

S: This I know. Of course the entropy increases. I can even compute by how much.

A: Aha, but now suppose that you were colorblind, you could not see the difference between green and red and you would see something pretty uniformly grey from beginning to end.

S: And the particles are be assumed to be individually taking each exactly the same paths as before?

A: Yes. The microsopics is identical.

S: Strange. Now I would say there is no increase in entropy. Is that not related to the Gibbs paradox?

A: Indeed, as is its solution. You see, there are many entropies. In one case you add the extra macrovariable giving you the red density-profile and if you do it colorblindly, you just inspect the overall density profile.

S: There is one thing I am now confused about. If there are these different entropies corresponding to what you and I decide to include as macrovariables, how can the measurement of entropy be objective, yielding the same result for all of us?

A: That is a good question. The solution is that you should think about what is relevant and what is not for the problem you are considering. The possibility of having different entropies is not at all harmful as long as you understand that it corresponds to different situations, e.g. are you interested in properties of the system which depend on the color of the particles or not? Of course in the former case you better have someone or some instrument which is color sensitive.

Joel L. Lebowitz, Christian Maes [224]

8.10   Dynamical Systems Vs Statistical Mechanics

In Chap. 4 we analyzed dynamical systems, in particular those that are in practice unpredictable, by statistical methods and we emphasized the importance of the
dynamical properties of those systems, such as ergodicity, mixing and sensitive
dependence on initial conditions.

In this chapter, we studied the dynamical evolution of systems characterized by a
large number of microscopic variables, but we, here, de-emphasized those dynamical
properties.

That may sound confusing. To clarify the issue, we can say, in a nutshell, that the
difference between the two kinds of systems is that:

Dynamical systems $= \lim_{t \to \infty}$,

Statistical mechanics $= \lim_{N \to \infty}$.

In the theory of dynamical systems, we dealt with few variables and therefore
a "small" phase space: an interval for the maps (4.1.3), (4.1.8), (4.1.9) or a solid
torus in $\mathbb{R}^3$ for the solenoid of Sect. 4.6.2.2. Therefore, it makes sense to study how
often small regions of phase space are visited, even over humanly accessible time
spans, which can be idealized as the limit $t \to \infty$. This study uses dynamical notions
like ergodicity or mixing and the averages times spent in regions of phase space are
similar to the averages in the law of large numbers, but averaged over time.

By contrast, in statistical mechanics, one studies systems with a very large number
of variables and therefore a "large" phase space and it would not make sense to study
the amount of time spent by the system in every region of that phase space. But, since
the number N of particles is large, we can obtain an almost deterministic evolution
for quantities that are averages over N (and those averages are of the type given by
the usual law of large numbers). This evolution holds over "finite" times, meaning
relatively short ones, or, at least, humanly accessible ones.

8.11   Summary

This chapter is the most central one in this book. In Sect. 8.1, we outline Boltzmann's
scheme to explain the approach to equilibrium: if a microstate starts in a region of
phase space corresponding to a non equilibrium value of the macrostate, one expects
it to move towards a region where the macrostate takes its equilibrium value, simply
because the latter region is so much larger than the former. This means that the
macrostate will evolve according to the relevant macroscopic law.

In Sect. 8.1.3 we answered the objection to that scheme based on the reversibility
of the equations of motion and in Sect. 8.1.4 we linked Boltzmann's scheme to our
notion of probabilistic explanation given in Sect. 2.5.

We also defined in Sect. 8.1.5 the time dependent Boltzmann entropy as proportional to the logarithm of the size of phase space corresponding to a given value
of the macrostate. Of course, given this definition, if the microstate evolves toward

larger and larger region of the phase space, the Boltzmann entropy will (trivially) increase and so the second law is an immediate consequence of Boltzmann's scheme and of Boltzmann's definition of entropy. On the other hand, we noticed that the Gibbs entropy defined in Sect. 6.6.2 is constant under the Hamiltonian evolution.

In Sect. 8.2, we answered several classical objections to Boltzmann's scheme: apart from the one of Loschmidt that was answered in Sect. 8.1.3, we discuss the one of Zermelo based on Poincaré's recurrence theorem as well as a less known objection due to Poincaré himself and philosophical objections to typicality arguments.

In Sect. 8.3 we discussed several misleading explanations (in our view) of the problem of irreversibility: ergodicity, mixing and the more radical Brussels-Austin school which wants to express the fundamental dynamics in terms of probability measures rather than microstates. We also discussed the argument based on the non isolation of real systems.

In Sect. 8.4 we turned our attention to the (in)famous Maxwell's demon which is again sometime presented as an objection to the Boltzmannian scheme or as requiring the introduction of the notion of "information" in the physical laws and we explained why this is not so.

Next, since the Boltzmannian scheme requires that the past was in a lower entropy state than the present, we tried to explain in Sect. 8.5 what this assumption involves and that brings us back to the origin of the Universe. This is actually the most subtle question concerning the explanation or irreversibility and the answer to it is open.

In the next two sections we turned to concrete applications of Boltzmann's ideas: first, with Boltzmann's equation that does justify his ideas for dilute gases and which is even partly proven mathematically (Sect. 8.6); and then, in several toy models that, even if they are not physically realistic, illustrate perfectly how Boltzmann's scheme works, are mathematically rigorous and show what one would like to be able to prove in more realistic contexts (Sect. 8.7).

In Sect. 8.8 we came back to the distinction between Boltzmann and Gibbs entropies, the first one evolving "correctly" in time and the second one being constant. Some people who like to focus on measures rather than on microstates view the constancy of the Gibbs entropy as a serious problem for the understanding of irreversibility. We suggested that one can modify the time evolution of measures in such a way that the Boltzmann and Gibbs entropies actually coincide also in non equilibrium situations. We argue that if probabilities are not viewed a physical objects but as mathematical tools then our modified time evolution of measures is defensible.

Finally, in Sect. 8.9 we discussed the sometimes alleged subjectivity of entropy, which is true only to the extent that the definition of entropies depend on the choice of the macroscopic variables but nothing more, and in Sect. 8.10 we contrasted the statistical approaches in the theory of dynamical systems and in statistical mechanics.

8.12   Exercises

Since this chapter is mostly theoretical, the exercises will be related to the models
of Sect. 8.7.

8.1. Check (8.7.5).

8.2. Prove the identity (8.7.6).

8.3. Use (8.7.9) to give an estimate on the return time, starting from the initial state
$n = N$.

8.4. Check that (8.7.1) implies that $\frac{P(n,t+1)}{G_N(n)}$ is a linear combination of the form
$\sum_m c_{m,n} \frac{P(m,t)}{G_N(m)}$ where the sum is over $m = n \pm 1$, with $c_{m,n} \geq 0, \sum_{m=n\pm 1} c_{m,n} = 1$, and $c_{m,n} G_N(n) = c_{n,m} G_N(m)$.

8.5. Derive (8.7.13). Hint: define a variable ξ_t that takes value $+1$ with probability
$1 - \frac{n_t}{N}$ and -1 with probability $\frac{n_t}{N}$. Then, write $n_{t+1} = n_t + \xi_t$ and write a rela-
tion between $< n(t+1)|n(t) >$ and n_t. Next, define $\delta_t = < n(t)|n(t-1) >$
$-\frac{N}{2}$ and use the previous relation to write recursion relation for δ_t and iterate
that recursion.

8.6. Consider the Kac ring model of Sect. 8.7.2 and assume that $N = 2^M$. Divide
the circle into 2^L equal subsets each containing 2^{M-L} sites and labelled $i = 1, \ldots, 2^L$. Let $N_+^i(t)$ be the number of $+$ particles at time t in the subset of index
i and associate a Boltzmann entropy to a given sequence of values $(N_+^i)_{i=1}^{2^L}$:

$$S_B((N_+^i)_{i=1}^{2^L}) = k \ln \Omega((N_+^i(t))_{i=1}^{2^L})$$

where $\Omega((N_+^i)_{i=1}^{2^L})$ is the number of configurations with given $N_+^1, \ldots N_+^{2^L}$.

Write explicitly $\Omega((N_+^i)_{i=1}^{2^L})$ in terms of factorials.

8.7. Derive bounds similar to (8.7.23) for each $N_+^i(t), i = 1, \ldots, 2^L$. What happens
if we set $L = M$? What is then the value of the entropy defined in exercise 8.6?
What happens if we set $L = M - m$ for fixed m?

8.8. Prove (8.7.31), using the conjugation of T_b with the shift map T_{shift} on symbol
sequences.

Appendix: Quotes

Although we quoted already in this chapter several reactions to Boltzmann's approach
to statistical mechanics, we will give here some longer quotes from famous scientists,
with rather different appreciations of Boltzmann.

8.A Boltzmannian Quotes

Einstein was very appreciative of Boltzmann's work:

> On the basis of the kinetic theory of gases Boltzmann had discovered that, aside from a constant factor, entropy is equivalent to the logarithm of the 'probability' of the [macro]state under consideration. Through this insight he recognized the nature of the course of events which, in the sense of thermodynamics, are 'irreversible'. Seen from the molecular-mechanical point of view, however, all courses of events are reversible. If one calls a molecular-theoretically defined state a microscopically described one, or, more briefly, micro-state, then an immensely large number (Z) of states belong to a macroscopic condition. Z then is a measure of the probability of a chosen macro-state. This idea appears to be of outstanding importance also because of the fact that its usefulness is not limited to microscopic description on the basis of mechanics.

Albert Einstein [122, p. 43]

Schrödinger was even more enthusiastic:

> The spontaneous transition from order to disorder is the quintessence of Boltzmann's theory [. . .]. This theory really grants an understanding and does not [. . .] reason away the dissymmetry of things by means of an a priori sense of direction of time [. . .]. No one who has once understood Boltzmann's theory will ever again have recourse to such expedients. It would be a scientific regression beside which a repudiation of Copernicus in favor of Ptolemy would seem trifling.

Erwin Schrödinger, [285, Sect. 6]

Schrödinger also wrote [286, p. 168]: "...no perception in physics has ever seemed more important to me than that of Boltzmann – despite Planck and Einstein."[41]

Let us also quote Boltzmann himself. Here he relates, in his reply to Zermelo, his explanation of irreversibility to typicality:

> Zermelo is therefore completely correct when he asserts that the motion is periodic in a mathematical sense; but, far from contradicting my theorem, this periodicity is in complete harmony with it. One should not forget that the Maxwell distribution is not a state in which each molecule has a definite position and velocity, and which is thereby attained when the position and velocity of each molecule approach these definite values asymptotically. For a finite number of molecules the Maxwell distribution can never be true exactly, but only to a high degree of approximation. It is in no way a special singular distribution which is to be contrasted to infinitely many more non-Maxwellian distributions; rather it is characterized by the fact that by far the largest number of possible velocity distributions have the characteristic properties of the Maxwell distribution, and compared to these there are only a relatively small number of possible distributions that deviate significantly from Maxwell's. Whereas Zermelo says that the number of states that finally lead to the Maxwellian state is small compared to all possible states, I assert on the contrary that by far the largest number of possible states are "Maxwellian" and that the number that deviate from the Maxwellian state is vanishingly small.

Ludwig Boltzmann, [35, p. 394–95], quoted by Goldstein in [151, p. 60].

---

[41] Both quotes of Schrödinger here are cited by Goldstein in [149].

Here is again Boltzmann about the reversibility objection[42]:

Ostwald concludes that, because of the complete time reversibility of the differential equations of mechanics, the mechanical conception of the world should not be able to explain why natural processes occur always in a privileged direction. It seems to me that in this way one neglects the fact that mechanical processes are defined not only by differential equations but also by initial conditions. In a manner directly opposed to Ostwald's thesis, I have shown that one of the most staggering confirmations of mechanicism is exactly that of giving an excellent picture of the dissipation of energy, if we concede that the world started from an initial state corresponding to certain initial conditions, which I designated as an improbable state.

Ludwig Boltzmann, [37] (quoted in [74, p. 207])

And about Poincaré's recurrence theorem:

Thus when Zermelo concludes, from the theoretical fact that the initial states in a gas must recur – without having calculated how long a time this will take – that the hypotheses of gas theory must be rejected or else fundamentally changed, he is just like a dice player who has calculated that the probability of a sequence of 1000 one's is not zero, and then concludes that his dice must be loaded since he has not yet observed such a sequence!

Ludwig Boltzmann, [35, p. 397]

Finally, an interesting analogy was made by the 19th century American philosopher Charles Sanders Peirce between the explanation of evolution of species by natural selection and the probabilistic explanation of the behavior of gases.[43]

Mr. Darwin proposed to apply the statistical method to biology. The same thing has been done in a widely different branch of science, the theory of gases. Though unable to say what the movements of any particular molecule of gas would be on a certain hypothesis regarding the constitution of this class of bodies, Clausius and Maxwell were yet able, eight years before the publication of Darwin's immortal work, by the application of the doctrine of probabilities, to predict that in the long run such and such a proportion of the molecules would, under given circumstances, acquire such and such velocities; that there would take place, every second, such and such a relative number of collisions, etc.; and from these propositions were able to deduce certain properties of gases, especially in regard to their heat-relations. In like manner, Darwin, while unable to say what the operation of variation and natural selection in any individual case will be, demonstrates that in the long run they will, or would, adapt animals to their circumstances.

Charles Sanders Peirce [252]

---

[42] Friedrich Wilhelm Ostwald was a German chemist and philosopher. He received the Nobel Prize in Chemistry in 1909 for his studies in catalysis and rates of reaction. His views, called "energetics" considered energy as more fundamental than matter and he rejected the atomistic conception of Boltzmann. He was followed, among others, by Ernst Mach, but, unlike Mach, Ostwald eventually accepted the existence of atoms in 1908.

[43] I learned that quote from Vincent Bauchau [20].

8.B   Quotes Critical of Boltzmann

We have already criticized the views of Ilya Prigogine (see Sect. 8.3.3). Here are some more quotes from this author and his collaborator the philosopher Isabelle Stengers:

> He (Boltzmann) was forced to conclude that the irreversibility postulated by thermodynamics was incompatible with the reversible laws of dynamics (Prigogine [273], p. 41).
>
> All attempts to construct an entropy function, describing the evolution of a set of trajectories in phase space, came up against Liouville's theorem, since the evolution of such a set cannot be described by a function that increases with time (Prigogine and Stengers [272], p. 104).
>
> According to the mechanical view of the world, the entropy of the universe is today identical to what it was at the origin of time (Prigogine [271], p. 160).

The philosopher Karl Popper is often considered, specially among scientists,[44] as the most important philosopher of science of the 20th century. We already discussed his views on probability and "propensity" in Sect. 2.4, which are maybe not unrelated to his misunderstanding of Boltzmann's ideas. He wrote:

"This law of the increase of disorder, interpreted as a cosmic principle, made the evolution of life incomprehensible, apparently even paradoxical" [266, p. 172].

This is forgetting the fact that the solar system and, indeed, the whole universe, starts from a low entropy state and that "life" may be complex from an organizational point of view, but uses a relatively small part of the low entropy energy sent to us by the sun.

Popper also used the following image to criticize what he thought was the "subjectivist" approach of Boltzmann to irreversibility: it "makes the catastrophe of Hiroshima an illusion" [268, p. 128].

But macroscopic events should not be viewed as "illusions" because they can be explained by microscopic ones, not more than tables and chairs should be considered as illusions because they are composed of atoms.

The great Soviet mathematician and one of the founders of the modern theory of probability Alexander Khinchin wrote, in his book on the *Mathematical Foundations of Statistical Mechanics*, that:

> [One often] states that because of thermal interaction of material bodies the entropy of the universe is constantly increasing. It is also stated that the entropy of a system "which is left to itself" must always increase; taking into account the probabilistic foundation of thermodynamics, one often ascribes to this statement a statistical rather than absolute character. This formulation is wrong if only because the entropy of an isolated system is a thermodynamic function – not a phase-function – which means that it cannot be considered as a random quantity; if E and all [external parameters] remain constant the entropy cannot change its value whereas by changing these parameters in an appropriate way we can make the entropy increase or decrease at will. Some authors[45] try to generalize the notion of entropy by considering it as being a phase function which, depending on the phase, can assume different

---

[44] But not necessarily among professional philosophers; for a critique of his general philosophy of science, see Stove [305] and [303, Chap. 4].

[45] Comp. Borel, Mécanique statistique classique, Paris 1925.

values for the same set of thermodynamical parameters, and try to prove that entropy so defined must increase, with overwhelming probability. However, such a proof has not yet been given, and it is not at all clear how such an artificial generalization of the notion of entropy could be useful to the science of thermodynamics.

Alexander Khinchin [191, p. 139]

This misses the point that, as we explained in Chaps. 6 and 8, the entropy, at least the Boltzmann one, is indeed "a phase function", i.e. a function depending on the microstate, and the value of that entropy does change with time, even for an isolated system.

We have mentioned and criticized in Sect. 8.9 the emphasis put on the role of "ignorance" by famous physicists (Heisenberg, Born, Gell-Mann); here is the view of another famous physicist, Wolfgang Pauli, which also gives a misleading role to our knowledge/ignorance:

> The first application of the calculus of probabilities in physics, which is fundamental for our understanding of the laws of nature, is the general statistical theory of heat, established by Boltzmann and Gibbs. This theory, as is well known, led necessarily to the interpretation of the entropy of a system as a function of its state, which, unlike the energy, depends on our knowledge about the system. If this knowledge is the maximal knowledge which is consistent with the laws of nature in general (micro-state), the entropy is always null. On the other hand thermodynamic concepts are applicable to a system only when the knowledge of the initial state of the system is inexact; the entropy is then appropriately measured by the logarithm of a volume in phase space.

Wolfgang Pauli [250, p. 45]

John von Neumann, who made profound contributions to many fields of mathematics, ranging from logic to computing science, along with mathematical foundations of quantum mechanics, gave also a special role to knowledge when he wrote:

> For a classical observer, who knows all coordinates and momenta, the entropy is therefore constant, and is in fact 0, since the Boltzmann "thermodynamic probability" is 1; just as in our theory for [pure quantum] states, since these again correspond to the highest possible state of knowledge of the observer, relative to the system. The time variations of entropy are based then on the fact that the observer does not know everything, that he cannot find out (measure) everything that is measurable in principle. His senses allow him to perceive only the so-called macroscopic quantities.

John von Neumann [325, p. 400–401]

It is interesting to notice that people associated with the "Copenhagen" interpretation of quantum mechanics (Heisenberg, Born, Pauli, von Neumann) tended to somewhat misrepresent Boltzmann, while those that were critical of that interpretation (Einstein, Schrödinger) were fully supporting of Boltzmann's approach. This in turn is related to the conflict between Boltzmann and the "energeticists" of his time (Mach, Ostwald) who rejected both Boltzmann's atomism and his somewhat realistic attitude towards science (see Cercignani [74, Chap. 10]).

The Copenhagen interpretation means different things to different people, but what is common to them and the energeticists, is a view of science as having to limit itself to the description of phenomena or to "save" them and of being skeptical of deeper microscopic explanations (see footnote 23 in Chap. 2 for a view of quantum mechanics that differs from the Copenhagen one and offers a microscopic explanation of why quantum mechanics works, that Boltzmann might have liked).

Sheldon Goldstein draws an interesting parallel between the Copenhagen interpretation of quantum mechanics and the approach to the foundations of statistical mechanics that considers ensembles and not microstates as fundamental:

> I began by mentioning a similarity between the foundations of statistical mechanics and the foundations of quantum mechanics: that they are both rather controversial. Here is another similarity. Most physicists don't have a clear understanding of what they themselves really think about these foundational issues. When it comes to quantum mechanics, by far most would say they accept the Copenhagen interpretation. For statistical mechanics, most would insist on the fundamentality of ensembles.
>
> One reason this might be so is this. Physicists tend to be busy people; they want and need to obtain practical results about experiments and observable phenomena, and in order to do so they use the best tools at their disposal. For quantum mechanics, the tools are those of the quantum formalism, i.e., what they think of as the Copenhagen interpretation. For statistical mechanics, the tools are ensemblist tools, pioneered by Gibbs and, some would argue, also by Boltzmann. So it is quite natural for physicists to imagine that they are Copenhagenists and ensemblists.
>
> It is of little practical value – for their research and for their professional advancement – for physicists to worry about why their tools work as well as they do and what those tools actually have to do with the reality that lies beneath observation. Nonetheless, I do think physicists should do a bit better in this regard. And philosophers of physics should as well: they should pay more careful attention to why physicists do in fact express the views that they do.

> Sheldon Goldstein [154]

Chapter 9
Phase Transitions

9.1 Phenomenology

It is well-known that water freezes at 0 degrees centigrade and boils at 100 degrees centigrade under normal conditions of pressure. This phenomenon of sudden transitions between one phase of matter and another has perplexed scientists as well as ordinary people for ages. Over the last century some progress has been made towards an understanding of these transitions, that are among the simplest examples of collective phenomena in nature. The freezing and boiling phenomena do not occur only under normal conditions of pressure but there are lines in a pressure-temperature diagram along which such transitions occur, see Fig. 9.1. The liquid-vapor transition line ends at a critical point (which, for water is equal to 373.946 degree centigrades and 217.75 atmospheres), beyond which one can pass continuously from the liquid to the vapor phase.

A related phenomenon is the one of ferromagnetic transitions. Some materials such as iron, nickel, cobalt, have the property that, at low temperatures, they remain magnetized when a magnetic field is applied to them and then switched off. This phenomenon however stops at sufficiently high temperatures, which is illustrated in Fig. 9.2.

This chapter aims only to provide a very elementary introduction to some of the basic concepts used in the study of phase transitions. There are so many books and review articles on that subject that we could not hope to make an adequate list. We will rely on the comprehensive study by Sacha Friedli and Yvan Velenik [131] to which we refer the reader interested in mathematical rigor, but presented in a pedagogical fashion. One of the first book on rigorous equilibrium statistical mechanics is the one of David Ruelle [280]. More recent books mathematical statistical mechanics dealing with phase transitions include the ones of Gallavotti [141], Ross Kinderman and Laurie Snell [193], of Yakov Sinai [297], of Barry Simon [292, 293], of Hans-Otto Georgii [145], and of Errico Presutti [269].

© Springer Nature Switzerland AG 2022
J. Bricmont, *Making Sense of Statistical Mechanics*, Undergraduate Lecture
Notes in Physics, https://doi.org/10.1007/978-3-030-91794-4_9

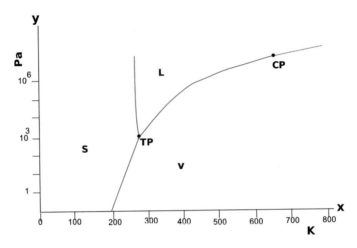

Fig. 9.1 A typical phase diagram with three phases: vapor, liquid and solid. We are only interested in the liquid vapor transition. The line of coexistence between the vapor an liquid phase ends at a critical point (noted CP in the figure). The point noted TP is the triple point where the vapor, liquid and solid phases coexist. Source: Eurico Zimbres, CC BY-SA 2.5 https://creativecommons. org/licenses/by-sa/2.5, via Wikimedia Commons

Fig. 9.2 A typical phase diagram of a ferromagnet with the dark line being the one where two phases with opposite magnetizations coexist

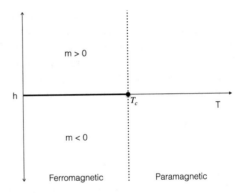

9.2 Lattice Models and Gibbs States

Since very little is known about phase transitions for continuum models (such as gases and liquids), while a great deal is known about lattice models, we will concentrate on the latter and focus at first on the most well known of them, the Ising model.[1] It is defined by:

1. A lattice usually taken to be $\mathbb{Z}^d$, with the Euclidean distance.

---

[1] The name comes from Ising who was student of Lenz. Ising solved the one-dimensional model in his thesis but his life in Germany was made impossible by the Nazi regime and he emigrated to the United States. For more information on Ising's life, see Kobe [195] and for the history of the Ising model, see Brush [60].

2. At each site of that lattice a random variable, called "spin" s_i that takes values in $\{-1, +1\}$.
3. A Hamiltonian or energy function

$$H = -J \sum_{<ij>} s_i s_j \qquad (9.2.1)$$

with $J > 0$ and the sum running over nearest neighbor pairs: $<ij>$ means that $|i - j| = \text{dist}(i, j) = 1$. As such, the sum is ill defined, but, for any finite set $\Lambda \subset \mathbb{Z}^d$, we can restrict the sum in (9.2.1) to $i, j \in \Lambda$ and then that sum is finite.

The phase space of the system is $\Omega = \{-1, +1\}^{\mathbb{Z}^d}$ or, in a finite volume Λ, $\Omega_\Lambda = \{-1, +1\}^\Lambda$. There are many simplifications in that model: our "spin" variables $s_i = \pm 1$ take only two values, there is no kinetic energy term in (9.2.1) and, in fact there is no Hamiltonian dynamics either in the Ising model.[2] The name "spin" comes from the fact that, in quantum mechanics, the spin (of "spin one-half particles") in a given direction takes two values: up or down. However the model here is purely classical.

It is easy to see that, for any finite set $\Lambda \subset \mathbb{Z}^d$, the sum in (9.2.1) reaches its minimum value (because $J > 0$) for two configurations: $s_i = +1, \forall i \in \Lambda$ and $s_i = -1, \forall i \in \Lambda$. That value equals $-J|B(\Lambda)|$ where $B(\Lambda)$ is the set of nearest neighbor pairs in Λ.[3] So, if we introduce the canonical probability distribution on Ω_Λ

$$\mu(\mathbf{s}_\Lambda) = \frac{\exp(-\beta H(\mathbf{s}_\Lambda))}{Z_\Lambda}, \qquad (9.2.2)$$

where $H(\mathbf{s}_\Lambda)$ is given by (9.2.1) with the sum restricted to $i, j \in \Lambda$ and

$$Z_\Lambda = \sum_{s_i = \pm 1, i \in \Lambda} \exp(-\beta H(\mathbf{s}_\Lambda)), \qquad (9.2.3)$$

we see that configurations with low energies are all the more probable that the temperature is lower (i.e. β is large). So we might expect two ferromagnetic phases at low temperatures: one where the spins $s_i = +1$ dominate and the other where the spins $s_i = -1$ dominate. One the other hand, there are many more "disordered" configurations where the values of the spin vary from site to site than "ordered" configurations where most spins are either all equal to $+1$ or all equal to -1.

This would suggest that there is a transition between a disordered regime at high temperatures (β small), where the energy difference between different configurations does not matter too much and those that are more numerous, namely the disordered configurations, "win" and an ordered regime at low temperatures where the energy differences "win" and the two ferromagnetic phases coexist. This is an example of a competition between entropy and energy: at high temperatures the large number of

---

[2] One can introduce a stochastic dynamics for that model, but we will not discuss that.
[3] Below we will use the same notation $B(\Lambda)$ for the set of nearest neighbor pairs under consideration, although the latter may vary between different formulas.

configurations, meaning large entropy, dominates while at low temperatures, energy dominates and it favors the relatively few ordered configurations.

One method to prove this conjecture, would be to add an magnetic field to the Hamiltonian (9.2.1)

$$H(\mathbf{s}_\Lambda) = -J \sum_{<ij>} s_i s_j - h \sum_i s_i \tag{9.2.4}$$

where the sums are restricted to $i, j \in \Lambda$. Then, try to prove that, as $h \to 0$, the average magnetization $\frac{\sum_{i\in\Lambda} s_i}{\Lambda}$, in typical configurations, remains strictly positive when $h \downarrow 0$ and strictly negative when $h \uparrow 0$. Since $\frac{\sum_{i\in\Lambda} s_i}{|\Lambda|}$ is a macroscopic observable for $|\Lambda|$ large, we can, because of the fundamental formula (6.6.21), study its value by studying its average $< \frac{\sum_{i\in\Lambda} s_i}{|\Lambda|} >_{\Lambda,\beta,h} = \frac{\sum_{i\in\Lambda} <s_i>_{\Lambda\beta,h}}{|\Lambda|}$ where the average $< \cdot >_{\Lambda,\beta,h}$ is taken with respect to the probability distribution $\mu_{\beta,h}$ of the form (9.2.2) with Hamiltonian (9.2.4).

An important property of the states $\mu_{\beta,h}$ is the flip spin symmetry: "flipping" the spins, namely changing all the variables $s_i \to -s_i, \forall i \in \mathbb{Z}^d$, is equivalent to changing $\mu_{\beta,h}$ into $\mu_{\beta,-h}$.

However, it is easy to see that $< s_i >_{\Lambda,\beta,h}$ is continuous (in fact even real analytic) in h and β and therefore vanishes at $h = 0$ because of the flip spin symmetry. So, in order to obtain a discontinuous behavior, we need to take the *thermodynamic or infinite volume limit*: $\Lambda \uparrow \mathbb{Z}^d$, which means any sequence of finite boxes $\Lambda_n, n = 1, 2, \dots$ such that, for any finite $\Lambda_0 \subset \mathbb{Z}^d, \exists n_0$ so that $\Lambda_0 \subset \Lambda_n, \forall n \geq n_0$. Indeed, in that limit, the functions that are continuous for Λ finite are no longer necessarily continuous.

Sometimes people are puzzled by this necessity: after all, every observation or experiment is made on sets of finite size; so, does this mean that there are no real phase transitions in nature? One answer is that experimentalists usually do not report the specific size of the substance on which the experiment is performed, so that they implicitly think of measuring a volume-independent number, which corresponds to the thermodynamic limit. In general, if F denotes the number being measured and if $F(\Lambda)$ depends in reality on the volume Λ, one writes:

$$F(\Lambda) = F(\infty) + \text{corrections vanishing as } \Lambda \uparrow \mathbb{Z}^d \tag{9.2.5}$$

where $F(\infty)$ is the thermodynamic limit and is the quantity being reported in actual experiments when the size of the substance on which the experiment is performed is not mentioned. Of course, one may also study the *"finite size corrections"* corresponding to the second term in (9.2.5), but the approximation given by the first term in (9.2.5) is usually sufficient.[4]

---

[4] One may also ask whether, in systems where there is a genuine time evolution, and there are several equilibrium states, the system might jump from one state to another. In a finite system, this will eventually happen but, for the usual macroscopic systems, it will take a time much longer than the one of the universe.

So, what we should do is:

(1) Prove the existence of the thermodynamic limit $\mu_{\beta,h} = \lim_{\Lambda \uparrow \mathbb{Z}^d} \mu_{\Lambda,\beta,h}$, at least for h close to zero. These limiting measures are often called "states".

(2) Show that, for β large, $\lim_{h \downarrow 0}(\lim_{\Lambda \uparrow \mathbb{Z}^d} \frac{<\sum_{i \in \Lambda} s_i >_{\Lambda,\beta,h}}{|\Lambda|}) > 0$ and $\lim_{h \uparrow 0}$ $(\lim_{\Lambda \uparrow \mathbb{Z}^d} \frac{<\sum_{i \in \Lambda} s_i >_{\Lambda,\beta,h}}{|\Lambda|}) < 0$.

The first statement is actually quite easy to prove, for all β, h, but we will not prove it.[5] The existence of that limit also shows that the states $\mu_{\beta,h}$ are invariant under lattice translations, which means: $\forall F \in L^1(\Omega)$ (in this situation with spins equal to ± 1, functions in $L^1(\Omega)$ are actually bounded), $\forall i \in \mathbb{Z}^d$,

$$< \tau_i F >_{\beta,h} = < F >_{\beta,h}, \tag{9.2.6}$$

where

$$\tau_i F(\mathbf{s}) = F(\tau_i(\mathbf{s})), \text{ with } \tau_i(\mathbf{s})_j = s_{i+j}.$$

Statement (2) can be reduced, since the states $\mu_{\beta,h}$ are invariant under lattice translations, to $\lim_{h \downarrow 0} \mu_{\beta,h}(s_0) > 0$, since, by the flip spin symmetry, we have then $\lim_{h \uparrow 0} \mu_{\beta,h}(s_0) = -\lim_{h \downarrow 0} \mu_{\beta,h}(s_0) < 0$.

This latter statement, although true for $d \geq 2$, is not so easy to prove directly. In fact, it would be easier if we could limit ourselves to studying the Hamiltonian at $h = 0$. Unfortunately, again by flip spin symmetry, $< s_0 >_{\beta,0} = 0$, $\forall \beta$.

However, there is a trick that allows us to detect phase transitions while studying only the model at $h = 0$ and that is based on the notion of *Gibbs states*.

Let us fix a configuration $\bar{\mathbf{s}} \in \Omega$ and define *the Hamiltonian in Λ with "boundary conditions"* $\bar{\mathbf{s}}_{\Lambda^c}$, namely a fixed configuration in Λ^c:

$$H(\mathbf{s}_\Lambda; \bar{\mathbf{s}}_{\Lambda^c}) = -J \sum_{<ij> \subset \Lambda} s_i s_j - J \sum_{<ij>, i \in \Lambda, j \in \Lambda^c} s_i \bar{s}_j$$

Associate to this Hamiltonian a state on Ω^Λ:

$$\mu_\Lambda(\mathbf{s}_\Lambda; \bar{\mathbf{s}}_{\Lambda^c}) = \frac{\exp(-\beta H(\mathbf{s}_\Lambda; \bar{\mathbf{s}}_{\Lambda^c}))}{Z_\Lambda(\bar{\mathbf{s}}_{\Lambda^c})}, \tag{9.2.7}$$

---

[5] This follows from Griffiths' *correlation inequalities* that prove that expectation values $< s_A >_{\Lambda,\beta,h}$ increase with Λ: $< s_A >_{\Lambda,\beta,h} \leq < s_A >_{\Lambda',\beta,h}$ if $\Lambda \subset \Lambda'$. This implies that the functions $< s_A >_{\Lambda,\beta,h}$ converge as $\Lambda \uparrow \mathbb{Z}^d$, for all finite subsets $A \subset \mathbb{Z}^d$, and from this one can deduce that the measures $\mu_{\Lambda,\beta,h}$ converge also, because linear combinations of functions of the form s_A form a dense set in $L^1(\Omega, \mu)$ or in $C(\Omega)$, in the corresponding topologies (a set E of functions is dense in $L^1(\Omega, \mu)$ if, $\forall f \in L^1(\Omega, \mu)$, $\forall \epsilon > 0$, $\exists g \in E$, with $\int_\Omega |f(x) - g(x)| d\mu(x) \leq \epsilon$. The same definition holds for dense sets in $C(\Omega)$, with the integral replaced by $\sup_{x \in \Omega} |f(x) - g(x)|$). See e.g. Griffiths [161, 162, Sect. VI], Kelley and Sherman [188], Glimm and Jaffe [148, p. 83], Friedli and Velenik [131, Sect. 3.6].

where

$$Z_\Lambda(\bar{\mathbf{s}}_{\Lambda^c}) = \sum_{s_i=\pm 1, i \in \Lambda} \exp(-\beta H(\mathbf{s}_\Lambda; \bar{\mathbf{s}}_{\Lambda^c})).$$

The state (9.2.7) is the Gibbs state in Λ with boundary conditions $\bar{\mathbf{s}}$.

Now, a Gibbs state μ on Ω is defined by a series of equations that its conditional probabilities must satisfy:

$$\mu(\mathbf{s}_\Lambda | \bar{\mathbf{s}}_{\Lambda^c}) = \frac{\exp(-\beta H(\mathbf{s}_\Lambda; \bar{\mathbf{s}}_{\Lambda^c}))}{Z_\Lambda(\bar{\mathbf{s}}_{\Lambda^c})}, \qquad (9.2.8)$$

where $\mu(\mathbf{s}_\Lambda | \bar{\mathbf{s}}_{\Lambda^c})$ is the conditional probability , for the distribution μ on $\Omega = \{-1, +1\}^{\mathbb{Z}^d}$, of the configuration $\mathbf{s}_\Lambda$ in Ω^Λ, given the configuration $\bar{\mathbf{s}}_{\Lambda^c}$ in Ω^{Λ^c}. To make this well defined, consider Λ as a subset of a larger box $\tilde{\Lambda}$, identify Λ^c with the finite set $\tilde{\Lambda} \setminus \Lambda$ and write (9.2.8) with that convention and the definition of conditional probabilities (2.2.6). Since all the interaction terms within $\tilde{\Lambda} \setminus \Lambda$ cancel between the numerator and the denominator in the right hand side of (9.2.8), we see that the dependence on $\bar{\mathbf{s}}_{\Lambda^c}$ is only through the finite set $\bar{\mathbf{s}}_{\partial\Lambda}$, where $\partial\Lambda = \{j \in \Lambda^c, \exists i \in \Lambda, |i - j| = 1\}$, and we can then take the $\tilde{\Lambda} \uparrow \mathbb{Z}^d$ limit to get (9.2.8).

The (9.2.8) are called the Dobrushin–Lanford–Ruelle equations, after their inventors [103, 204]. They provide a very natural characterization of equilibrium or Gibbs states without making any reference to external parameters such as the magnetic field in (9.2.4). There is a nice mathematical theory of those Gibbs states, which extends to a larger class of interactions, for example:

$$H = -\sum_{i,j} J_{ij} s_i s_j \qquad (9.2.9)$$

where $J_{ij} \geq 0$ if we want ferromagnetic interactions and the sum is restricted as above to finite sets Λ. For the model to be well defined as $\Lambda \uparrow \mathbb{Z}^d$, the interaction energy between one spin and all the others has to be bounded: $\forall i \in \mathbb{Z}^d$,

$$\sum_{j \in \mathbb{Z}^d} |J_{ij}| < \infty \qquad (9.2.10)$$

and one usually considers interactions that are translation invariant and also invariant under lattice rotations, namely of the form: $J_{ij} = J(|i - j|)$, for a function $J : \mathbb{N} \to \mathbb{R}_+$. Then, the condition (9.2.10) becomes $\sum_{n \in \mathbb{N}} |J(n)| < \infty$.

One may also include many-body interactions (meaning more than two body). Define

$$H = -\sum_A J_A s_A, \qquad (9.2.11)$$

with $s_A = \prod_{i \in A} s_i$ and where $J_A \geq 0$ if we want ferromagnetic interactions. Again, for the model to be well defined, as $\Lambda \uparrow \mathbb{Z}^d$, the interaction between one spin and all the others has to be bounded:

$$\sum_{0 \in A} |J_A| < \infty. \tag{9.2.12}$$

And we will consider here only translation invariant interactions: $J_A = J_{A+i}, \forall i \in \mathbb{Z}^d$.

We will only need here certain consequences of the theory of Gibbs states:

(1) The set of Gibbs states is the closed convex hull of all the limits $\lim_{\Lambda \uparrow \mathbb{Z}^d} \mu(s_\Lambda; \bar{s}_{\Lambda^c})$, for all $\bar{s} \in \Omega$. Concretely, it means that one takes all the states μ obtained as limits of the form: $\lim_{n \to \infty} \mu(s_{\Lambda_n}; \bar{s}_{\Lambda_n^c})$, for some sequence $\Lambda_n \uparrow \mathbb{Z}^d, \bar{s} \in \Omega$, then take all convex combinations of those limiting states and, if necessary, also take all possible limits of sequences of the states thus obtained. This may sound complicated, but, in many examples, we will need only to consider limits for some special $\bar{s} \in \Omega$, as we will see below.

(2) The set of translation invariant (see (9.2.6)) Gibbs states[6] for Hamiltonians of the form (9.2.11) is also the set of limits $\lim_{\lambda \to 0} \mu_\lambda$, where μ_λ is a translation invariant Gibbs state for a Hamiltonian of the form

$$H_\lambda = H + \lambda \tilde{H}$$

where $\tilde{H}$ is also of the form (9.2.11). This means that any state obtained as limits of Gibbs states for Hamiltonians of the form (9.2.4) when $h \to 0$ are also Gibbs states obtained by using boundary conditions as in (1) above.

(3) The set of Gibbs states is a (Choquet)[7] *simplex*, which means that it is a convex set (that is true by construction) but such that every Gibbs state can be written *uniquely* as a combination of extreme points of that set (i.e. of points that are not convex combinations of other points of the set).[8] If there are finitely many or countably many extreme points $\mu_1, \mu_2, \ldots$, then, for any Gibbs state μ there are positive numbers $a_1, a_2, \ldots$ with $\mu = \sum_{\gamma=1}^{\infty} a_\gamma \mu_\gamma$. In case the set of extreme Gibbs states in continuous, the sum here is replaced by an integral. The set of translation invariant Gibbs states is also a Choquet simplex. See e.g. Friedli and Velenik [131, p. 291] for a proof.

4 Extreme translation invariant Gibbs states satisfy a law of large numbers: if μ is such a state then, for any $F \in L^1(\Omega, \mu)$, and for μ-almost all s,

$$\lim_{\Lambda \uparrow \mathbb{Z}^d} \frac{\sum_{i \in \Lambda} F(\tau_i(s))}{|\Lambda|} = < F >_\mu \tag{9.2.13}$$

---

[6] See Sect. 9.7.6 for a discussion of non translation Gibbs states.

[7] The French mathematician Gustave Choquet extended the notion and theory of simplexes to infinite dimensional spaces.

[8] For example, an interval in one dimension, a triangle in two dimensions, a tetrahedron in three dimensions etc.

with $\tau_i(\mathbf{s})_j = s_{i+j}$.

This is analogous to the law of large numbers (2.3.9), but with the sum over $n = 1, \ldots, N$ replaced by a sum over $i \in \Lambda$. It can also be seen as a generalized form of ergodicity, see (4.3.8), with T^n, $n \in \mathbb{N}$ replaced by the maps τ_i, $i \in \mathbb{Z}^d$. Extreme translation invariant Gibbs states are often referred to as "pure phases" or phases of the model.

As we explained in footnote 9.5, instead of considering all functions $F \in L^1(\Omega, \mu)$, it is often enough to consider the set of correlation functions $\mu(s_A) = < s_A >$, for all functions in the set $\{ s_A = \prod_{i \in A} s_i \mid A \subset \mathbb{Z}^d, |A| < \infty \}$. The following formula, with $< \cdot >$ being the expectation value with respect to a finite volume Gibbs state with a Hamiltonian of the form (9.2.11), is very useful:

$$\frac{\partial < s_A >}{\partial \beta J_B} = < s_A s_B > - < s_A >< s_B > . \tag{9.2.14}$$

The proof of (9.2.14) is left as an exercise.

It may be useful to explain the connection between the theory of Gibbs states and the general theory of equilibrium explained in Chap. 6. When there are several extreme Gibbs states (and there might be infinitely many of them, as we will see in Sects. 9.7.5 and 9.7.6), it means that the "large" set Ω_{eq} in Fig. 6.2 is divided into as many subsets as there are extreme Gibbs states, and, in each such subset, the macroscopic variables take values determined by that state.[9] For extreme translation invariant Gibbs states, formula (9.2.13) illustrates this: for a given such state μ, the macroscopic variables given by the left hand side of (9.2.13) take a value determined by that state, almost everywhere with respect to it. Since there is no dynamics in these models, the issue of convergence to equilibrium cannot be formulated, but, if one considers more realistic models for which a dynamics exists, one expects convergence from a non equilibrium initial condition of the system to a subset of Ω_0 of measure one with respect to a given Gibbs state, the latter being determined by those initial non equilibrium conditions.[10]

For more information on the theory of Gibbs states, see e.g. Georgi [145], Friedli and Velenik [131], Malyshev and Minlos [235], Ruelle [282], Simon [292], van Enter, Fernandez and Sokal [318].

---

[9] From that point of view, non extremal states, namely states that are convex combinations of extremal ones are not physical although they are useful mathematically.

[10] David Wallace thinks that this non-uniqueness of the equilibrium state is a problem for the Botzmannian approach, see [326, Sect. 5], because it does not explain why one has the same probability to arrive at any given Gibbs state. But that can be answered, again, by a typicality argument, see [155, Sect. 7].

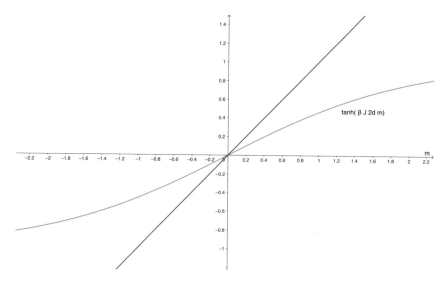

Fig. 9.3 Graphical solution of equation (9.3.1) at $h = 0$ for $\beta J2d < 1$

9.3 Mean Field Theory

When confronted with apparently complicated models such as the Ising model, it is tempting to try to "guess" the behavior of the model by making some approximations. The simplest such approximation is the mean field one and it is worth considering it both because it is a method often used in other contexts and because here we can analyze the model in more detail and thus see the limits of that approximation. Let us compute $m = <s_0>$ by replacing the variables s_i, with $|i| = 1$, namely the nearest neighbors of the site 0, by their average value m. Since the interactions between s_0 and the rest of the spin variables goes only through the interaction between s_0 and the variables s_i, with $|i| = 1$, this replacement effectively reduces the computation of m to:

$$m(\beta, h) = \frac{\sum_{s_0 = \pm 1} s_0 \exp(\beta J 2 d m(\beta, h) s_0 + \beta h s_0)}{z}$$

where $z = \sum_{s_0 = \pm 1} \exp(\beta J 2 d m(\beta, h) s_0 + \beta h s_0) = 2 \cosh(\beta J 2 d m(\beta, h) + \beta h)$ and $2d$ is the number of nearest neighbors of a given site in d dimensions. Doing the sums, we get:

$$m(\beta, h) = \tanh(\beta J 2 d m(\beta, h) + \beta h) \tag{9.3.1}$$

which is an implicit equation for $m(\beta, h)$.

It is easy to see from the graphs in Figs. 9.3, 9.4 and 9.5 that:

1. For $\beta J 2 d \leq 1$, $\lim_{h \to 0} m(\beta, h) = 0$, with $\frac{dm(\beta,h)}{dh}|_{h=0} < \infty$ when $\beta J 2 d < 1$ and $\frac{dm(\beta,h)}{dh}|_{h=0} = \infty$ when $\beta J 2 d = 1$, because we get from (9.3.1) at $h = 0$ and

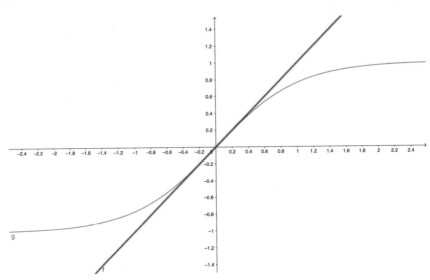

Fig. 9.4 Graphical solution of equation (9.3.1) at $h = 0$ for $\beta J 2d = 1$

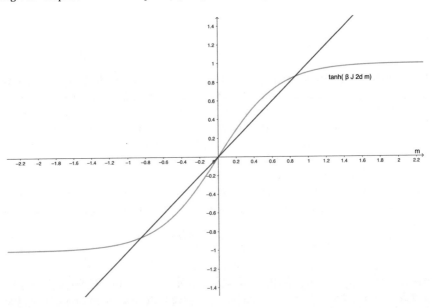

Fig. 9.5 Graphical solution of equation (9.3.1) at $h = 0$ for $\beta J 2d > 1$. By analyzing the right hand side of (9.3.1) for $h \neq 0$, one sees that the solution of (9.3.1) for $h \neq 0$ converges as $h \to 0$ to the non zero solutions of (9.3.1) at $h = 0$

$\beta J2d < 1$, $\frac{dm(\beta,h)}{dh}(1 - \beta J2d) = \beta$, which implies that $\frac{dm(\beta,h)}{dh}$ tends to ∞ as $\beta J2d \to 1$.

2. For $\beta J2d > 1$, $\lim_{h\downarrow 0} m(\beta, h) = -\lim_{h\uparrow 0} m(\beta, h) \equiv m_0(\beta) > 0$.

Thus this approximation predicts a ferromagnetic phase transition *in any dimension*. We will see in the next sections to what extent this prediction is correct (see also Sect. 9.9.2 for other qualities and defects of this approximation).

Another way to derive the mean field aproximation is to consider an interaction of the form (9.2.9), but with $J_{ij} = \frac{2dJ}{|\Lambda|}$, $\forall i, j \in \Lambda$ (all spins interact equally with each other), so the Hamiltonian is:

$$H_\Lambda(s_\Lambda) = -\frac{2dJ}{|\Lambda|} \sum_{i,j,\in\Lambda,i\neq j} s_i s_j - h \sum_{i\in\Lambda} s_i = -\frac{2dJ}{2|\Lambda|}\left(\left(\sum_{i\in\Lambda} s_i\right)^2 - |\Lambda|\right) - h\sum_{i\in\Lambda} s_i,$$

where, in the second equality, we use $s_i^2 = 1$ for the diagonal terms in $(\sum_{i\in\Lambda} s_i)^2$. Then, the sums over s_i, $i \in \Lambda$ in $Z_\Lambda = \sum_{s_i,i\in\Lambda} \exp(-\beta H_\Lambda(s_\Lambda))$ are easy to perform, using the identity

$$\exp\left(\frac{a^2}{2}\right) - \frac{1}{\sqrt{2\pi}}\int_{\mathbb{R}} \exp\left(-\frac{x^2}{2} + ax\right) dx,$$

which follows by completing the square: $-\frac{x^2}{2} + ax = -\frac{(x-a)^2}{2} + \frac{a^2}{2}$ and using the identity:

$$\frac{1}{\sqrt{2\pi}}\int_{\mathbb{R}} \exp\left(-\frac{y^2}{2}\right) dy = 1,$$

with $y = x - a$. Indeed, letting $a = \left(\frac{\beta 2dJ}{|\Lambda|}\right)^{\frac{1}{2}} (\sum_{i\in\Lambda} s_i)$, we get:

$$Z_\Lambda = \sum_{s_i,i\in\Lambda} \exp\left(-\beta H_\Lambda(s_\Lambda)\right)$$

$$= \exp\left(\frac{\beta 2dJ}{2}\right) \sum_{s_i,i\in\Lambda} \frac{1}{\sqrt{2\pi}}\int_{\mathbb{R}} \exp\left(-\frac{x^2}{2} + \left(\left(\frac{\beta 2dJ}{|\Lambda|}\right)^{\frac{1}{2}} x + \beta h\right)\sum_{i\in\Lambda} s_i\right) dx$$

Since

$$\exp\left(\left(\left(\frac{\beta 2dJ}{|\Lambda|}\right)^{\frac{1}{2}} x + \beta h\right)\sum_{i\in\Lambda} s_i\right) = \prod_{i\in\Lambda} \exp\left(\left(\left(\frac{\beta 2dJ}{|\Lambda|}\right)^{\frac{1}{2}} x + \beta h\right) s_i\right),$$

we can perform the sum over each s_i and obtain:

$$Z_\Lambda = \frac{\exp\left(\frac{\beta 2dJ}{2}\right)}{\sqrt{2\pi}} \int_\mathbb{R} \exp\left(-\frac{x^2}{2}\right)\left[2\cosh\left(\left(\frac{\beta 2dJ}{|\Lambda|}\right)^{\frac{1}{2}} x + \beta h\right)\right]^{|\Lambda|} dx$$

Changing variables $m = \left(\frac{1}{\beta 2dJ|\Lambda|}\right)^{\frac{1}{2}} x$, we get:

$$Z_\Lambda = \left(\frac{\beta 2dJ|\Lambda|}{2\pi}\right)^{\frac{1}{2}} \exp\left(\frac{\beta 2dJ}{2}\right) \int_\mathbb{R} \exp(|\Lambda|h(m))dm, \qquad (9.3.2)$$

with $h(m) = -\frac{\beta 2dJm^2}{2} + \ln(2\cosh(\beta 2dJm + \beta h))$. From Laplace's method (see Appendix 6.A.1), we get:

$$f(\beta, h) = -kT \lim_{|\Lambda|\to\infty} \frac{\ln Z_\Lambda}{|\Lambda|} = -kT \sup h(m),$$

and differentiating with respect to m we get that the sup is reached for $m = m(\beta, h)$ a solution of (9.3.1). For that m, we get

$$f(\beta, h) = -kT \lim_{|\Lambda|\to\infty} \frac{\ln Z_\Lambda}{|\Lambda|} = \frac{2dJm^2(\beta, h)}{2} - kT \ln(2\cosh(\beta 2dJm(\beta, h) + \beta h))$$

$$(9.3.3)$$

where $m(\beta, h)$ is the solution of (9.3.1) (equal to 0 for $h = 0$ and $\beta J2d \le 1$).

The fact that one recovers the mean field approximation by setting $J_{ij} = \frac{2dJ}{|\Lambda|}$, $\forall i, j \in \Lambda$ suggests that one might recover the same approximation by taking long range interactions and letting the range of the interaction go to infinity while letting the strength of these interactions go to 0. This can be shown, but is more difficult than the previous calculation. More precisely, one considers Kac interactions; let $J(x) : \mathbb{R}^d \to \mathbb{R}^+$ but such that $\int_\mathbb{R} J(x)dx = J$ and define $J_{ij}^\gamma = \gamma^d J(\gamma|i - j|)$. Then $\lim_{\gamma\to 0} \sum_{j\in\mathbb{Z}^d} J_{ij}^\gamma = J$ (to see that, write the sum over $j \in \mathbb{Z}^d$ as a Riemann sum over $j \in (\gamma\mathbb{Z})^d$).

One can show that, in the *van der Waals limit* $\gamma \to 0$, both the free energy of the model and its correlation functions converge to the corresponding quantities in the mean field model, see Lebowitz and Penrose [216], Thompson [308, Chap. 4], Presutti [269], Friedli and Velenik [131, Chap. 4].

For proofs of phase transitions in this model for γ small but not zero, see Cassandro and Presutti [71], Bovier and Zahradník [44], Lebowitz, Mazel and Presutti [222, 269].

9.4   One Dimension

The one dimensional Ising model can be solved analytically for all values of β and h, see (9.4.20 below). But we will be concerned here only with the issue of phase transitions at $h = 0$. It will be convenient to use the elementary identity (which will be used also in Sect. 9.5)

$$\exp(\pm x) = \cosh(x) \pm \sinh(x) = \cosh(x)(1 \pm \tanh(x)), \qquad (9.4.1)$$

to rewrite:

$$\exp(-\beta H(\mathbf{s}_\Lambda; \bar{\mathbf{s}}_{\Lambda^c})) = \prod_{<ij> \in \Lambda} \exp(\beta J s_i s_j) \prod_{<ij>, i \in \Lambda, j \in \Lambda^c} \exp(\beta J s_i \bar{s}_j)$$

$$= \cosh(\beta J)^{B(\Lambda)} \prod_{<ij> \in \Lambda} (1 + s_i s_j \tanh(\beta J)) \prod_{<ij>, i \in \Lambda, j \in \Lambda^c} (1 + s_i \bar{s}_j \tanh(\beta J)) \quad (9.4.2)$$

where $\Lambda = [-L, L] \cap \mathbb{Z}$, and $B(\Lambda)$ is the set of nearest neighbor pairs $< ij >$ in the products. Expanding the products, we get:

$$\exp(-\beta H(\mathbf{s}_\Lambda; \bar{\mathbf{s}}_{\Lambda^c})) = \cosh(\beta J)^{|B(\Lambda)|} \sum_{G \subset B(\Lambda)} \tanh(\beta J)^{|G|} \prod_{<ij> \in G \cap \Lambda} s_i s_j \prod_{<ij> \in G; j \in \partial \Lambda} s_i \bar{s}_j,$$
$$(9.4.3)$$

where the sum runs over all subsets G of $B(\Lambda)$, and $\partial \Lambda = \{j \in \Lambda^c, \text{ with } \exists i \in \Lambda, |i - j| = 1\}$. We can rewrite

$$\prod_{<ij> \in G \cap \Lambda} s_i s_j \prod_{<ij> \in G \cap \partial \Lambda} s_i \bar{s}_j = \prod_{i \in \Lambda} s_i^{n_i(G)} \prod_{j \in \partial \Lambda} \bar{s}_j. \qquad (9.4.4)$$

where $n_i(G) = |\{i | \exists j, \text{ with } < ij > \in G\}|$. Now comes the basic observation:

1. If $n_i(G)$ is odd,

$$\sum_{s_i = \pm 1} s_i^{n_i(G)} = 0. \qquad (9.4.5)$$

2. If $n_i(G)$ is even (including 0),

$$\sum_{s_i = \pm 1} s_i^{n_i(G)} = 2. \qquad (9.4.6)$$

We define ∂G, the boundary of G as:

$$\partial G = \{i | \exists j, \text{ with } < ij > \in G, \text{ and } n_i(G) \text{ odd}\}. \qquad (9.4.7)$$

To see the consequences of this observation, we shall proceed slowly. Consider first the partition function of the one dimensional Ising model without a boundary condition term:

$$Z(\Lambda, \beta) = \sum_{s_i = \pm 1, i \in \Lambda} \exp(-\beta J \sum_{<ij> \subset \Lambda} s_i s_j) \qquad (9.4.8)$$

Using (9.4.1)–(9.4.4) we obtain:

$$Z(\Lambda, \beta) = \cosh(\beta J)^{|B(\Lambda)|} \sum_{s_i = \pm 1, i \in \Lambda} \sum_{G \subset B(\Lambda)} \tanh(\beta J)^{|G|} \prod_{i \in \Lambda} s_i^{n_i(G)} \qquad (9.4.9)$$

Now, since we are in one dimension, it is easy to see that any G in (9.4.9) is a union of intervals and that, on the boundaries of those intervals, $n_i(G) = 1$, namely is odd and by (9.4.5) the sum over $s_i = \pm 1$ vanishes. Thus, all the sums $\sum_{s_i = \pm 1, i \in \Lambda}$ in (9.4.9) vanish, except for $G = \emptyset$, for which, by (9.4.6), the corresponding sum equals $2^{|\Lambda|} = 2^{2L+1}$, since $\Lambda = [-L, L] \cap \mathbb{Z}$. And, for that Λ, $|B(\Lambda)| = 2L$; so, we obtain, in the thermodynamic limit $L \to \infty$,

$$f(\beta) = -kT \lim_{L \to \infty} \frac{1}{2L + 1} \ln Z(\Lambda, \beta)$$

$$= -kT \lim_{L \to \infty} \frac{1}{2L + 1} \ln(2^{2L+1} \cosh(\beta J)^{2L}) = -kT \ln(2 \cosh(\beta J)) \qquad (9.4.10)$$

The next step is to compute the partition function with boundary conditions:

$$Z(\Lambda, \beta; \bar{s}_{-L-1}, \bar{s}_{L+1})$$

$$= \sum_{s_i = \pm 1, i \in \Lambda} \exp\left(-\beta J \left(\sum_{<ij> \subset \Lambda} s_i s_j + s_{-L} \bar{s}_{-L-1} + s_L \bar{s}_{L+1} \right) \right)$$

Proceeding as in the derivation of (9.4.9), we get:

$$Z(\Lambda, \beta; \bar{s}_{-L-1}, \bar{s}_{L+1})$$

$$= \cosh(\beta J)^{|B(\Lambda)|} \sum_{s_i = \pm 1, i \in \Lambda} \sum_{G \subset B(\Lambda)} \tanh(\beta J)^{|G|} \prod_{i \in \Lambda} s_i^{n_i(G)} \prod_{j=-L-1, L+1} \bar{s}_j^{n_j(G)}.$$

$$(9.4.11)$$

Again, if G contains an interval strictly inside Λ, its contribution will vanish when we sum over $s_i = \pm 1$, $i \in \Lambda$, for the same reason as before. We are left with the $G = \emptyset$ term, as before, and a term where G includes all the pairs $<ij> \in B(\Lambda)$: $< -L - 1, -L >$, $< -L, -L + 1 >$, $\cdots < L, L + 1 >$. For that term, all $n_i(G) = 2$ for $i \in \Lambda$. Thus, its contribution, after summing over $s_i = \pm 1$, $i \in \Lambda$, is $2^{|\Lambda|} \tanh(\beta J)^{|B(\Lambda)|} \alpha = 2^{2L+1} \tanh(\beta J)^{2L+2} \alpha$, where $\alpha = \prod_{j=-L-1, L+1} \bar{s}_j$ ($n_j(G) = 1$ for $j = -L - 1, L + 1$ for that graph) and $2L + 2$ comes from the inclusion of the boundary terms. So:

$$Z(\Lambda, \beta; \bar{s}_{-L-1}, \bar{s}_{L+1}) = 2^{2L+1} \cosh(\beta J)^{2L+2} (1 + \tanh(\beta J)^{2L+2} \alpha) \qquad (9.4.12)$$

and since $\tanh(\beta J) < 1$ for $\beta < \infty$, we get $\tanh(\beta J)^{2L+2} \to 0$ as $L \to \infty$, and

$$f(\beta) = -kT \lim_{L \to \infty} \frac{1}{2L+1} \ln Z(\Lambda, \beta; \bar{s}_{-L-1}, \bar{s}_{L+1}) \tag{9.4.13}$$

$$= -kT \lim_{L \to \infty} \frac{1}{2L+1} \ln[2^{2L+1} \cosh(\beta J)^{2L+2}(1 + \tanh(\beta J)^{2L+2}\alpha)]$$

$$= -kT \ln(2 \cosh(\beta J))$$

As an aside, let us remark that the fact that the free energy is independent of the boundary conditions in the thermodynamic limit, as in (9.4.14) is a very general fact. Indeed, it is easy to show that, for the Hamiltonian (9.2.4) and for any pair of boundary conditions $\bar{s}^1, \bar{s}^2$, one has

$$|H(\mathbf{s}_\Lambda; \bar{\mathbf{s}}^1_{\Lambda^c}) - H(\mathbf{s}_\Lambda; \bar{\mathbf{s}}^2_{\Lambda^c})| \leq C(|\partial \Lambda|), \tag{9.4.14}$$

for some constant $0 < C < \infty$, because the only terms in the left hand side of (9.4.14) that differ between $H(\mathbf{s}_\Lambda; \bar{\mathbf{s}}^1_{\Lambda^c})$ and $H(\mathbf{s}_\Lambda; \bar{\mathbf{s}}^2_{\Lambda^c})$, are those coming from the nearest neighbor pairs $< ij >$ with $i \in \partial \Lambda$.

From (9.4.14), we get:

$$\exp(-C|\partial \Lambda|)) \leq \frac{Z_\Lambda(\bar{\mathbf{s}}^1_{\Lambda^c})}{Z_\Lambda(\bar{\mathbf{s}}^2_{\Lambda^c})} \leq \exp(C|\partial \Lambda|)). \tag{9.4.15}$$

Since, for a box of size L, $|\Lambda| \approx L^d$ and $|\partial \Lambda| \approx L^{d-1}$, taking the logarithm of the ratio in 9.4.15), dividing by $|\Lambda|$ and letting $|\Lambda| \to \mathbb{Z}^d$ shows that the free energy is independent of the boundary conditions in the thermodynamic limit.

The reader is invited to extend this result to Hamiltonians of the form (9.2.11), with translation invariant interactions.

Let us now compute the correlation functions, starting with $< s_0 >_{\Lambda, \beta, \bar{s}_{-L-1}, \bar{s}_{L+1}}$. Proceeding as in the derivation of (9.4.11), we get:

$$< s_0 >_{\Lambda, \beta, \bar{s}_{-L-1}, \bar{s}_{L+1}} =$$

$$\frac{\cosh(\beta J)^{|B(\Lambda)|} \sum_{s_i = \pm 1, i \in \Lambda} \sum_{G \subset B(\Lambda)} \tanh(\beta J)^{|G|} s_0 \prod_{i \in \Lambda} s_i^{n_i(G)} \prod_{j=-L-1, L+1} \bar{s}_j^{n_j(G)}}{Z(\Lambda, \beta; \bar{s}_{-L-1}, \bar{s}_{L+1})}.$$

$$\tag{9.4.16}$$

If we examine the graphs in the numerator that may yield a non zero contribution, we see that ∂G (defined in (9.4.7)) must include 0, otherwise the sum over $s_0 = \pm 1$ would vanish. It must also include at least one of the endpoints $-L - 1, L + 1$; but it cannot include both of these endpoints and 0, otherwise we would have a sum over $s_0 = \pm 1$ of s_0^3, which vanishes also. So, we are left with two terms: one where $G = \{< -L - 1, -L >, \cdots < -1, 0 >\}$ and the other with $G = \{< 0, 1 >, \cdots < L, L + 1 >\}$. After summing over $s_i = \pm 1, i \in \Lambda$, both terms will equal in absolute value $2^{|\Lambda|} \tanh(\beta J)^{L+1} = 2^{2L+1} \tanh(\beta J)^{L+1}$. Using (9.4.12) for the denominator in (9.4.16), we get:

$$| < s_0 >_{\Lambda,\beta,\bar{s}_{-L-1},\bar{s}_{L+1}} | \leq \frac{2(\tanh(\beta J))^{L+1}}{1 + \tanh(\beta J)^{2L+2}\alpha},$$

since $|\alpha| = |\prod_{j=-L-1,L+1} \bar{s}_j^{n_j(G)}| \leq 1$ and the factor $2^{2L+1} \cosh(\beta J)^{|B(\Lambda)|}$ appears both in the numerator and the denominator.

Therefore,

$$| < s_0 >_{\Lambda,\beta,\bar{s}_{-L-1},\bar{s}_{L+1}} | \to 0$$

independently of the boundary conditions $\forall \beta < \infty$.

The reader is invited to extend this argument to all correlation functions: we get, given $i_1, \ldots, i_m$,

1. If m is odd, for all $\bar{s} \in \Omega$,

$$\lim_{L \to \infty} < \prod_{l=1}^{m} s_{i_l} >_{\Lambda,\beta,\bar{s}_{-L-1},\bar{s}_{L+1}} = 0.$$

2. If m is even, for all $\bar{s} \in \Omega$,

$$\lim_{L \to \infty} < \prod_{l=1}^{m} s_{i_l} >_{\Lambda,\beta,\bar{s}_{-L-1},\bar{s}_{L+1}} = \prod_{l=1}^{m/2} \tanh(\beta J)^{|i_{2l} - i_{2l} - 1|} \tag{9.4.17}$$

As an exercise, the reader can initiate herself to the transfer matrix formalism. Consider the Hamiltonian in one dimension with nearest neighbor interactions (of the form (9.2.4)) and what are called *periodic boundary conditions*:

$$H(s_\Lambda) = -J \sum_{i=-L,\ldots,L} s_i s_{i+1} - h \sum_{i=-L,\ldots,L} s_i, \tag{9.4.18}$$

where we set $s_{L+1} = s_{-L}$, namely we consider the model to be defined on a circle (this is why one speaks of periodic boundary conditions).

Now, consider the 2×2 matrix T:

$$T = \begin{pmatrix} \exp(\beta J + \beta h) & \exp(-\beta J) \\ \exp(-\beta J) & \exp(\beta J - \beta h) \end{pmatrix}, \tag{9.4.19}$$

and check that the partition function (9.2.3) corresponding to the Hamiltonian (9.4.18) equals:

$$Z_\Lambda = \text{Trace } T^{2L+2} = \lambda_1(\beta, h)^{2L+2} + \lambda_2(\beta, h)^{2L+2},$$

where $\lambda_1(\beta, h), \lambda_2(\beta, h)$ (with, by convention, $\lambda_1 > \lambda_2$), are the eigenvalues of T; they are analytic in β, h (exercise: compute those eigenvalues and check that). From this, we get, with $|\Lambda| = 2L + 1$:

$$f(\beta, h) = -kT \lim_{\Lambda \to \mathbb{Z}} \frac{\ln Z_\Lambda}{|\Lambda|} = -kT \ln \lambda_1(\beta, h). \qquad (9.4.20)$$

Further exercises include: write the partition function with other boundary conditions in term of the transfer matrix, do the same for the correlation functions and compute their thermodynamic limit.

Although in more than one dimension we do not have solutions for the correlation functions as explicit as (9.4.17), one can analyze qualitatively the Ising model at high and low temperatures.

9.5 High Temperature Expansions

We will first use formulas (9.4.8)–(9.4.9) to analyze the partition function without boundary conditions in any dimension. After summing over $s_i = \pm 1, i \in \Lambda$, and using (9.4.5)–(9.4.6), we get:

$$Z(\Lambda, \beta) = \cosh(\beta J)^{|B(\Lambda)|} 2^{|\Lambda|} \sum_{G \subset B(\Lambda), \partial G = \emptyset} \tanh(\beta J)^{|G|}, \qquad (9.5.1)$$

since any graph G with $\partial G \neq \emptyset$ will vanish after summing over $s_i = \pm 1$, for $i \in \partial G$, because of (9.4.5). So, the partition function without boundary conditions is expressed as a sum over closed graphs, i.e. with $\partial G = \emptyset$.

For the partition function with boundary conditions, we get from (9.4.11):

$$Z(\Lambda, \beta; \bar{s}_{\Lambda^c}) = \cosh(\beta J)^{|B(\Lambda)|} 2^{|\Lambda|} \sum_{G \subset B(\Lambda), \partial G \subset \Lambda} \tanh(\beta J)^{|G|} \prod_{j \in \partial G} \bar{s}_j^{n_j(G)}, \qquad (9.5.2)$$

since again graphs G with $\partial G \not\subset \partial \Lambda$ will vanish after summing over $s_i = \pm 1$, for $i \in \partial G \setminus \partial \Lambda$. If $\partial G \subset \partial \Lambda$, since there is no sum over $\bar{s}_j, j \in \partial \Lambda, \bar{s}_j$ being fixed, the corresponding term in the sum does not vanish.

To simplify the notation, we will consider now only $+$ boundary conditions: $\bar{s}_j = +1, \forall j \in \Lambda^c$, so that $\prod_{j \in \partial G} \bar{s}_j^{n_j(G)} = 1$.

If we want to compute $< s_0 >_{\Lambda, \beta; \bar{s}_{\Lambda^c}}$, we get the analogue of (9.4.16):

$$< s_0 >_{\Lambda, \beta; \bar{s}_{\Lambda^c}} = \frac{\cosh(\beta J)^{|B(\Lambda)|} \sum_{s_i = \pm 1, i \in \Lambda} \sum_{G \subset B(\Lambda)} \tanh(\beta J)^{|G|} s_0 \prod_{i \in \Lambda} s_i^{n_i(G)}}{Z(\Lambda, \beta; \bar{s}_{\Lambda^c})}. \qquad (9.5.3)$$

Again, the sum over $s_i = \pm 1, i \in \Lambda$ makes the terms where $\partial G \not\subset \partial \Lambda \cup 0$ vanish and we get, using (9.5.2):

$$| < s_0 >_{\Lambda, \beta; \bar{s}_{\Lambda^c}} | \leq \frac{\sum_{G \subset B(\Lambda), \partial G \subset \partial \Lambda \cup 0} \tanh(\beta J)^{|G|}}{\sum_{G \subset B(\Lambda), \partial G \subset \partial \Lambda} \tanh(\beta J)^{|G|}} \qquad (9.5.4)$$

So far we just wrote down identities in order to derive (9.5.4). Now, we will use the representation (9.5.4) to obtain estimates on $< s_0 >_{\Lambda,\beta;\bar{s}_{\Lambda^c}}$. Observe that, if $\partial G \subset \partial \Lambda \cup 0$, there must exist a path $P \subset G$, made of nearest neighbor pairs, with $P = < 0\, i_i >, < i_1\, i_2 >, \cdots < i_{k-1}\, i_k >$ and $i_k \in \partial \Lambda$. We denote such paths as $P : 0 \to \partial \Lambda$. Indeed, if there was no such path, the part of ∂G containing 0 would have another boundary point not in $\partial \Lambda$ and its contribution to the numerator of (9.5.4) would vanish. So, "choose" such a path $P : 0 \to \partial \Lambda$ for each G with $\partial G \subset \partial \Lambda \cup 0$; we can then write, for each G in the numerator of (9.5.4) $G = G' \cup P$ where $\partial G' \subset \partial \Lambda$ and $G' \cap P = \emptyset$. Let us write (9.5.4) as:

$$| < s_0 >_{\Lambda,\beta;\bar{s}_{\Lambda^c}} | \leq \frac{\sum_{P:0\to\partial\Lambda} \sum_{G'\subset B(\Lambda),\partial G'\subset\partial\Lambda,G'\cap P=\emptyset} \tanh(\beta J)^{|P|} \tanh(\beta J)^{|G'|}}{\sum_{G\subset B(\Lambda),\partial G\subset\partial\Lambda} \tanh(\beta J)^{|G|}}$$

$$(9.5.5)$$

For every P in the numerator of (9.5.5), we have

$$\sum_{G'\subset B(\Lambda),\partial G'\subset\partial\Lambda,G'\cap P=\emptyset} \tanh(\beta J)^{|G'|} \leq \sum_{G\subset B(\Lambda),\partial G\subset\partial\Lambda} \tanh(\beta J)^{|G|}, \qquad (9.5.6)$$

since the constraint $G' \cap P = \emptyset$ is imposed on the summands of the left hand side but not of the right hand side and all the terms are positive. Since the right hand side of (9.5.6) equals the denominator of (9.5.5), we bound (9.5.4) by:

$$| < s_0 >_{\Lambda,\beta;\bar{s}_{\Lambda^c}} | \leq \sum_{P:0\to\partial\Lambda} \tanh(\beta J)^{|P|} \qquad (9.5.7)$$

It is easy to show that the upper bound in (9.5.7) tends to 0 as $\Lambda \uparrow \mathbb{Z}^d$ for β small. First, let us count the number of paths P of length $L = |P|$ starting from a given point 0: since each point has $2d$ nearest neighbors in d dimensions, at each step along the path one has $2d$ "choices" of where to take the next step; actually, that number is only $2d - 1$ for all steps except the first one since there is no "turning back" of the path. So, we get

$$|\{\text{paths } P, |P| = L\}| \leq 2d(2d - 1)^{L-1}$$

Now, let $L_0 = \text{dist}(0, \partial\Lambda)$; we get an upper bound on the right hand side of (9.5.7):

$$\sum_{P:0\to\partial\Lambda} \tanh(\beta J)^{|P|}$$

$$\leq \frac{2d}{2d - 1} \sum_{L\geq L_0} ((2d - 1) \tanh(\beta J))^L \leq C((2d - 1) \tanh(\beta J))^{L_0}$$

for β small so that $(2d - 1) \tanh(\beta J) < 1$ with $C = \frac{2d}{2d-1} \frac{1}{1-(2d-1)\tanh(\beta J)}$. Inserting this in (9.5.7), we get:

$$| < s_0 >_{\Lambda,\beta;\bar{s}_{\Lambda^c}} | \leq C((2d - 1)\tanh(\beta J))^{L_0} \qquad (9.5.8)$$

which tends to 0 as $L_0 \to \infty$ or $\Lambda \uparrow \mathbb{Z}^d$.

The bound $(2d - 1)\tanh(\beta J) < 1$ is approximately equal to the value obtained in the mean field model. This actually gives the correct asymptotics ($\beta_c \approx \frac{1}{d}$) as $d \to \infty$ for the value of β for which a phase transition occurs, see Driessler, Landau and Perez [110] or Lebowitz and Mazel [220].

It is left as an exercise for the reader to prove a similar bound on $| < s_A >_{\Lambda,\beta;\bar{s}_{\Lambda^c}} |$, where $s_A = \prod_{i \in A} s_i$, for any $A \subset \mathbb{Z}^d$, $|A| < \infty$, and $|A|$ odd. One can also prove that $\lim_{\Lambda \to \mathbb{Z}^d} < s_A >_{\Lambda,\beta;\bar{s}_{\Lambda^c}}$ is independent of the boundary conditions $\bar{s}_{\Lambda^c}$ for $|A|$ even, see [63, 139, 235, 292] and exercise 9.5.

One can also use such expansions to show that spin variables are asymptotically independent when they are spatially far apart: for β small,

$$| < s_A s_{B+i} > - < s_A >< s_B > | \leq C \exp(-c|i|) \qquad (9.5.9)$$

for $i \in \mathbb{Z}^d$, see exercise 9.5.

Let us mention that these high temperature expansions can be extended in many directions: to non nearest neighbor interactions as well as many body ones (9.2.11) and to "spin" variables that do not have only ± 1 as values. The only place where we used this restriction to $s_i = \pm 1$ in the proof above was in the identity (9.4.1).

To explain how to avoid this limitation, write first the Hamiltonian in (9.2.1) as

$$H(\mathbf{s}_\Lambda) = -J \sum_{<ij> \subset \Lambda} (s_i s_j + 1) + J|B(\Lambda)| = \tilde{H}(\mathbf{s}_\Lambda) + J|B(\Lambda)|,$$

so that $(s_i s_j + 1)$ is always positive (we assume here, for simplicity, that $|s_i| \leq 1$, but not necessarily that $s_i = \pm 1$).[11] Redefine μ and Z in (9.2.2), (9.2.3) with $\tilde{H}$ instead of H. This changes the original Z by a multiplicative factor $\exp(\beta J|B(\Lambda)|)$ (and the free energy $F(\beta) = -kT \lim_{\Lambda \uparrow \mathbb{Z}^d} \frac{1}{|\Lambda|} \ln Z(\Lambda, \beta)$ by a constant $\beta J d$ since $\lim_{\Lambda \to \mathbb{Z}^d} \frac{|B(\Lambda)|}{|\Lambda|} = d$) but does not change anything to the expectation values $< s_A >_{\Lambda,\beta;\bar{s}_{\Lambda^c}}$, since one has the same multiplicative factor $\exp(\beta J|B(\Lambda)|)$ in the numerator and the denominator.

Then, instead of (9.4.1), write

$$\exp(\beta J(s_i s_j + 1)) = 1 + (\exp(\beta J(s_i s_j + 1)) - 1) \equiv 1 + g(s_i s_j) \qquad (9.5.10)$$

where $0 \leq g(s_i s_j) \leq C\beta$ for some constant C that can be taken as $C = 2J \exp(2\beta J)$ (using $e^x - 1 \leq xe^x$, for $x > 0$).

Using (9.5.10) and expanding the product over $< ij >\in \Lambda$ we get, instead of (9.5.1):

---

[11] If we had $|s_i| \leq C$ instead of $|s_i| \leq 1$, we would replace $s_i s_j + 1$ by $s_i s_j + C^2$ and the arguments below would still be valid.

$$Z(\Lambda, \beta) = \sum_{s_i = \pm 1, i \in \Lambda} \sum_{G \subset B(\Lambda)} \prod_{<ij> \in G} g(s_i s_j),$$

where the sum is over all graphs, not only closed ones, since the sum over $s_i = \pm 1, i \in \Lambda$ does not make $\prod_{<ij> \in G} g(s_i s_j)$ vanish for $\partial G \neq \emptyset$. And, instead of (9.5.3), we get:

$$< s_0 >_{\Lambda, \beta; \bar{s}_{\Lambda^c}} = \frac{\sum_{s_i = \pm 1, i \in \Lambda} \sum_{G \subset B(\Lambda)} s_0 \prod_{<ij> \in G \cap \Lambda} g(s_i s_j) \prod_{<ij> \in G, i \in \Lambda, j \in \Lambda^c} g(s_i \bar{s}_j)}{Z(\Lambda, \beta; \bar{s}_{\Lambda^c})}.$$

$$(9.5.11)$$

One can show that the contribution to the sum in the numerator of (9.5.11) of any graphs not containing a path connecting 0 to the boundary of Λ vanishes. Indeed, consider the connected subgraph G_0 of G containing 0; if it does not intersect $\partial \Lambda$, then, since $g(s_i s_j)$ is even and s_0 is odd under the exchange $s_i \to -s_i, \forall i \in G_0$, after summing over $s_i = \pm 1$, that term vanishes. Then the arguments proving a bound like (9.5.8) (with $\tanh(\beta J)$ replaced by $C\beta$, since $0 \leq g(s_i s_j) \leq C\beta$) can proceed as above.

For a rather general application of such expansions to many body and to long range interactions, see [54].

One can use expansions similar to (9.5.11), called *cluster expansions* to prove analyticity of the free energy and of the correlations functions for β small. See Brydges [63], Friedli and Velenik [131, Chap. 5] for pedagogical introductions to the cluster expansion formalism.

Besides, an entirely different approach to spin models at high temperatures was developed first by Dobrushin and then extended by Dobrushin and Shlosman see [108] or Simon [292, Chap. 5].

9.6   Low Temperatures and the Peierls Argument

The behavior at high temperatures is not very surprising: when β is small the coupling between the spins is weak and they behave more or less like independent random variables; in (9.5.9), we have an exponential decay of correlations instead of a product, as would be the case for independent random variables when $A \cap (B + i) = \emptyset$.

What happens at low temperatures is somewhat more surprising and the question of whether statistical mechanics is able to describe phase transitions was open before the proof by Rudolf Peierls in 1935 [251] that the Ising model in two dimensions or more does exhibit different phases at low temperatures.[12] This argument has been extended in many directions (see Sects. 9.7 and 9.8) and is one of the two main arguments used to prove the existence of phase transitions (the other being the

---

[12] Peierls' argument was not completely rigorous. The first rigorous proofs of a phase transition in the Ising model were given by Dobrushin in [102] and by Griffiths in [160].

infrared bounds, see Fröhlich, Simon and Spencer [136], that lead to proofs of phase transitions for models with a continuous symmetry discussed in Sect. 9.7.5).

To prove the coexistence of two phases at low temperatures, it will be enough to prove:

Proposition 9.1 *There exists β_0 so that, $\forall \beta \geq \beta_0$, $\exists \delta > 0$ so that, $\forall \Lambda \subset \mathbb{Z}^d$, with $d \geq 2$,*

$$< s_0 >_{\Lambda,\beta,+} \geq \delta, \tag{9.6.1}$$

where the subscript $+$ indicates the "plus" boundary conditions: $\mathbf{s}^+ = (s_i = +1)_{\forall i \in \mathbb{Z}^d}$.

Indeed that proposition implies, by the spin flip symmetry, that $< s_0 >_{\Lambda,\beta;-} \leq -\delta$, where $-$ indicates the "minus" boundary conditions: $\bar{s}_i = -1$, $\forall i \in \mathbb{Z}^d$. Then, for any limiting states $\mu_+ = \lim_{\Lambda_n \to \mathbb{Z}^d} \mu_{\Lambda_n,\beta;+}$ and $\mu_- = \lim_{\Lambda_n \to \mathbb{Z}^d} \mu_{\Lambda_n,\beta;-}$, we have $\mu_+(s_0) \neq \mu_-(s_0)$; hence there exists a least two Gibbs states.[13] The bound (9.6.1) is equivalent to $\mu_{\Lambda,\beta;+}(s_0 = -1) \leq \frac{1-\delta}{2}$ since $< s_0 >_{\Lambda,\beta;+} = \mu_{\Lambda,\beta;+}(s_0 = +1) - \mu_{\Lambda,\beta;+}(s_0 = -1) = 1 - 2\mu_{\Lambda,\beta;+}(s_0 = -1)$. We will prove (9.6.1) for $d = 2$; the extension to all $d > 2$ is left as an exercise.

Proof To prove (9.6.1), we need the notion of *contour*. Consider a configuration with $+$ boundary conditions, $\mathbf{s}_\Lambda \vee \mathbf{s}_{\Lambda^c}^+$ (meaning a configuration equal to $\mathbf{s}$ in Λ and to $\mathbf{s}^+$ in Λ^c), as in Fig. 9.6, and draw a line perpendicular to each bond $< ij >$ with $|i - j| = 1$ and $s_i s_j = -1$ and going through the middle of that bond; we then obtain a set of connected lines, called *contours*.

Next, consider a configuration with $+$ boundary conditions and $s_0 = -1$. Let $V_-(\mathbf{s}_\Lambda)$ be the largest connected set (meaning that there for any two sites of the set, there is a path going from one site to the other) containing 0 with $s_i = -1$, $\forall i \in V_-(\mathbf{s}_\Lambda)$. The exterior boundary $\partial^e V$ of a connected subset $V \subset \Lambda$ is defined as: $\{< ij >, i \in V, j \in V^c, \exists \text{ a } path\ P : j \to \partial\Lambda, P \subset \Lambda \setminus V\}$.

A *contour* $\gamma(\mathbf{s}_\Lambda)$ *associated to a configuration* $\mathbf{s}_\Lambda$ *with* $s_0 = -1$ is the unique contour $\gamma(\mathbf{s}_\Lambda)$ associated to the exterior boundary $\partial^e V_-(\mathbf{s}_\Lambda)$. For each $< ij > \in \partial^e V_-(\mathbf{s}_\Lambda)$, we have $s_i s_j = -1$.

We write $\text{Int}\gamma = V_-(\mathbf{s}_\Lambda)$ and $\text{Ext}\gamma = V_-(\mathbf{s}_\Lambda)^c$.

So, we have:

$$\mu_{\Lambda,\beta;+}(s_0 = -1) = \sum_\gamma \mu_{\Lambda,\beta;+}(\gamma), \tag{9.6.2}$$

where $\mu_{\Lambda,\beta;+}(\gamma) = \sum_{\mathbf{s}_\Lambda | \gamma(\mathbf{s}_\Lambda) = \gamma} \mu_{\Lambda,\beta;+}(\mathbf{s}_\Lambda)$. Let us estimate $\mu_{\Lambda,\beta;+}(\gamma)$. We know that, for each bond $< ij > \in \gamma$, $s_i s_j = -1$. Now, associate to each configuration $\mathbf{s}_\Lambda$ with $\gamma(\mathbf{s}_\Lambda) = \gamma$, a configuration $\tilde{\mathbf{s}}_\Lambda$ with $\tilde{s}_i = -s_i$ for $i \in \text{Int}\gamma$, $\tilde{s}_i = s_i$ for $i \in \text{Ext}\gamma$. Then, for all $< ij > \notin \gamma$ we have $s_i s_j = \tilde{s}_i \tilde{s}_j$ and, for all $< ij > \in \gamma$ we have $s_i s_j =$

---

[13] Actually, using correlation inequalities, one can show that the limits $\Lambda \uparrow \mathbb{Z}^d$ of the states with "plus" or "minus" boundary conditions exist, because expectation values $< s_A >_{\Lambda,\beta,+}$ decrease with Λ: $< s_A >_{\Lambda,\beta,+} \geq < s_A >_{\Lambda',\beta,+}$ if $\Lambda \subset \Lambda'$ (see Griffiths [162, Sect. VI] or Friedli and Velenik [131, Sect. 3.6]).

Fig. 9.6 A spin
configuration with +
boundary conditions, and the
lines made of the bonds
perpendicular to each
$< ij >$ with $|i - j| = 1$ and
$s_i s_j = -1$ and going through
the middle of that bond

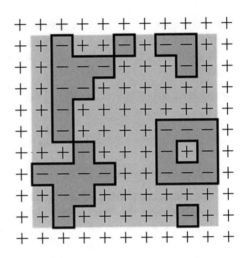

$-\tilde{s}_i \tilde{s}_j$. So, $H(\mathbf{s}_\Lambda; \mathbf{s}^+_{\Lambda^c}) = H(\tilde{\mathbf{s}}_\Lambda; \mathbf{s}^+_{\Lambda^c}) + 2J|\gamma|$, with $|\gamma|$ being the number of bonds in γ.

We can bound from below the partition function $Z_{\Lambda,\beta;+}$ by

$$Z_{\Lambda,\beta;+} \geq \sum_{\tilde{\mathbf{s}}_\Lambda;\gamma} \exp(-\beta H(\tilde{\mathbf{s}}_\Lambda; \mathbf{s}^+_{\Lambda^c})) \qquad (9.6.3)$$

where the sum runs over configurations $\tilde{\mathbf{s}}_\Lambda$ such that, for the corresponding configuration $\mathbf{s}_\Lambda$, $\gamma(\mathbf{s}_\Lambda) = \gamma$. Now, we have a one-to-one correspondance between terms with $\mathbf{s}_\Lambda$ in the numerator of $\mu_{\Lambda,\beta;+}(\gamma)$ and terms with $\tilde{\mathbf{s}}_\Lambda$ in the right hand side of (9.6.3). Since for each such pair of terms we have $H(\mathbf{s}_\Lambda; \mathbf{s}^+_{\Lambda^c}) = H(\tilde{\mathbf{s}}_\Lambda; \mathbf{s}^+_{\Lambda^c}) + 2J|\gamma|$, we get:

$$\mu_{\Lambda,\beta;+}(\gamma) \leq \exp(-2\beta J|\gamma|). \qquad (9.6.4)$$

Let us insert this bound in (9.6.2). We know that, since the contours are connected, the number of contours of length L is bounded by C^L for some $C < \infty$. It is easy to see geometrically that, if $\gamma \ni < ij >$ and $\mathrm{Int}\gamma \ni 0$, then $|\gamma| \geq 2|i|$. Using those two bounds, we get, for β large enough (so that $C \exp(-2\beta J) < 1$),

$$\mu_{\Lambda,\beta;+}(s_0 = -1) \leq \sum_{i \in \Lambda, i \neq 0} \sum_{L \geq 2|i|} C^L \exp(-2\beta J L) \leq C' \sum_{i \in \Lambda i \neq 0} \exp(-4\beta J|i|) \leq C'' \exp(-4\beta J),$$
$$(9.6.5)$$

for some constants C', C''. Since the right hand side of (9.6.5) goes to 0 as $\beta \to \infty$, we get that $\mu_{\Lambda,\beta;+}(s_0 = -1) \leq \frac{1-\delta}{2}$ holds for β large and this proves the Proposition for $d = 2$.

The proof for $d > 2$ is left as an exercise. $\qquad \square$

This proves that, for low temperatures there are least two distinct translation invariant Gibbs states, while at high temperature, there is a unique one. Thus, there must be a transition between the two regimes and one can show that the transition temperature, called the *critical temperature* T_c is unique (see Aizenman et al. [7] and Friedli and Velenik [131, Theorem 3.25]).

9.7 Other Models

There are many other models in which phase transitions occur, where the existence of a phase transition is proven by a version of the Peierls argument at low temperatures (their behavior at high temperatures is similar to the one of the Ising model, as we saw in Sect. 9.5). We give here a short list of such models, together with a guide to the literature.

9.7.1 Trivial Extensions

There are two models that are isomorphic to the Ising one and have therefore the same behavior at all temperatures:

1. Define the variables $\rho_i = \frac{s_i+1}{2}$; these variables take values 0 and 1 and can be considered as occupation numbers: $\rho_i = 1$ if site i is occupied by a particle and $\rho_i = 0$ if site i is empty. We have $s_i = 2\rho_i - 1$ and we can thus trivially rewrite the Hamiltonian (9.2.1) as well as all the measures considered in the previous sections in terms of the ρ_i variables and translate all the results obtained on the Ising model in terms of those variables.

   Models with these variables are often called "lattice gases" which is quite an oxymoron since gases don't like to sit on lattices, but if we forget about that, what was the "plus" phase in the Ising model becomes a dense or liquid phase in terms of the occupation variables and the "minus" phase becomes a rarified or gas phase.

   More realistically, one can interpret $\rho_i = 1$ as meaning that site i is occupied by an A particle and $\rho_i = 0$ as meaning that site i is occupied by a B particle, where A and B are two types of molecules present in an alloy. The two phases at low temperatures become then an A-rich phase and a B-rich one.

2. One may also consider the antiferromagnetic Ising model, with $J < 0$ in (9.2.1). It is called antiferromagnetic because the lowest energy state are now those where $s_i s_j = -1$, $\forall <ij>$. This is realized for the configurations $s_i = (-1)^{|i|}$ and $s_i = (-1)^{|i|+1}$, $\forall i \in \mathbb{Z}^d$, namely configurations where sites with $s_i = +1$ and $s_i = -1$ alternate.

If one changes variables $s_i \to (-1)^{|i|} s_i$, $\forall i \in \mathbb{Z}^d$, then the Hamiltonian (9.2.1) with $J > 0$ is transformed into the same Hamiltonian but with $J < 0$ since this transformation changes the sign of $s_i s_j$ for every pair $< ij >$. That transformation extends trivially to measures and we get at low temperatures two phases, one where $s_i = (-1)^{|i|}$ for most i's and one where $s_i = (-1)^{|i|+1}$ for most i's.

9.7.2   Long Range Interactions

Let us consider interactions of the form (9.2.9) and satisfying (9.2.10). If the interaction J_{ij} is such that $J_{ij} > 0$ for $|i - j| = 1$, namely if the nearest neighbor interaction is non zero, then it is easy to prove the existence of two phases at low temperatures:

Indeed, let J^1 and J^2 be two sets of ferromagnetic interactions. If $J_{ij}^1 \geq J_{ij}^2 \geq 0$, $\forall i, j \in \mathbb{Z}^d$, then, $\forall A \subset \Lambda \subset \mathbb{Z}^d$,

$$< s_A >_{\Lambda, \beta, J^1; +} \geq < s_A >_{\Lambda, \beta, J^2; +} \tag{9.7.1}$$

This is another application of Griffiths correlation inequalities, see Griffiths [162] or Friedli and Velenik [131, Sect. 3.6]. Then, letting $J^1 = J$ and letting J_2 to be the restriction of J to its nearest neighbor interactions, the combination of (9.7.1) and Proposition 9.1 proves the existence of two phases at low temperatures for the model with interaction J.

If $J_{ij} = 0$ for $|i - j| = 1$ then the existence of at least two phases at low temperatures for $d \geq 2$ is not immediate, although I am not aware of a non-trivial example without such a coexistence (a trivial example would be J_{ij} identically zero). But if we define the even and the odd sub-lattice by the parity of $|i|$, and we have $J_{ij} = 0$ for $|i - j| = 1$, but $J_{ij} > 0$ for nearest neighbors on each sub-lattice (so, in two dimensions $J_{ij} > 0$ for $|i - j| = \sqrt{2}$), then we can have two phases on each sub-lattice and thus four phases for the system on $\mathbb{Z}^d$.

If we let the interactions have a long range, then one might ask whether the uniqueness of the Gibbs state still holds in one dimension. The answer is that it does hold, provided that the total interaction between two half lines is bounded, meaning (since we consider always translation invariant interactions $J_{ij} = J(|i - j|)$) that the sum of the interaction between spins on the sites with $i \leq 0$ and $j > 0$ is bounded[14]:

$$\sum_{i \leq 0, j > 0} |J_{ij}| = \sum_n n |J(n)| < \infty$$

For $|J(n)| \approx n^{-\alpha}$, it means $\alpha > 2$. On the other hand there are two phases at low temperatures in one dimension for ferromagnetic interactions, $J_{ij} \geq 0$, and $|J(n)| \approx n^{-\alpha}$, with $1 < \alpha \leq 2$ ($\alpha > 1$ is needed so that (9.2.10) holds), see Dyson [116], Fröhlich and Spencer [138] and Aizenman et al. [8].

---

[14] See Bricmont, Lebowitz and Pfister [47] for a simple and general proof.

9.7.3 Many Body Interactions

The theory of thermodynamic limits and of Gibbs states can be extended to Hamiltonians of the form (9.2.11) with J_A satisfying (9.2.12). If $J_{ij} > 0$ for $|i - j| = 1$ and $J_A \geq 0$, for all other $A \subset \mathbb{Z}^d$ and $J_A = 0$ for $|A|$ odd (so that the model still possesses the spin flip symmetry) then we can use inequality (9.7.1) to prove the existence of two ferromagnetic phases at low temperatures. But one might wonder what happens if $J_A = 0$ unless $|A| = 4$ for example or any other choice of J not including nearest neighbor interactions. There is actually a nice algebraic theory characterizing the low temperature phases for this class of models, see Holsztynski and Slawny [173], Miekisz [241], Bricmont et al. [48], and, for a review, Slawny [299].

9.7.4 Continuous Spins

Another extension consists of freeing ourselves of the fact that the spins take only two values: ± 1. Nothing would change qualitatively if the spin took values $-m, -m + 1, \ldots, m$, with $m \in \mathbb{N}$. We introduce a *single spin measure* ν, which is an even measure on $\mathbb{R}$; assume first that its support is bounded. Then instead of (9.2.2), we define:

$$d\mu(\mathbf{s}_\Lambda) = \frac{\exp(-\beta H(\mathbf{s}_\Lambda))d\nu(\mathbf{s}_\Lambda)}{Z_\Lambda}, \tag{9.7.2}$$

where $H(\mathbf{s}_\Lambda)$ is given by (9.2.1), (9.2.9) or (9.2.11) with the sums restricted to interactions in Λ, $d\nu(\mathbf{s}_\Lambda) = \prod_{i \in \Lambda} d\nu(s_i)$, and

$$Z_\Lambda = \int_{\mathbb{R}^\Lambda} \exp(-\beta H(\mathbf{s}_\Lambda))d\nu(\mathbf{s}_\Lambda). \tag{9.7.3}$$

The theory of Gibbs states extends to that framework, as well as the one of phase transitions, for ν of bounded support, see Friedli and Velenik [131, Sect. 6.10.2].

A more interesting but more delicate set of models is given by (9.7.2) with $d\nu(s)$ not having a bounded support; to simplify matters, let us limit ourselves to the Hamiltonian with nearest neighbor interactions (9.2.2). Since the Hamiltonian may grow quadratically in the s variables, we need that $\nu(s)$ decreases as $s \to \infty$ faster than $\exp(-\mathcal{O}(s^2))$ for the integrals (9.7.3) to converge.

In fact if $d\nu(s) = \exp(-as^{2k} + P(s))ds$ with $a > 0$, $k \geq 2$, and $P(s)$ a polynomial of degree less than $2k$, it is easy to see that the integrals (9.7.3) converge. An important example that has been very much studied is the one with

$$d\nu(s) = \exp(-as^4 + bs^2)ds, \tag{9.7.4}$$

often called the lattice ϕ^4 model, because it is a lattice model of a quantum field theory (whose variables are usually denoted by the letter ϕ). Since one can scale the s variable

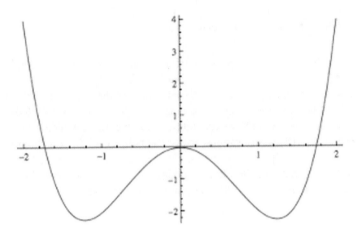

Fig. 9.7 A double well potential of the form $as^4 - bs^2$

by $\sqrt{\beta J}$ and redefine accordingly a and b in (9.7.4), let us set $\beta J = 1$. This model is also called the Ginzburg-Landau model. One can write $dv(s) = \exp(-V(s))ds$, with $V(s) = as^4 - bs^2$. The function $as^4 - bs^2$, with $b > 0$, is called a double well, name whose meaning is obvious if one looks at its graph, see Fig. 9.7. Note that, if $b = 2a\beta$ one can write $V(s) = a(s^2 - \beta)^2 - \beta^2$, and neglecting the constant β^2 (the factor e^{β^2} cancels between numerator and denominator in the correlation functions), we see that $c_a \exp(-V(s)) \to \delta(s^2 - \beta)$ as $a \to \infty$, for c_a's so that $c_a \int_{\mathbb{R}} \exp(-V(s))ds = 1$. Rescaling s by $\sqrt{\beta}$, one obtains the usual Ising model ($\delta(s^2 - 1)$ means $s = \pm 1$) at inverse temperature β.

If the wells are deep enough (taking b large for a fixed in (9.7.4)), one can again show by a Peierls-type argument that there are two phases, while if they are shallow (b small or negative), there is a unique phase.

The interest of this model is that in a suitable limit where $a \to 0$, with $b = b(a) \to 0$, in dimensions two and three, one obtains a quantum field theory[15] defined on $\mathbb{R}^d$ after a rescaling of the lattice variables: roughly speaking, replace $\mathbb{Z}^d$ by $\epsilon\mathbb{Z}^d$, which is only a relabelling of the indices of the s_i variables and let $\epsilon \to 0$ as a function of $a \to 0$. Then, in that limit, the model is defined formally on $\{s_x, x \in \mathbb{R}^d\}$, which makes sense as a measure defined on a space of distributions, see [148].

One can show that there are phase transitions in that limit: depending on the choice of the function $b = b(a) \to 0$ we can have field theories with either one or two phases, see Glimm, Jaffe and Spencer [147], Imbrie [175], Borgs and Imbrie [41].

---

[15] To be precise, the Euclidean version of such a theory, see Glimm and Jaffe [148].

9.7.5 Models with a Continuous Symmetry

Yet another extension of our "spin models" is when the spin variables take values in $\mathbb{R}^n$, with $n \geq 2$. Since all our previous models had a $s \to -s$ symmetry, we will be interested in models in $\mathbb{R}^n$ that are rotationally symmetric. Consider the single-spin measure:

$$d\nu(\mathbf{s}) = \delta(|\mathbf{s}| = 1)d\mathbf{s},$$

with $d\mathbf{s}$ the Lebesgue measure on $\mathbb{R}^n$ and, for simplicity, the Hamiltonian (9.2.1) but with $s_i s_j$ replaced by $\mathbf{s}_i \cdot \mathbf{s}_j$. The Ising model can be viewed as the $n = 1$ version of that model.

The behavior of those models is that there is a unique Gibbs state in one *and* *two* dimensions, see Dobrushin and Shlosman [105], but infinitely many at low temperatures in three or more dimensions, see Fröhlich, Simon and Spencer [136]. The latter are parametrized by the average $\mu(\mathbf{s}_0)$ which can take any value in the unit sphere S^{n-1} in $\mathbb{R}^n$. The existence of such phases is called a breakdown of a continuous symmetry (the rotation symmetry of the model), a concept which is important in quantum field theory.

Moreover, for $d = 2$ and $n = 2$ there is, at low temperatures, a phase with a power law decay of correlations: while in all our previous models the correlation functions decay exponentially at low temperatures in the extremal Gibbs states, as they did at high temperatures, see (9.5.9), here we have:

$$< \mathbf{s}_0 \cdot \mathbf{s}_i > \approx |i|^{-p(\beta)}$$

where $p(\beta) < 2$ and $p(\beta) \to 0$ as $\beta \to \infty$, see Fröhlich and Spencer [137]. It is expected, but not proven that, for $d = 2$ and $n \geq 3$ the correlation functions decay exponentially at all temperatures.[16]

9.7.6 Non Translation Invariant Gibbs States

So far, we have only discussed translation invariant Gibbs states, but could there exist also non translation invariant ones? Consider again the Ising model with the simplest Hamiltonian (9.2.1). A natural candidate for such non translation invariant states would be states with a spatial coexistence of phases: for $i \in \mathbb{Z}^d$ with $i_1 \geq 0$, we would have predominantly $s_i = +1$, and, for $i_1 < 0$, we would have predominantly $s_i = -1$. Of course, if such states exist, then we have infinitely many of them by replacing 0 by any $l \in \mathbb{Z}$.

It would also seem natural, in order to produce such states, to consider the following boundary conditions $\mathbf{s}$ with $s_i = +1$ for $i_1 \geq 0$ and $s_i = -1$ for $i_1 < 0$, called the $\pm$ boundary conditions. It turns out that, in three dimensions or more, these boundary

---

[16] This is considered one of the hardest open problem in the field of lattice spin models.

conditions do lead to non translation Gibbs invariant states in the thermodynamic limit for $\Lambda = [-L, L]^d$, with $L \to \infty$, at low temperatures, see Dobrushin [104], van Beijeren [317], with an "interface" separating a region ($i_1 \geq 0$) where the spins are predominantly "plus" and another region ($i_1 < 0$) where they are predominantly "minus". Thus, we get infinitely many such states, since any translation of a non translation invariant Gibbs state in the direction perpendicular to its interface produces another such state. Moreover, we can replace i_1 by any i_l, $l = 2, \ldots, d$ and obtain new non translation invariant Gibbs states.

On the other hand, in two dimensions, under very general conditions, there are no non translation invariant states at all temperatures, see Aizenman [3], and see Friedli and Velenik [131, Theorem 3.61] for a review of more recent results.

9.8   The Pirogov–Sinai Theory

A final extension concerns model without spin flip symmetries. What if we consider Ising models with Hamiltonians of the form (9.2.11) but including terms with $J_A \neq 0$ for $|A|$ odd?

9.8.1   A Simple Example

Let us start with a simple example with three body interactions:

$$H(s_\Lambda) = -J \sum_{<ij>} s_i s_j - \epsilon \sum_{<ijk>} s_i s_j s_k - h \sum_i s_i, \qquad (9.8.1)$$

where the sum over $< ijk >$ is over triangles with $|i - j| = 1$, $|j - k| = 1$ and $|i - k| = \sqrt{2}$ and we will assume ϵ small. Since there is no spin flip symmetry $s_i \to -s_i$ in the Hamiltonian (9.8.1), there is no reason to expect a $+$ and a $-$ phase to coexist at $h = 0$. Let us compute first the average energy per site of the $+$ configuration:

$$\lim_{\Lambda \to \mathbb{Z}^d} \frac{H(s_\Lambda^+)}{|\Lambda|} = -Jd - \epsilon \frac{2d(2d-2)}{3} - h,$$

since each site belongs on average to d nearest neighbor pairs,[17] and to $\frac{2d(2d-2)}{3}$ triangles.[18] Similarly, for the $-$ configuration:

---

[17] Each site has $2d$ nearest neighbor, but each pair contains 2 sites.

[18] Given a site i one has $2d$ choices of a site j with $|i - j| = 1$ and then $2d - 2$ choices of a site $k \neq i$ with $|j - k| = 1$ and $|i - k| = \sqrt{2}$. We divide by 3 because each triangles contains three sites.

$$\lim_{\Lambda \to \mathbb{Z}^d} \frac{H(s_\Lambda^-)}{|\Lambda|} = -Jd + \epsilon \frac{2d(2d-2)}{3} + h,$$

so that the energy per site of these two configurations are equal provided that:

$$-\epsilon \frac{2d(2d-2)}{3} - h = +\epsilon \frac{2d(2d-2)}{3} + h,$$

or $h = -\epsilon \frac{2d(2d-2)}{3}$.

So, one can find a value of h so that the $+$ and the $-$ configurations have equal energies. But, again because of the lack of spin flip symmetry, there is no reason to expect the $+$ and the $-$ phases to coexist at low temperatures for that particular value of h. In 1975, Sergei Pirogov and Yasha Sinai managed to prove [257, 297]:

Proposition 9.2 *For β large and ϵ small enough, there exists a function $h(T, \epsilon)$, with $h(T, \epsilon) \to -\epsilon \frac{2d(2d-2)}{3}$ as $T \to 0$, such that the Hamiltonian (9.8.1) with $h = h(T, \epsilon)$ has a $+$ and a $-$ phase.*

The proof is far more complicated than the usual Peierls argument, because of has to "construct" the function $h(T, \epsilon)$ as the same time as one proves that phase coexistence occurs for that value of h.

9.8.2 The General Pirogov–Sinai Theory

In fact, Pirogov and Sinai proved a far more general proposition than the one above. Their theory is the most beautiful and subtle extension of the 1935 Peierls argument. To express their result, we need several concepts: we consider Hamiltonians of the form (9.2.11) with finite range interactions and J_A not necessarily positive, and define:

1. *Ground states.* They are configurations such that any configuration obtained by modifying that configuration on a finite number of sites has a higher energy. We assume that we have a Hamiltonian H_0 with a finite number m of periodic ground states (meaning periodic under the set of shifts τ_i, $i \in \mathbb{Z}^d$), all having the same energy per site.
2. We assume that H_0 satisfies a *Peierls condition*, which means that one associates to each configuration that differs from a ground state on a finite number of sites a contour (similar to what we did in Sect. 9.6), i.e. a set of lattice boxes (of a fixed size) on which the configuration does not coincide with any ground state of H_0. The Peierls condition assumes that the energy (relative to the ground state energy) associated to any contour is proportional to its size, as in Sect. 9.6.
3. A set of perturbations that *lifts the degeneracy of the ground states*. If H_0 has m periodic ground states $s_1, \ldots, s_m$ one can introduce $m - 1$ Hamiltonians

$H_1, \ldots H_{m-1}$ and parameters $\lambda = (\lambda_1, \ldots \lambda_{m-1})$, such that the set of Hamiltonians $H_\lambda = H_0 + \sum_{q=1}^{m-1} \lambda_q H_q$ has the following property: the origin 0 is the intersection of m lines in the parameter space λ for which the Hamiltonians H_λ have $m - 1$ periodic ground states. Each of those lines is an intersection of $(m - 1)$ two-dimensional surfaces in the parameter space λ, with boundaries on those lines, for which the Hamiltonians H_λ have $m - 2$ periodic ground states, etc., up to $(m - 1)$ dimensional subsets in the parameter space λ for which the Hamiltonians H_λ have a unique ground state.

These properties of the set of ground states is called the *Gibbs phase rule*. In the example of Hamiltonians given by (9.8.1), H_0 corresponds to $h = -\epsilon \frac{2d(2d-2)}{3}$, there are two ground states and the perturbation h lifts the degeneracy of the ground states. The Peierls condition will be satisfied in this model for ϵ small, because the first term in (9.8.1) will be dominant and does satisfy that condition.

To take a less trivial model, consider the Blume-Capel model: $s_i = -1, 0, 1$ at each $i \in \mathbb{Z}^d$ with Hamiltonian $H_0(s_\Lambda) = \sum_{<ij>}(s_i - s_j)^2$ which has three translation invariant ground states (s^+, with all $s_i = 1$, s^0 with all $s_i = 0$ and s^- with all $s_i = -1$) and no other periodic one. It is easy to see that H_0 satisfies Peierls condition.

To lift the degeneracy of the ground states, introduce the family of Hamiltonians (with $\lambda_1 = \lambda$, $\lambda_2 = h$ in our previous notation):

$$H_{\lambda,h}(s_\Lambda) = \sum_{<ij> \subset \Lambda} (s_i - s_j)^2 + \lambda \sum_{i \in \Lambda} s_i^2 + h \sum_{i \in \Lambda} s_i.$$

We get:

$$\lim_{\Lambda \to \mathbb{Z}^d} \frac{H(s_\Lambda^\pm)}{|\Lambda|} = \lambda \pm h,$$

and

$$\lim_{\Lambda \to \mathbb{Z}^d} \frac{H(s_\Lambda^0)}{|\Lambda|} = 0$$

We get three ground states $+1$, 0 and -1 when $\lambda = h = 0$; we get the two ground states $+1$ and 0 when $\lambda + h = 0$, $\lambda > 0$, the two ground states -1 and 0 when $\lambda - h = 0$, $\lambda < 0$ and the two ground states $+$ and $-$ when $h = 0$, $\lambda < 0$. We get one ground state for the other parameter values.

Stated informally, the result of Pirogov and Sinai is that, at low temperatures, the phase diagram is a small perturbation of the zero temperature one: there exists a curve $\lambda(T)$, defined for T small, with m periodic Gibbs states at temperature T for $H_{\lambda(T)}$; each of these Gibbs states has configurations that coincide with one of the ground states of H_0 at most sites, and $\lambda(T) \to 0$ as $T \to 0$. For each T small, there exists also $m - 1$ curves intersecting at $\lambda(T)$ on which there are $m - 1$ periodic Gibbs states, $m - 2$ surfaces intersecting two by two on these curves with $m - 2$ periodic Gibbs states etc.

For a pedagogical introduction to that theory, see Sinai [297], Slawny [299], Fernández [127], Friedli and Velenik [131, Chap. 7], and a lecture by Kotecký available on http://www.cts.cuni.cz/~kotecky/publ/K-PS-EMP06.pdf.

This remarkable Pirogov–Sinai theory has been extended in many directions and new proofs of it have been found, see Zahradnìk [332] and Kotecký and Preiss [198]. One of these is based on a renormalization group approach and is due to Gawedzki, Kotecký, and Kupiainen [143]. The theory has been extended to quantum field theories, see Borgs and Imbrie [41]or Borgs and Waxler [42], to some systems with infinitely many ground states, see Dinaburg and Sinai [100] or Bricmont and Slawny [50, 52], to disordered systems, see Bricmont and Kupiainen [51], to some continuum fluids models, see Bricmont et al. [49], specially those with Kac-type interactions (defined in Sect. 9.3) see Lebowitz et al. [222]. For more recent results, see Mazel et al. [240] and references therein.

9.9 Critical Points

How do we pass from a high-temperature disordered phase to a low-temperature ordered one? The answer is: in general, through a critical point (see Figs. 9.1 and 9.2), or critical temperature whose existence and uniqueness is known for Ising models see Friedli and Velenik [131, Theorem 3.25]. Those critical points have been the subject of considerable interest at least since the 1960s and their (partial) understanding is one of the great successes of modern theoretical physics. But, as for the rest of this chapter, one would need a whole book to treat adequately that subject, hence we will have to limit ourselves to an elementary introduction.

9.9.1 Phenomenology

As one can see from Figs. 9.4 and 9.5, there a critical value (in the mean field theory but it is also true in more realistic models) β_c such that for $\beta \leq \beta_c$, the magnetization vanishes at zero field $\lim_{h \to 0} m(\beta, h) = 0$, while for $\beta > \beta_c$, $\lim_{h \to 0} m(\beta, h) \neq 0$. One sees that, at β_c, the slope of the curve $m(\beta_c, h)$ becomes infinite at $h = 0$: $\lim_{h \to 0} \frac{dm(\beta_c, h)}{dh} = \infty$. One also sees in Fig. 9.3 that this slope increases as β increases towards its critical value $\frac{dm(\beta, h)}{dh} |_{h=0} \to \infty$ as $\beta \uparrow \beta_c$.

The way it increases as a function of β allows us to define a first *critical exponent*:

$$\chi(\beta, h) |_{h=0} \equiv \frac{dm(\beta, h)}{dh} |_{h=0} \approx (T - T_c)^{-\gamma}, \tag{9.9.1}$$

as $T \downarrow T_c$, where we introduce the notation $\chi(\beta, h) = \frac{dm(\beta, h)}{dh}$, which is called the magnetic susceptibility and used the variable T instead of β. To give a precise meaning to (9.9.1), let us define the notation $f(x) \approx x^a$, as $x \downarrow 0$ to mean

$$\lim_{x \downarrow 0} \frac{\ln f(x)}{\ln x} = a$$

Using (9.2.14) and an interchange of limits between the derivative with respect to h and the limit $\Lambda \uparrow \mathbb{Z}^d$ (which can be justified but we will not do it here), it is easy to express $\chi(\beta, h)$ in terms of correlation functions, up to a factor β,

$$\chi(\beta, h) = \sum_{i \in \mathbb{Z}^d} < s_0 s_i > - < s_0 > < s_i > .$$

For $T \neq T_c$, we have exponential decay of correlations, see (9.5.9), and so $\chi < \infty$, but not at $T = T_c$.

Before introducing other critical exponents, let us explain why they attract a great deal of interest. While the value of the critical point or the dependence in β, h of the magnetization or of the other correlation functions varies from one model to another, the value of the exponent γ, as well as the other critical exponents defined below, are "universal" meaning that they take the same value for a whole class of models, but not for all of them. The set of models can then be divided into "universality classes", for which the critical exponents are the same.

One can also define a critical exponent characterizing the way $m(\beta, 0)$, which is strictly positive for $T < T_c$ vanishes as $T \uparrow T_c$:

$$m(\beta, 0) \approx |T - T_c|^{\beta},$$

where the choice of the letter β is probably one of the worst ones in the history of physics! For examples of values taken by the exponent β, see (9.9.6) and (9.9.9) below.

When $h \to 0$ for $\beta < \beta_c$, the curve $m(\beta, h)$ is smooth, meaning that it vanishes proportionally to h. For $\beta > \beta_c$, $m(\beta, h = 0)$ is strictly positive and the difference $|m(\beta, h) - m(\beta, h = 0)| \approx |h|$ as $h \to 0$. But at $\beta = \beta_c$, the way $m(\beta_c, h)$ vanishes is singular:

$$|m(\beta_c, h)| \approx |h|^{\frac{1}{\delta}},$$

as $h \to 0$, and that defines the exponent δ.

Since we have exponents β, γ, δ, there must some exponent α. And that governs the behavior of the second derivative of the free energy with respect to β:

$$C \equiv \frac{d^2 f(\beta, 0)}{d\beta^2} \approx |T - T_c|^{-\alpha}, \tag{9.9.2}$$

as $T \downarrow T_c$; C is the *specific heat* (up to some T dependent factor).[19] Here we take the second derivative, because the first derivative is just proportional to $< s_i s_j >$ and is thus continuous and bounded at β_c.

Again, using (9.2.14) and an interchange of limits, one can write, up to a T dependent factor:

$$C = \sum_{<ij>} < s_0 s_1 s_i s_j > - < s_0 s_1 >< s_i s_j >, \qquad (9.9.3)$$

where the sum runs over all nearest neighbor pairs in $\mathbb{Z}^d$. Because of the exponential decay of correlations, see (9.5.9), $C < \infty$, for $T \neq T_c$, but not at $T = T_c$.

There is also an exponent η, defined by the decay of the spin-spin correlation function at T_c:

$$< s_0 s_i > \approx \frac{1}{|i|^{d-2+\eta}},$$

as $i \to \infty$.

Finally, there is an exponent governing the way the rate of decay of the spin-spin correlation function goes to zero as $T \downarrow T_c$, or rather how the inverse of this rate, called the *correlation length* ξ goes to infinity. The reason for that expression is that one can write $< s_0 s_i > \approx \exp(-\frac{|i|}{\xi})$, which means that $< s_0 s_i > \approx 0$ when $|i|$ is much larger that ξ. Define

$$\lim_{i \to \infty} \frac{|\ln < s_0 s_i >_\beta|}{|i|} = \frac{1}{\xi(T)}. \qquad (9.9.4)$$

Then one defines the critical exponent ν

$$\xi(T) \approx (T - T_c)^{-\nu}, \qquad (9.9.5)$$

as $T \downarrow T_c$.

9.9.2 Mean Field Theory

Let us compute the critical exponents in the mean field approximation, by analyzing (9.3.1): if $m_0 = \lim_{h \downarrow 0} m(\beta, h) > 0$ is the positive solution of (9.3.1) for $T < T_c$, we have, by using $\tanh x = x - \frac{x^3}{3} + \mathcal{O}(x^5)$, for $x \to 0$,

---

[19] Strictly speaking, we should distinguish here between exponents α and α', depending on whether $T \to T_c$ from above or from below, but we will not do that distinction, since in the examples discussed here both exponents are equal. The same remark holds for the exponent ν defined in (9.9.5) below, and could have been made about the exponent γ in (9.9.1), where, for $T < T_c$, it would be defined through the one sided derivative of $m(\beta, h)$ with respect to h at $h = 0$.

$$m_0 = \beta J 2 d m_0 - \frac{(\beta J 2 d m_0)^3}{3} + \mathcal{O}(m_0^5),$$

as $T \uparrow T_c$. This implies, since $\beta_c J 2 d = 1$ in the mean field approximation, that

$$m_0 \approx |T - T_c|^{\frac{1}{2}} \tag{9.9.6}$$

and thus the mean field exponent $\beta = \frac{1}{2}$.

Using the same approximation for the function $\tanh x = x - \frac{x^3}{3} + \mathcal{O}(x^5)$, when $\beta_c J 2 d = 1$, we get, from (9.3.1),

$$m(\beta_c, h) = (m(\beta_c, h) + \beta_c h) - \frac{(m(\beta_c, h) + \beta_c h)^3}{3} + \mathcal{O}(m(\beta_c, h) + \beta_c h)^5$$

which implies that, when $h, m \to 0$,

$$h \approx m(\beta_c, h)^3$$

i.e. that the mean field exponent $\delta = 3$. Consider now the susceptibility: $\chi = \frac{dm(\beta, h)}{dh}$. From (9.3.1) and $\frac{d \tanh x}{dx} = 1 - (\tanh x)^2$, we get, for $h = 0$:

$$\frac{dm(\beta, h)}{dh} = \frac{\beta(1 - m^2(\beta, h))}{1 - 2d\beta J(1 - m^2(\beta, h))}$$

For $T > T_c$, $m(\beta, 0) = 0$ and we get

$$\frac{dm(\beta, h)}{dh} \Big|_{h=0} = \frac{\beta}{1 - 2d\beta J} \approx \frac{1}{|T - T_c|},$$

since $2d\beta_c J = 1$ in the mean field model. Thus, the mean field exponent $\gamma = 1$.

To compute α, we get from (9.3.3)

$$f(\beta, 0) = -kT \ln 2, \quad \text{for} \quad T \geq T_c$$
$$f(\beta, 0) = dJm_0^2 - kT \ln(2 \cosh \beta 2 d J m_0), \quad \text{for} \quad T < T_c. \tag{9.9.7}$$

Thus, the first and second derivatives of $f(\beta, 0)$ with respect to β vanish for $T \geq T_c$, while for $T < T_c$ the second derivative tends to a non-zero constant as $T \uparrow T_c$, since $f(\beta, 0) \approx m_0(T)^2 \approx |T - T_c|$ by (9.9.6). Thus, the specific heat C has a jump discontinuity at T_c, which is, by convention, denoted as $\alpha = 0$.

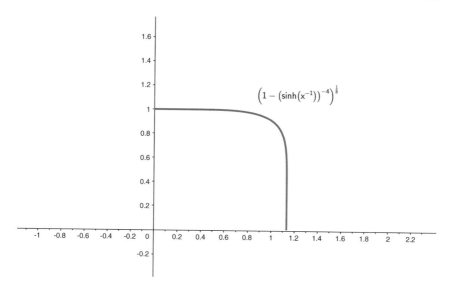

Fig. 9.8 The graph of the spontaneous magnetization of the two dimensional Ising model with $x = \frac{kT}{2J}$

9.9.3 More Realistic Models

In 1944 Lars Onsager [249] was able to compute exactly the value of the free energy for the two-dimensional Ising model, using the transfer matrix formalism (see Sect. 9.4) and from that and subsequent work [68, 331], one knows the values of the critical exponents in that model[20]:

$$\alpha = 0 \tag{9.9.8}$$
$$\beta = \frac{1}{8}$$
$$\gamma = \frac{7}{4}$$
$$\delta = 15$$

The value $\beta = \frac{1}{8}$ comes from the explicit formula for the spontaneous magnetization due to Yang [331]: for $T < T_c$, see Fig. 9.8:

$$\lim_{h \downarrow 0} m(\beta, h) = [1 - (\sinh(2\beta J))^{-4}]^{\frac{1}{8}} \approx |T - T_c|^{\frac{1}{8}}, \tag{9.9.9}$$

with T_c given by $\sinh(2\beta_c J) = 1$.

---

[20] Here $\alpha = 0$ means a logarithmic singularity rather than a discontinuity as in the mean field model, see (9.9.7) The proof of $\delta = 15$ is in [68].

On the other hand, for $d > 4$, the exponents are known to equal their mean field values ($\alpha = 0$, $\beta = \frac{1}{2}$, $\gamma = 1$, $\delta = 3$) see Fernández, Fröhlich and Sokal [126] and are expected to equal their mean field values up to logarithmic corrections in $d = 4$; this is partially but not completely proven, see [4–6, 302].

For the most physically interesting case of $d = 3$, only estimates based on numerical methods are known, see [254] for a review.

In order to study the critical exponents, a very important field of research based on the idea of the renormalization group has been developed. This idea has had applications not only in statistical physics, but also in quantum field theories, dynamical systems, partial differential equations, and areas of pure mathematics, such as the convergence of Fourier series. But dealing, even superficially, with those ideas would truly require another book.

9.10 Summary

In this chapter we tried to understand phase transitions and phase coexistence, for example between liquid and vapor or between different states of spontaneous magnetization of a ferromagnet, whose phenomenology is described in Sect. 9.1. We concentrate ourselves on lattice models, that are the best studied ones.

But since there may be more than one equilibrium state, we cannot simply use the formalism of Chap. 6. In Sect. 9.2, we introduce a class of lattice models, like the Ising one, and briefly explain the notion of Gibbs states that allows us to discuss situations where many (even infinitely many) equilibrium states coexist for given thermodynamic parameters.

Since the study of the set of Gibbs states, even for the most elementary models is obviously complicated, physicists have invented a simple but powerful approximation, the mean field theory, which is explained in Sect. 9.3.

If one wants to go beyond that approximation, the easiest example are the one-dimensional models discussed in Sect. 9.4. One can write explicit solutions for some of them and show very generally that there are no phase transitions in one dimension, with some exceptions for long range interactions, see Sect. 9.7.2.

The next simplest situations are models at high temperatures: there one can use graphical expansions to prove that there is a unique Gibbs state, again under very general assumptions, see Sect. 9.5.

Low temperature models are more subtle but an argument due to Peierls allows to prove that there are at least two translation invariant Gibbs states at low temperatures for the Ising model, see Sect. 9.6.

In Sect. 9.7 we discuss various extensions of what was done before: long range interactions (Sect. 9.7.2), many body interactions (Sect. 9.7.3), continuous spins (Sect. 9.7.4), models with a continuous symmetry, i.e. models where the spin takes value on a circle or a sphere (Sect. 9.7.5) and non translation invariant Gibbs states (Sect. 9.7.6).

In Sect. 9.8 we sketch a far reaching extension of Peierls' argument, that applies to models that, unlike the Ising one, do not have a spin flip symmetry.

Finally Sect. 9.9 is devoted to what happens at and near the critical temperature separating the high and the low temperature phases. At that point, various quantities like the magnetization or their derivatives diverge, and the way they do so is characterized by critical exponents, defined in Sect. 9.9.1. Those exponents are easy to compute within the mean field approximation (see Sect. 9.9.2) but very hard to compute in more realistic model, except in two dimensions (see Sect. 9.9.3).

9.11 Exercises

Some of the exercises in this section are taken from Thompson [308].

9.1. Prove (9.2.14).

9.2. Check all the steps in the derivation of (9.3.2).

9.3. Using the transfer matrix (9.4.19), write the partition function with other boundary conditions in term of the transfer matrix, do the same for the correlation functions and compute their thermodynamic limit.

9.4. Prove for β small a bound of the form (9.5.8) for $| < s_A >_{\Lambda,\beta;\bar{s}_{\Lambda^c}} |$, for any $A \subset \mathbb{Z}^d$, $|A| < \infty$, and $|A|$ odd.

9.5. Prove that, for β small $\lim_{\Lambda \uparrow \mathbb{Z}^d} < s_A >_{\Lambda,\beta;\bar{s}_{\Lambda^c}}$, exists for any $A \subset \mathbb{Z}^d$, $|A| < \infty$, and is independent of $\bar{s}_{\Lambda^c}$. Hint: use "duplicate variables" given two boundary conditions $\bar{s}$, $\bar{s}'$, write $< s_A >_{\Lambda,\beta;\bar{s}_{\Lambda^c}} - < s_A >_{\Lambda,\beta;\bar{s}'_{\Lambda^c}} = \sum_{s_i, s'_i \in \Lambda} (s_A - s'_A) \mu_{\bar{s}_{\Lambda^c}}(s_\Lambda) \mu(s'_\Lambda)_{\bar{s}'_{\Lambda^c}}$.
Prove also (9.5.9) using duplicate variables.

9.6. Extend the Peierls argument to the three dimensional Ising model. Hint: consider "plaquettes" i.e. surfaces of area one perpendicular to the bonds $< i, j >$ with $s_i s_j = -1$.

9.7. Check the power $\frac{1}{8}$ in the right hand side of formula (9.9.9).

9.8. Check that, for the mean field free energy given by (9.3.3), $f(\beta, 0) = \lim_{h \to 0} f(\beta, h)$ is given by (9.9.7).

9.9 Using the transfer matrix formalism, compute the thermodynamic free energy of a one-dimensional Ising model with Hamiltonian (9.2.1) and the spins taking values $s_i = -1, 0, 1$.

9.10. Using the transfer matrix formalism, compute the grand-canonical partition function $\mathcal{Z}_{gc}(V, T, z)$ for the one-dimensional Ising model with Hamiltonian (9.2.1), but expressed in terms of the variables $\rho_i = \frac{s_i+1}{2}$, with periodic boundary conditions, $\rho_{-L} = \rho_{-L+1}$.

Chapter 10
Conclusion: Statistical Mechanics and Reductionism

Reductionism is a scientifico-philosophical attitude that tries to explain everything in terms of a few simple principles or laws. Recently, reductionism has acquired a rather bad reputation and words like "emergence" or expressions like "the whole is more than the sum of its parts" have become fashionable, so it may be worthwhile to discuss that issue, in particular in relation with thermodynamics and statistical mechanics.[1]

This opposition to reductionism is rather common among philosophers or religiously inclined people, but is has also opposed two famous physicists, both Nobel Prize winners, Steven Weinberg, who defends reductionism in *Dreams of a Final Theory* [328], and Philip Anderson who criticizes it in "More is Different" [13]. However, when concepts are clearly defined, the opposition between Weinberg and Anderson is not as radical as it may seem.

One could define the notion of reductionism by saying that biology is "nothing more" than chemistry, that chemistry is "nothing more" than physics and that the various branches of macroscopic physics are "nothing more" than atomic and molecular physics, the latter being in turn "nothing more" than particle physics. One could go further and say that psychology is "nothing more" than biology.

Stated like that, reductionism may sound simplistic and indeed, it all depends on what one means by "nothing more". Obviously, and that is one of Anderson's criticisms, if one did know about things such as superconductivity or phase transitions, one would probably not have discovered them by doing only computations "from bottom up", starting with atoms and molecules.

But the reductionist answer is that indeed, "higher level" sciences like macroscopic physics, chemistry or biology cannot be *deduced* from fundamental physics, but nevertheless the *explanations* of the phenomena studied by those sciences do not require principles independent of the fundamental physical ones. Or, as Weinberg puts it, the explanatory arrows always point towards the fundamental physical laws

---

[1] For a detailed discussion of reductionism made by statistical physicists, see Chibarro et al. [78].

© Springer Nature Switzerland AG 2022
J. Bricmont, *Making Sense of Statistical Mechanics*, Undergraduate Lecture
Notes in Physics, https://doi.org/10.1007/978-3-030-91794-4_10

and microscopic physics. One does not explain the structure of atoms by appealing to the concept of turbulence, but one does explain chemical reactions by invoking atoms and electrons.

The question of whether we need "something more" was rather colorfully expressed by Richard Feynman:

> The next great era of awakening of human intellect may well produce a method of understanding the qualitative content of equations. Today we cannot. Today we cannot see that the water flow equations contains such things as the barber pole structure of turbulence that one sees between rotating cylinders. Today we cannot see whether Schrödinger's equation contains frogs, musical composers, or morality – or whether it does not. We cannot say whether something beyond it like God is needed, or not. And so we can all hold strong opinions either way.[2]

> Richard Feynman [129, Sect. 41]

To clarify further the notion of reductionism, consider the following historical examples:

1. In the nineteenth century, there was a fierce discussion between reductionists and vitalists: do we need a new principle to account for life or is it explainable in principle by physics and chemistry? In that discussion, it is fair to say that the reductionists won. In psychology, there has been a similar debate as to whether we need a separate "substance" to account for the functioning of the mind; that substance used to be called the soul, but, although here reductionism has not triumphed, the idea of a substance separate from the body is not very popular among scientists.

2. Still in the nineteenth century, when electric and magnetic phenomena were discovered and analyzed and when the laws governing those phenomena were unified by Maxwell, they were not reduced to the laws of classical mechanics but were considered as new fundamental laws of physics. The same thing happened in the twentieth century with the weak and strong forces of particle physics.

3. Chemistry and the theory of chemical binding of atoms into molecules was also developed in the nineteenth century. This is now considered as being "reduced" to physics, but that reduction needed the development of quantum physics in the twentieth century, so that nineteenth century chemistry was indeed reduced to physics, but not to the physics of that time.

So, in the first example, the "nothing more is needed" prevailed, in the second one, "more" was added—new laws of physics—and in the third one, "more" meant radically changing those laws of physics.

In view of what precedes one should see reductionism as a program, somewhat similar to the search for deterministic laws, rather than a fixed doctrine.

---

[2] Of course, "God" is such an ill-defined notion that one cannot expect it to enter our scientific explanations. Music and morality are also not well defined, but these subjects are studied in the framework of evolutionary psychology, which is an example of a reductionist approach. (Note by J.B.).

Let us now turn to what is often considered as the prime example of reductionism, the one of thermodynamics to statistical mechanics. It is even expressed in a sort of slogan: "heat is molecular motion." And, in this book, we saw reductionism in action: all the thermodynamic formulas are expressed in terms of the microstates. Moreover, Boltzmann's approach was grounded in his strong "trust" in the existence of atoms (see Cercignani [74]) and further developments in statistical mechanics not covered here (like Einstein's analysis of Brownian motion) strengthened the case in favor of their existence.

But this example also indicates some caveats to keep in mind when discussing reductionism.

First of all, not all microstates in a surface of given energy in phase space are equilibrium ones, only the vast majority of them. This was even more relevant in Chap. 8 when we discussed how irreversible macroscopic equations can be, in some sense, derived from microscopic laws. The derivation was not straightforward and did not take the form of a simple deduction:

$$\text{microscopic laws} \longrightarrow \text{macroscopic laws.}$$

Some philosophical views on reductionism defines the reduction of a theory T_1 to a theory T_2, if T_1 is a *logical consequence* of T_2,[3] but our apparently typical example of theoretical reduction, the one of thermodynamics to statistical physics, shows that things are not so simple (see Lazarovici and Reichert [212, Sect. 4] for a discussion similar to the one given here).

The other caveat is that we can only derive macroscopic laws when there are such laws, which requires the macroscopic system to behave in a predictable and reproducible way. If one takes any complex system, it is possible that its behavior exhibits sensitive dependence on initial conditions (in the sense of Sect. 4.5) at the microscopic level, but with effects at the macroscopic level, in which case there is simply no macroscopic law to be derived from the microscopic ones: one can think, among many examples, of any cardiovascular accident that will affect the whole brain, or of the butterfly effect discovered by Lorenz, or, speaking about history, of Pascal's remark on Cleopatra's nose.

And if one adopts such a nuanced view of reductionism (as a goal of science rather than a fixed doctrine) then, there is no big conflict between Anderson and Weinberg, only a difference of emphasis: Anderson is right in the sense that we would not be able to discover the facts of macroscopic physics, chemistry, biology, etc. if we only knew the fundamental laws of physics; so that those higher level science have a certain degree of autonomy (their own laws and concepts) with respect to fundamental physics; but Weinberg is also right in the sense that whenever new phenomena are discovered, we try to see if they can at least in principle be explained on the basis of the fundamental laws or if their explanation requires new physical laws (like

---

[3] This simplistic view is not accepted nowadays; see [319] for a discussion of various philosophical definitions of reductionism.

for electromagnetic phenomena) or modified physical laws (like the reduction of chemistry to physics).

The people who are genuinely antireductionist (and they compose the vast majority of mankind although not of scientists) are those who believe in magic, interventions of deities or direct actions of the mind on the body. I can illustrate this difference by an anecdote: some time ago, I was speaking, at a school in Trieste for students in scientific journalism, and Anderson was also there. He gave an antireductionist talk in the spirit of "More is different", but a little later I heard students who had (mis)understood him as supporting things like holistic medicine. The students viewed that positively, but, of course, Anderson had done no such thing.

The real division between reductionists and antireductionists opposes friends and enemies of the scientific world-view, and, in that opposition, Anderson and Weinberg are on the same side of the barricade.

Chapter 11
Hints and Solutions for the Exercises

11.1 Exercises of Chap. 2

2.1. $= \dfrac{1}{2^6} \left(\dfrac{6!}{4!2!} + \dfrac{6!}{5!1!} + \dfrac{6!}{6!0!} \right)$

2.2. That probability equals 1 minus the probability $\bar{P}(N)$ that all birthdays are different. And that latter probability can be computed by labelling people $1, \dots, N$ and multiplying the probabilities that the kth person's birthday is different from the ones of the previous $k - 1$ persons, for $k = 2, \dots, N$, so, with $l = k - 1$,

$$\bar{P}(N) = \prod_{l=1}^{N-1} \frac{365 - l}{365} = \frac{364!}{365^{N-1}(365 - N)!}$$

Numerical evaluation shows that $\bar{P}(N) < \frac{1}{2}$ (thus the probability that two people have the same birthday exceeds $\frac{1}{2}$) for $N \geq 23$.

2.3. 1.

$$\left(\int_{\mathbb{R}} \exp\left(-\frac{x^2}{2} \right) dx \right)^2 = \int_{\mathbb{R}^2} \exp\left(-\frac{x^2 + y^2}{2} \right) dxdy$$
$$= \int_{\mathbb{R}_+} \int_0^{2\pi} \exp\left(-\frac{r^2}{2} \right) rdrd\phi = 2\pi \qquad (11.1.1)$$

by setting $\frac{r^2}{2} = u$ and $\int_{\mathbb{R}_+} \exp(-\frac{r^2}{2})rdr = \int_{\mathbb{R}_+} \exp(-u)du = 1$.

3. We get:

$$\mathbb{E}_g(\exp(ikx)) = \sum_{l=0}^{\infty} \frac{(ik)^l}{l!} \mathbb{E}_g(x^l),$$

© Springer Nature Switzerland AG 2022
J. Bricmont, *Making Sense of Statistical Mechanics*, Undergraduate Lecture Notes in Physics, https://doi.org/10.1007/978-3-030-91794-4_11

and

$$\exp\left(-\frac{\sigma^2 k^2}{2}\right) = \sum_{m=0}^{\infty} \frac{(-\sigma^2 k^2)^m}{2^m m!}$$

By identifying the coefficients of the powers of k $(l = 2m)$, we get

$$\mathbb{E}_g(x^{2m}) = \frac{(2m)!}{2^m m!}\sigma^{2m} \quad \forall m \in \mathbb{N} \tag{11.1.2}$$

(and $\mathbb{E}_g(x^{2m+1}) = 0, \forall m \in \mathbb{N}$).

2.4. Parts 1–3 of are a straightforward extension of exercise 2.3, but some care is needed to derive (2.8.2) For part 4, we put $x_1, \ldots, x_{2m} = x$ and note that the number of partitions of $2m$ objects into pairs is $\frac{(2m)!}{m!2^m}$: arrange the $2m$ objects in some order and take them two by two to construct a partition into pairs. There are $(2m)!$ such ordering but nothing changes in the final partition if we permute the m pairs or the elements in each pairs. So, one has to divide $(2m)!$ by $m!$ and by 2^m.

2.5.

$$\mathbb{E}(X) = \sum_{k=1}^{\infty} k \frac{\lambda^k}{k!} \exp(-\lambda) = \lambda \sum_{k=1}^{\infty} \frac{\lambda^{k-1}}{(k-1)!} \exp(-\lambda) = \lambda \tag{11.1.3}$$

since $\sum_{k=1}^{\infty} \frac{\lambda^{k-1}}{(k-1)!} = \sum_{k=0}^{\infty} \frac{\lambda^k}{k!} = \exp(\lambda)$.

$$\mathbb{E}(X^2) = \sum_{k=1}^{\infty} k^2 \frac{\lambda^k}{k!} \exp(-\lambda) = \lambda \exp(-\lambda) \sum_{k=1}^{\infty} k \frac{\lambda^{k-1}}{(k-1)!}$$

$$= \lambda \exp(-\lambda) \frac{d}{d\lambda} \sum_{k=1}^{\infty} \frac{\lambda^k}{(k-1)!} = \lambda(\lambda + 1) \tag{11.1.4}$$

since $\sum_{k=1}^{\infty} \frac{\lambda^k}{(k-1)!} = \lambda \sum_{k=0}^{\infty} \frac{\lambda^k}{k!} = \lambda \exp(\lambda)$ and $\frac{d}{d\lambda}\lambda \exp(\lambda) = (\lambda + 1)\exp(\lambda)$.
So, the variance: $\mathbb{E}(X^2) - \mathbb{E}(X)^2 = \lambda$.

2.6. We need to show that

$$\lim_{n\to\infty} \frac{n!}{(n-k)!k!} p_n^k (1 - p_n)^{n-k} = \frac{\lambda^k}{k!} \exp(-\lambda) \tag{11.1.5}$$

We have, setting $p_n = \frac{\lambda}{n}$

$$\frac{n!}{(n-k)!k!} p_n^k (1-p_n)^{n-k} = \frac{\prod_{j=0}^{k-1}(n-j)}{k!n^k}\lambda^k\left(1 - \frac{\lambda}{n}\right)^{n-k} \tag{11.1.6}$$

Now use $\frac{\prod_{j=0}^{k-1}(n-j)}{n^k} = \frac{n^k + \mathcal{O}(n^{k-1})}{n^k} \to 1$, as $n \to \infty$ for fixed k, and $(1 - \frac{\lambda}{n})^{n-k} \to \exp(-\lambda)$ in the same limit.

2.7. The distribution of X is the binomial one (2.3.5), with $P(1) = p$, $P(0) = 1 - p$, and $P(X = k) = \frac{N!}{k!(N-k)!} p^k (1 - p)^{N-k}$. Use the fact that the expected value and the variance of a sum of independent random variables are the sum, respectively, of the expected value and of the variance of each variable. Since the probability of A is p, we get p for the expected value of each variable and thus Np for the expected value of the sum; for the variance of each variable we get $p - p^2 = p(1 - p)$ and thus $Np(1 - p)$ for the variance of the sum.

2.8. The probability that a given person is the most unlucky among N persons is, by symmetry, $\frac{1}{N}$. So, the probability that the first person who is more unlucky that you is the Nth one is the probability that you are the most unlucky person among $N - 1$ persons, minus the probability that the most unlucky person among N persons is the Nth one.

So, the probability that one has to meet N persons before finding someone more unlucky than oneself is $C(\frac{1}{N-1} - \frac{1}{N}) = C \frac{1}{(N-1)N}$, where $C = \frac{1}{\sum_{N=2}^{\infty} \frac{1}{(N-1)N}}$ is the normalization factor. It is easy to see that the expected value of that variable, $C \sum_{N=2}^{\infty} \frac{N}{(N+1)N} = \infty$, so that the expected time that you have to wait before you meet someone more unlucky than you is infinite. Of course, the same reasoning works for the waiting time before you meet someone more lucky than you.

2.9. The probability that the first time that the coin falls heads is on the nth flip is $\frac{1}{2^n}$ (that is the probability of a sequence of $n - 1$ tails followed by a heads). So, the expected gain $\mathbb{E}$ of Peter equals:

$$\mathbb{E} = \sum_{n=1}^{\infty} \frac{2^{n-1}}{2^n} = \infty. \tag{11.1.7}$$

Thus, the amount of money that Peter should be ready to give in order to play the game should be arbitrarily large, since his expected gain is infinite.

The paradox is that nobody will be willing to do that, the reason being that the probability of winning a large sum 2^{n-1} (for n large) is very small ($\frac{1}{2^n}$). Maudlin in [236] suggests to solve that paradox by reasoning in terms of typicality: a typical game (if played only once) will stop after a few tosses of the coin, so one should not give a too large sum in order to enter the game.

2.10. The probability that we have N_0 heads and $N - N_0$ tails is

$$\frac{N!}{N_0!(N - N_0)!} \left(\frac{1}{3}\right)^{N_0} \left(\frac{2}{3}\right)^{N-N_0} \tag{11.1.8}$$

where N_0 is the number of heads; for $N_0 \approx \frac{N}{2}$, we can approximate $\frac{N!}{N_0!(N-N_0)!} \approx 2^N$, using Stirling's formula for $N_0 = \frac{N}{2}$. So (11.1.8) equals approximately $2^N (\frac{1}{3}\frac{2}{3})^{\frac{N}{2}} \leq \exp(-cN)$ with $c = \frac{1}{2} \ln \frac{9}{8}$, which goes to 0 exponentially in N as $N \to \infty$.

2.11. Stirling's formula gives

$$\ln \frac{N!}{(\frac{N}{2} + \sqrt{N}n)!(\frac{N}{2} - \sqrt{N}n)!}$$
$$\approx N \ln N - N - \left(\frac{N}{2} + \sqrt{N}n\right) \ln \left(\frac{N}{2} + \sqrt{N}n\right) + \left(\frac{N}{2} + \sqrt{N}n\right)$$
$$- \left(\frac{N}{2} - \sqrt{N}n\right) \ln \left(\frac{N}{2} - \sqrt{N}n\right) + \left(\frac{N}{2} - \sqrt{N}n\right) \qquad (11.1.9)$$

Write

$$\ln \left(\frac{N}{2} \pm \sqrt{N}n\right) = \ln N - \ln 2 + \ln \left(1 \pm \frac{2n}{\sqrt{N}}\right)$$
$$= \ln N - \ln 2 \pm \frac{2n}{\sqrt{N}} + \mathcal{O}\left(\frac{n^2}{N}\right),$$

so that

$$-\sqrt{N}n \ln \left(\frac{N}{2} + \sqrt{N}n\right) + \sqrt{N}n \ln \left(\frac{N}{2} - \sqrt{N}n\right)$$
$$= -2\sqrt{N}n\frac{2n}{\sqrt{N}} + \mathcal{O}\left(\frac{n^3}{\sqrt{N}}\right) = -4n^2 + \mathcal{O}\left(\frac{n^3}{\sqrt{N}}\right).$$

We also have:

$$-\frac{N}{2}\left(\ln\left(1 + \frac{2n}{\sqrt{N}}\right) + \ln\left(1 - \frac{2n}{\sqrt{N}}\right)\right)$$
$$= -\frac{N}{2}\left(\ln\left(1 - \frac{4n^2}{N}\right)\right) = 2n^2 + \mathcal{O}\left(\frac{n^4}{N}\right),$$

which implies:

$$-\frac{N}{2}\left(\ln\left(\frac{N}{2} + \sqrt{N}n\right) + \ln\left(\frac{N}{2} - \sqrt{N}n\right)\right)$$
$$= -N \ln N + N \ln 2 + 2n^2 + \mathcal{O}\left(\frac{n^4}{N}\right).$$

Insert all this in (11.1.9):

$$\ln \frac{N!}{\left(\frac{N}{2} + \sqrt{Nn}\right)! \left(\frac{N}{2} - \sqrt{Nn}\right)!}$$

$$\approx N \ln 2 - 4n^2 + 2n^2 + \mathcal{O}\left(\frac{n^3}{\sqrt{N}}\right) + \mathcal{O}\left(\frac{n^4}{N}\right) \qquad (11.1.10)$$

$$= N \ln 2 - 2n^2 + \mathcal{O}\left(\frac{n^3}{\sqrt{N}}\right) + \mathcal{O}\left(\frac{n^4}{N}\right),$$

which shows that the distribution of n approaches a Gaussian one as $N \to \infty$ (the term $N \ln 2$ is compensated by the factor 2^{-N} in the formula (2.3.5) with $p = \frac{1}{2}$).

2.12. The probability equals $\prod_{i=1}^{L} p_i^{n_i}$ times the number of ways to choose $n_1, n_2, \ldots n_L$ balls among N balls, which is given by (6.2.8). So the probability equals:

$$\frac{N!}{\prod_{i=1}^{L} n_i!} \prod_{i=1}^{L} p_i^{n_i} \qquad (11.1.11)$$

2.13. Each step $s_n = \pm 1$ with equal probability $\frac{1}{2}$ and thus $x(T) = \sum_{n=1}^{T} s_n$ is a sum of independent random variables.

Since $\mathbb{E}(s_n^2) = 1$, we get $\mathbb{E}(x(T)^2) = T$ and for the limiting behavior when $T \to \infty$ of $\frac{x(T)}{\sqrt{T}}$, use the central limit Theorem 2.1.

2.14. 1. Let us denote the boxes GG GS and SS and compute the conditional probability of the chosen box being GG (the next coin will be a golden one if the chosen box is GG) given that the coin withdrawn from it is a golden one. We need to use (2.2.7), with three hypotheses, GG GS and SS, all with prior probability $\frac{1}{3}$, with data G and with $P(G \mid GG) = 1$, $P(G \mid GS) = \frac{1}{2}$, $P(G \mid SS) = 0$. So, we need to compute:

$$P(GG \mid G) = \frac{\frac{1}{3} P(G \mid GG)}{\frac{1}{3}(1 + \frac{1}{2} + 0)} = \frac{2}{3}, \qquad (11.1.12)$$

which is more than $\frac{1}{2}$, $\frac{1}{2}$ being the answer that might be given intuitively, since there are an equal number of silver and gold coins (but that is if we ignore the data that the coin withdrawn from the chosen box is a golden one).

2. You should open the door that you have not chosen, since the one you have chosen has probability $\frac{1}{3}$ and thus the two others combined have probability $\frac{2}{3}$, but, since the host gave you information about one of those doors (there is nothing behind it), the probability that the treasure is behind the remaining door is also $\frac{2}{3}$.

This problem has led to intense discussions, see e.g. the Monty Hall problem on wikipedia.

2.15. Let A, B, and C be the event that the chosen coin was type A, type B, and type C. We have three hypotheses ($n = 3$), and our hypotheses are: $H_1 =$ the chosen coin is of type A, $H_2 =$ the chosen coin is of type B and $H_3 =$ the chosen coin is of type C.

Let D be the data, namely that the toss is heads. We have to find $P(A \mid D)$, $P(B \mid D)$, $P(C \mid D)$.

Since the drawer has 2 coins of type A, 2 of type B and 1 or type C we have for the prior probabilities $P(A) = 0.4$, $P(B) = 0.4$, $P(C) = 0.2$. We have that $P(D \mid A) =$ probability of heads if the coin is type A equals 0.5 and $P(D \mid B) = 0.6$, $P(D \mid C) = 0.9$.

For $P(D)$ we get:

$$P(D) = P(D \mid A)P(A) + P(D \mid B)P(B) + P(D \mid C)P(C)$$
$$= 0.5 \cdot 0.4 + 0.6 \cdot 0.4 + 0.9 \cdot 0.2 = 0.62. \tag{11.1.13}$$

Now each of the three posterior probabilities can be computed:

$$P(A \mid D) = \frac{P(D \mid A)P(A)}{P(D)} = \frac{0.5 \cdot 0.4}{0.62} = \frac{0.2}{0.62}. \tag{11.1.14}$$

$$P(B \mid D) = \frac{P(D \mid B)P(B)}{P(D)} = \frac{0.6 \cdot 0.4}{0.62} = \frac{0.24}{0.62}. \tag{11.1.15}$$

$$P(C \mid D) = \frac{P(D \mid C)P(C)}{P(D)} = \frac{0.9 \cdot 0.2}{0.62} = \frac{0.18}{0.62}. \tag{11.1.16}$$

2.16. The set $\mathbb{Q}$ of rational numbers is not closed since irrational number are limits of rational ones and it is not open either since every open set containing a rational number contains irrational ones. A point is a closed set, i.e. it is a Borel set and $\mathbb{Q}$ is a countable union of points, hence it is also a Borel set.

2.17. It is obvious that $I \cap J$, where I and J are intervals is also an interval. And it is easy to see that $I \setminus J$ is a disjoint union of intervals. The extension to rectangles is immediate since the two properties defining a semi-algebra are valid for each factor in products of the form $I_1 \times \cdots \times I_n$.

2.18. As in exercise 2.17, we use the fact that the two properties defining a semi-algebra are valid for each factor in products of the form (2.A.5).

2.19. Since $\mathbb{Q}$ is a Borel set, its indicator function is measurable; since each point is of measure 0, and since a countable union of sets of measure 0 is of measure 0, $\mathbb{Q}$ is of measure 0 and the integral of its indicator function also equals 0.

2.20. The proof is similar to the one of (2.3.9); we have only to check that $E_a = \{\mathbf{x} \mid \lim_{N \to \infty} n_\alpha(\mathbf{x}) = P_a\}$, for $a \in \mathbb{R}$, is invariant under the shift map T_{shift}, which is obvious, and then proceed as in the proof of (2.3.9).

11.2 Exercises of Chap. 3

3.1. The conservation of energy gives $E = \frac{1}{2}v_0^2 + V(x_0) = \frac{1}{2}(\frac{dx(t)}{dt})^2 + V(x(t))$, $\forall t$. This means that $\frac{dx(t)}{dt} = \sqrt{2(E - V(x(t)))}$ or (formally) $dt = \frac{dx(t)}{\sqrt{2(E - V(x(t)))}}$. Integrate both sides of that equation over t and replace $x(t)$ by x in the resulting right hand side and we get (3.7.1).

3.2. The solution is obviously $x(t) = at$ mod 1, $y(t) = bt$ mod 1. The motion is periodic if there exist $n, m \in \mathbb{N}$ with $at = n, bt = m$. If $\frac{a}{b} \in \mathbb{Q}$, $\frac{a}{b} = \frac{p}{q}$, $p, q \in \mathbb{N}$, we can choose $n = p, m = q$ and $t = \frac{n}{a} = \frac{m}{b}$.

If $\frac{a}{b} \notin \mathbb{Q}$, consider the circle in the torus given by $y = 0$. The orbit crosses this circle at times t_n, with $bt_n = n, n \in \mathbb{N}$. The x coordinate at these times equals $at_n = \frac{a}{b}n$. Since $\frac{a}{b} \notin \mathbb{Q}$, the sequence $(\frac{a}{b}n)_{n\in\mathbb{N}}$ is the orbit of an irrational rotation of the circle; but those orbits are dense in the circle, since irrational rotations of the circle are ergodic, see Sect. 4.3.1 (hence, by definition of ergodicity, the orbit must visit every interval of the circle). We can repeat the same argument for the circle in the torus given by $y = y_0$ for any $y_0 \in [0, 1[$, which proves that the curve $x(t) = at$ mod 1, $y(t) = bt$ mod 1 is dense in the torus $\mathbb{T}^2$.

Another, more direct way to see that irrational rotations of the circle have dense orbits is to, first, notice that an orbit $x + \alpha n, n \in \mathbb{N}$, with $\alpha \notin \mathbb{Q}$ is composed of infinitely many distinct points: $x + \alpha n = x + \alpha m$ mod 1 implies $n - m$. Then, if there exists an interval I of length ϵ that the orbit does not visit (which must occur if the orbit is not dense), then $|\alpha n - \alpha m| \geq \epsilon$, $\forall n \neq m$. Indeed, otherwise, the set of rotations by integer multiples of $\alpha n - \alpha m$ must be made of points at a distance strictly less than ϵ from each other; but then, one of these points must be in I. If $|\alpha n - \alpha m| \geq \epsilon$, the orbit $x + \alpha n, n \in \mathbb{N}$ contains at most $\frac{1}{\epsilon}$ distinct points, which contradicts the fact that the orbit $x + \alpha n$ contains of infinitely many distinct points.

3.3. The solution of (3.7.2) with the given initial conditions is:

$$x(t) = \cos(\omega_1 t)$$
$$y(t) = \cos(\omega_2 t) \qquad\qquad (11.2.1)$$

Since the cosine function is periodic, the functions $\omega_1 t, \omega_2 t$ are similar to the functions $x(t), y(t)$ of the solution of the previous exercise and we can then argue as we did there.

3.4. From (3.3.6) and (3.2.5), we get:

$$\frac{d}{dt}\mathbf{P}(t) = \sum_{i=1}^{N} \frac{d}{dt}\mathbf{p}_i(t) = \sum_{k=1}^{N} \nabla_{\mathbf{q}_k} \sum_{i<j=1}^{N} \tilde{V}_{ij}(\mathbf{q}_i - \mathbf{q}_j) = 0, \qquad (11.2.2)$$

because the terms with index i and j in the second sum have opposite signs for $k = i$ and $k = j$ in the first sum (see (3.2.7)).

3.5. From the definition of $\mathbf{M}(t)$, we have

$$\frac{d}{dt}\mathbf{M}(t) = \sum_{i=1}^{N} \frac{d}{dt}(\mathbf{r}_i(t) \wedge \mathbf{p}_i(t)) \tag{11.2.3}$$

We have also: $\frac{d}{dt}(\mathbf{r}_i(t) \wedge \mathbf{p}_i(t)) = (\frac{d}{dt}\mathbf{r}_i(t) \wedge \mathbf{p}_i(t)) + (\mathbf{r}_i(t) \wedge \frac{d}{dt}\mathbf{p}_i(t))$. Since $\frac{d}{dt}$ $\mathbf{r}_i(t)$ is proportional to $\mathbf{p}_i(t)$, the first term vanishes. The sum $\sum_{i=1}^{N}$ of second term vanishes by (3.2.7), as in the previous exercise.

3.6. From the formula for the determinant, we see that every non diagonal element of the matrix $1 + tA$ will be multiplied by other non diagonal elements and thus will be of order at least t^2. And for the term with only diagonal elements, we have: $\prod_i (1 + tA_{ii}) = 1 + t\,\text{trace}\,A + \mathcal{O}(t^2)$, where the second term corresponds to choosing one A_{ii} in the product and 1 for all the other factors. Choosing two or more A_{ii}'s gives a term $\mathcal{O}(t^2)$.

3.7. In Newton's equation (3.2.1), with $i = 1$, change variables $\tilde{t} = \sqrt{\frac{m'}{m}}t$ and write the equation for $\mathbf{q}(\tilde{t})$. We get the same equation as for $\mathbf{q}(t)$, but with m replaced by m', which proves the claim.

3.8. (1) $V(x, y, z) = -\frac{1}{2}(x^2 + y^2 + z^2)$
(2) $V(x, y, z) = -\frac{c}{2}(x^2 + y^2)$
(3) $V(x, y, z) = -xyz$
(4) Using the hint, we see that there is no potential: $\partial_y F_x = 0 \neq \partial_x F_y = 1$.
(5) $V(x, y, z) = mgz$

3.9. From (3.7.1), we have to integrate $\int_0^{x_1} \frac{dx}{\sqrt{2(E - \frac{x^2}{2})}}$, with $x_0 = 0$. Set $x = a\sin\phi$, with $a^2 = 2E$, we get $2(E - \frac{x^2}{2}) = 2E(1 - \sin^2\phi)$, $dx = \sqrt{2E}\cos\phi\,d\phi$ and

$$\int_0^{x_1} \frac{dx}{\sqrt{2(E - \frac{x^2}{2})}} = \int_0^{\phi_1} d\phi = \phi_1, \tag{11.2.4}$$

with $\phi_1 = \arcsin\frac{x_1}{a}$. Since this integral equals t, by setting $x_1 = x(t)$, we get $x(t) = \sqrt{2E}\sin t$.

11.3 Exercises of Chap. 4

4.1. The analogue of (4.3.10) for the n-dimensional torus is:

$$F(\mathbf{x}) = \sum_{l \in \mathbb{Z}^n} c_l \exp(2\pi i \mathbf{l} \cdot \mathbf{x}), \tag{11.3.1}$$

and the equation $F(x) = F(T_\alpha x)$ implies:

$$c_l(e^{2\pi i \mathbf{l} \cdot \alpha} - 1) = 0 \quad \forall \mathbf{l} \in \mathbb{Z}^n, \tag{11.3.2}$$

which implies $c_l = 0$ for $l \neq 0$, if and only if $l \cdot \alpha = \sum_{i=1}^{n} l_i \alpha_i = m$ has no solution except $l_i = 0$, $\forall i$, and $m = 0$.

4.2. For the map T defined by (4.1.7),we just follow the proof given for the map (4.1.3) and get, instead of (4.3.14), the equation $c_n = c_{pn}$, which leads to the conclusion that $c_n = 0$, for $n \neq 0$.

4.3. Using the Fourier series on the n-dimensional torus as in (11.3.1) and the equation $F(x) = F(Tx)$ for T as in (4.1.10), we get, instead of (4.3.15), $c_l = c_{A^t l}$, where A^t is the transpose of A. Iterating, we get that the coefficients $c_{(A^t)^k l}$ are equal $\forall k \in \mathbb{N}$, which implies, since $\sum_{l \in \mathbb{Z}^n} |c_l|^2 < \infty$, that $c_l = 0 \; \forall l \in$

$\mathbb{Z}^n$, unless $\exists k, l$ with $(A^t)^k l = l$, which means that A^t (and therefore also A) has an eigenvalue that is a root of unity.

On the other hand, if A has an eigenvalue that is a root of unity, i.e. a vector l with $(A^t)^k l = l$, the function $F(x) = \sum_{j=0}^{k-1} \exp(2\pi i l \cdot A^j x) = \sum_{j=0}^{k-1} \exp(2\pi i (A^t)^j l \cdot x)$ satisfies $F(x) = F(Tx)$ and is not constant.

4.4. The mixing property follows by writing (4.6.3) as $T(x, y) = \sigma^{-1} T_{shift} \sigma(x, y)$, which implies $T^n = \sigma^{-1} T^n_{shift} \sigma$, and then, inserting this in the definition 4.10 of mixing and using the property of mixing for the shift map, see the end of Sect. 4.4.

For the sensitive dependence on initial conditions, we need to check that the conjugation σ in (4.6.3) and its inverse satisfy (4.10.1). Here, $\omega = (x, y)$, $\omega' = (x', y')$ and $\Phi(\omega) = \sigma(x, y) = \mathbf{a} = (a_n)_{n \in \mathbb{Z}}$, $\Phi(\omega') = \sigma(x', y') = \mathbf{a}' = (a'_n)_{n \in \mathbb{Z}}$. The first inequality in (4.10.1) is immediate: dist $((x, y), (x', y')) \leq \sum_{n \in \mathbb{Z}} \frac{|a_n - a'_n|}{2^n}$.

For the second inequality, let N be the smallest integer such that $\frac{1}{2^N} < $ dist $((x, y), (x', y'))$, which means that dist $((x, y), (x', y')) \leq \frac{1}{2^{N-1}}$. Then, we must have, for some i, with $|i| \leq N + 1$, $a_i \neq a'_i$, otherwise dist $(\mathbf{a}, \mathbf{a}') \leq \sum_{n \in \mathbb{Z}, |n| \geq N+2} \frac{1}{2^n} \leq 2 \cdot 2^{-N-1}$ and dist $((x, y), (x', y')) \leq \sum_{n \in \mathbb{Z}} \frac{|a_n - a'_n|}{2^n}$ would then imply dist $((x, y), (x', y')) \leq \frac{1}{2^N}$. But, if $a_i \neq a'_i$ with $|i| \leq N + 1$, we have dist $(\mathbf{a}, \mathbf{a}') \geq 2^{-|i|} \geq 2^{-(N+1)} \geq \frac{1}{4}$ dist $((x, y), (x', y'))$. So, (4.10.1) holds with $L = 4$.

4.5. Choose (p, q) so that $\frac{1}{q} \leq \epsilon$, and so that $|\alpha - \frac{p}{q}| \leq \frac{1}{q^2}$. With $n = q$, we have $|n\alpha - p| \leq \epsilon$, so n can be chosen as $n = [\frac{1}{\epsilon}] + 1$.

4.6. With $x = C(y) = \sin \pi y$, $1 - 2x^2 = 1 - 2C(y)^2 = 1 - 2 \sin^2 \pi y = \cos 2\pi y$, so that the map $x \to 1 - 2x^2$ is conjugated by $C(y) = \sin \pi y$ to the map $x \to 2x \mod 1$. Then, reason as in the derivation of (4.6.10) using the fact that $C'(y) = \pi \cos \pi y = \pi \sqrt{1 - x^2}$.

4.7. The reasoning that led to (4.3.18) here gives:

$$n \log_{10} 3 \in J_p \equiv [\log_{10} p, \log_{10}(p + 1)[\quad \mod 1. \qquad (11.3.3)$$

and, by ergodicity, the frequency of a digit p is $\log_{10}(1 + \frac{1}{p})$, provided that $\log_{10} 3 \notin \mathbb{Q}$, which is easy to check. The same result holds for all sequences

$(a^n)_{n=1}^\infty$, $a \in N$, provided that $\log_{10} a \notin \mathbb{Q}$, which is true for all a's that are not powers of 10 (because $\log_b a$, for $a, b \in \mathbb{N}$ is either an integer or an irrational number).

4.8. Write x and y in base 4 and then proceed as for the baker's map.

4.9. Because of the conjugation with the shift map, it is the same number as the number of points that have a periodic orbit of length n under the shift map, i.e. 2^n (take any sequence of symbols of length n and repeat it periodically)

4.10. A subset E of $[0, 1[$ is dense if, for any finite sequence $(b_n)_{n=0}^M$, $\exists x \in E$, $x = \sum_{n=1}^\infty \frac{a_n}{2^n}$, with $a_n = b_n$, $\forall n = 0, \ldots, M$, because then, $\forall y \in [0, 1[$, $y = \sum_{n=1}^\infty \frac{b_n}{2^n}$, and $\forall \epsilon > 0$, $\exists x \in E$ with $|x - y| \le \epsilon$, by choosing M such that $2^{-M} \le \epsilon$.

Thus, $x \in [0, 1[$ has a dense orbit under the map T (4.1.3) if every finite sequence of 0's and 1's appears somewhere in the binary expansion of $x = \sum_{n=0}^\infty \frac{a_n}{2^n}$, because then, since $Tx = \sum_{n=0}^\infty \frac{a_{n+1}}{2^n}$, that sequence will appear, under the iteration of T, at the first places of the binary expansion of $T^k x$ for some k.

Any given finite sequence of 0's and 1's of length M appears somewhere in the binary expansion of almost all $x \in [0, 1[$, because it has a non-zero probability of appearing between $n = j$ and $n = j + M$ for each $j = 1, \ldots, \infty$. Thus it appears with probability one between $n = j$ and $n = j + M$ for some j (see footnote 21 in Chap. 2) and thus, the set of x's where the given finite sequence of 0's and 1's of length M does appear in the binary expansion of x is of measure 0. Now use the fact that the set of finite sequence of 0's and 1's is countable and the fact that a countable union of sets of measure 0 is of measure 0 (by (2.A.1)) to conclude the argument.

4.11. Write x in base 3: $x = \sum_{n=1}^\infty \frac{a_n}{3^n}$ and choose $a_n = 0$ if n is a prime number or $n = 0, 1$, and $a_n = 2$ if n is not a prime number.
This defines a unique x.

4.12. That follows from the conjugation (4.6.3) and the fact that the set of periodic points for the shift map is dense. This is because, for any $\epsilon > 0$, choose N so that $2^{2-N} \le \epsilon$; then, note that any sequence $(a_n)_{n=-\infty}^\infty$ can be approximated by a periodic sequence $(b_n)_{n=-\infty}^\infty$ with $a_n = b_n$, $|n| \le N$ (repeat periodically the symbols b_n for $|n| > N$) and we have $\sum_{|n| \ge N+1} \frac{|a_n - b_n|}{2^n} \le 2^{2-N} \le \epsilon$.

4.13. We know that the transformation given by the map $f(x) = 4x(1 - x)$ is ergodic with respect to the measure whose density is $\frac{1}{\pi \sqrt{x(1-x)}}$, see (4.6.10). So, by Birkhoff's ergodic theorem, we get that:

$$\lim_{N \to \infty} \frac{1}{N} \sum_{n=0}^{N-1} (x_n(1 - x_n))^{3/2} = \int_0^1 \frac{(x(1-x))^{3/2}}{\pi \sqrt{x(1-x)}} dx = \int_0^1 \frac{(x(1-x))}{\pi} dx = \frac{1}{6\pi}.$$

$$(11.3.4)$$

4.14. This follows immediately from (4.7.2) and (4.7.4) with $\mu(\Omega_1) = p_1$ and $\mu(\Omega_2) = p_2 + p_3$.

4.15. It is easy to check that the measure (4.10.3) on cylinder sets satisfies the properties of definition (2.2) and therefore can be extended to a measure μ on Ω using Proposition (2.4) and that this extension is invariant under the shift T_{shift}.

For the entropy, use (4.7.2), $\ln(p_{a_0} \prod_{i=1}^{N-1} P_{a_i a_{i+1}}) = \ln p_{a_0} + \sum_{i=1}^{N-1} \ln P_{a_i a_{i+1}}$ and then use repeatedly $\sum_{x=1}^{l} p_x P_{xy} = p_y$ and $\sum_{y=1}^{l} P_{xy} = 1$ to get:

$$S(\mu, T_{\text{shift}}) = -\sum_{x,y=1}^{l} p_x P_{xy} \ln P_{xy}. \qquad (11.3.5)$$

11.4　Exercises of Chap. 5

5.1. We get (5.7.8) from (5.7.4), by identifying the derivatives of S or directly from (5.7.5); (5.7.9) follows from differentiating both sides of (5.7.8) and using (5.6.17). From (5.6.21) and (5.6.18), we get:

$$d\Phi = -SdT - PdV + \sum_{i=1}^{l} \mu_i dN_i - \sum_{i=1}^{l} N_i d\mu_i - \sum_{i=1}^{l} \mu_i dN_i$$

$$= -SdT - PdV - \sum_{i=1}^{l} N_i d\mu_i, \qquad (11.4.1)$$

which is (5.6.22).

5.2. Using (5.3.9) and (5.7.2), we get

$$\delta Q = NkcdT + PdV = Nk(cdT + \frac{T}{V}dV) \qquad (11.4.2)$$

So, we have:

(1) for constant temperature

$$\int \delta Q = \int_{V_i}^{V_f} Nk\frac{T}{V}dV = NkT \ln \frac{V_f}{V_i}.$$

(2) For constant volume,

$$\int \delta Q = \int_{T_1}^{T_2} NkcdT = Nkc(T_2 - T_1).$$

(3) For constant pressure, in (11.4.2), can use $PdV = NkdT$ and get, if we integrate over T,

$$\int \delta Q = \int_{T_1}^{T_2} Nk(1+c)dT = Nk(1+c)(T_2 - T_1),$$

and, if we integrate over V:

$$\int \delta Q = \int_{V_1}^{V_2} P(1+c)dV = P(1+c)(V_2 - V_1),$$

that are equivalent formulas.

5.3. (1) From the previous exercise $\int_{\text{Cycle}} \delta Q = Nk(T_2 - T_1) - T_1 \ln(\frac{V_2}{V_1})$, where V_2 is the volume at the end of the transformation at constant pressure. Indeed, we get a term $Nk(1+c)(T_2 - T_1)$ for the transformation at constant pressure, a term $-Nkc(T_2 - T_1)$ for the transformation at constant volume and a term $-T_1 \ln(\frac{V_2}{V_1})$ for the transformation at constant temperature.

If the pressure is constant when going from (T_1, V_1) to (T_2, V_2), we can use $PV = NkT$ to write $\int_{\text{Cycle}} \delta Q = NK(T_2 - T_1) - T_1 \ln(\frac{T_2}{T_1}) = NkT_1(\frac{T_2}{T_1} - 1 - \ln(\frac{T_2}{T_1}))$. One can check that $x - 1 - \ln x$ is positive for $x \geq 1$, with a minimum equal to 0 at $x = 1$. So, $\int_{\text{Cycle}} \delta Q > 0$ for $T_2 > T_1$.

(2) We get the opposite result.

(3) The integral would vanish since one integrates over a cycle.

(4) $-\delta W = \delta Q$, by (5.3.9), with the work done by the gas being $-\delta W$ in that formula and using the fact that the integral of the differential dE over a cycle vanishes.

5.4. Starting from (5.7.20):

(1) $\Delta S = Nk \ln \frac{V_2}{V_1}$, because the temperature is constant and where V_1, V_2 are the volumes at both ends of the process..

(2) $\Delta S = cNk \ln \frac{T_2}{T_1}$, because the volume is constant and where T_1, T_2 are the temperatures at both ends of the process.

(3) We need $dS = 0$, hence $\delta Q = 0$, which implies that $TV^{\gamma-1}$ is constant, see Sect. 5.7.4.

5.5. (1) Using (5.6.11), $S(T, V) = -\frac{\partial}{\partial T} F(T, V)$, we get: $S(T, V) = 4CVT^3$.

(2) By (5.6.9), $P(T, V) = -\frac{\partial}{\partial V} F(T, V)$, and we get: $P(T, V) = CT^4$.

(3) $E(T, V) = F(T, V) + TS(T, V) = 3CVT^4$

(4) Since $F(T, V)$ is independent of N, $\mu = 0$.

(5) We have, from (2) and (3) above: $PV = \frac{1}{3}E$, and so, $dE = 3PdV + 3VdP$. In an adiabatic transformation $\delta Q = 0$ and $-dE = PdV$; adding those relations, we get $4PdV + 3VdP = 0$, or $\frac{4}{3}\frac{dV}{V} + \frac{dP}{P} = 0$; which gives, after integration $PV^{\frac{4}{3}} = $ constant.

5.6. Use (5.6.22) and (5.8.1), we get:

$(\frac{\partial P}{\partial \mu})_{T,V} = (\frac{\partial N}{\partial V})_{T,\mu}$,

$(\frac{\partial S}{\partial \mu})_{T,V} = (\frac{\partial N}{\partial T})_{V,\mu}$,

$(\frac{\partial S}{\partial V})_{T,\mu} = (\frac{\partial P}{\partial T})_{V,\mu}$, where on both sides, the function are functions of T, V, μ, and the variables indicated as indices of the parentheses are kept constant.

5.7. (1) ω_1 not exact: $\frac{\partial}{\partial y}a(x, y) = x \neq \frac{\partial}{\partial x}b(x, y) = 0$

(2) $\omega_2 = dF$, with $F = -\cos x \cos y$.

(3) $\omega_3 = dF$, with $F = \frac{x^3}{3} + \frac{y^3}{3}$.

(4) ω_4 not exact: $\frac{\partial}{\partial y}a(x, y) = 2y \neq \frac{\partial}{\partial x}b(x, y) = 2x$.

5.8. From (5.4.4), (5.7.1), which gives $\alpha = Nk$ in (5.4.4), and $\frac{V_B}{V_A} = 4, T_1 = 200\,\text{K}$, we get $Q_1 = Nk200 \ln 4$. From (5.4.5), (5.4.6) that implies $\frac{V_C}{V_D} = 4$, and $T_2 = 100\,\text{K}$, we get $Q_2 = Nk100 \ln 4$. By (5.4.1), $\Delta W = Nk200 \ln 4 - Nk100 \ln 4 = Nk100 \ln 4$.

5.9. For $f(x) = x^\alpha, \alpha > 1, f^*(p) = C_\alpha p^{\frac{\alpha}{\alpha-1}}$ for some constant C_α, and for $f(x) = e^x, f^*(p) = p(\ln p - 1)$.

5.10. We get: $\frac{1}{T} = a^{-\frac{1}{2}}e^{-\frac{3}{4}}v^{\frac{1}{2}}$, and
$\frac{P}{T} = 2a^{-\frac{1}{2}}e^{\frac{1}{4}}v^{-\frac{1}{2}}$.

Now, insert this in (5.6.4), we get: $ds = a^{-\frac{1}{2}}\left(e^{-\frac{3}{4}}v^{\frac{1}{2}}de + 2e^{\frac{1}{4}}v^{-\frac{1}{2}}dv\right)$, which

equals $4a^{-\frac{1}{2}}d\left(e^{\frac{1}{4}}v^{\frac{1}{2}}\right)$, so that

$s = 4a^{-\frac{1}{2}}e^{\frac{1}{4}}v^{\frac{1}{2}} + s_0$, or $S = 4a^{-\frac{1}{2}}E^{\frac{1}{4}}V^{\frac{1}{2}}N^{\frac{1}{4}} + Ns_0$.

5.11. (1) We have $e(s, v) = a\frac{s^3}{v}$; so $T(s, v) = \frac{\partial e}{\partial s} = 3a\frac{s^2}{v}$ and $P(s, v) = -\frac{\partial e}{\partial v} = a\frac{s^3}{v^2}$.

(2) We have $e(s, v) = as^2 - bv^2$; so $T(s, v) = 2as$, $P(s, v) = 2bv$.

(3) We have $e(s, v) = u\frac{s^2}{v}\exp(bs)$; so $T = 2a\frac{s}{v}\exp(bs) + ab\frac{s^2}{v}\exp(bs)$, $P = a\frac{s^2}{v^2}\exp(bs)$.

5.12. Write $\frac{1}{T} = (a\frac{V}{E})^{\frac{1}{4}}$ and $\frac{P}{T} = \frac{1}{3}a^{\frac{1}{4}}(\frac{E}{V})^{\frac{3}{4}}$.

The fundamental relation can be obtained from the Euler relation (5.7.5): $S = \frac{E}{T} + \frac{PV}{T} = \frac{4}{3}a^{\frac{1}{4}}E^{\frac{3}{4}}V^{\frac{1}{4}}$.

11.5 Exercises of Chap. 6

6.1. By expanding $s(\mathbf{x})$ to third order in $\mathbf{x}$ around $\mathbf{x}_0$, we get $s(\mathbf{x}) = s(\mathbf{x}_0) - c_L \sum_{i=1}^{L}(n_i - \frac{1}{L})^2(1 + \mathcal{O}(\epsilon))$, for some constant $c_L > 0$. Then we can bound from below $|\{\mathbf{n} \in \Omega | |\mathbf{n} - \mathbf{x}_0| \leq \epsilon\}|$ by $|\{\mathbf{n} \in \Omega | |\mathbf{n} - \mathbf{x}_0| \leq \frac{\epsilon}{2}\}|$ and proceed as in the proof of (6.2.6).

6.2. For the relative probabilities of two configurations $\mathbf{x}_1, \mathbf{x}_2$ of the subsystem of energies $E(\mathbf{x}_1) = \tilde{E}_1$ and $E(\mathbf{x}_2) = \tilde{E}_2$, and number of particles $N(\mathbf{x}_1) = \tilde{N}_1$ and $N(\mathbf{x}_2) = \tilde{N}_2$, we get from (6.6.11):

$$\frac{|\Omega(E - \tilde{E}_1, V - \tilde{V}, N - \tilde{N}_1)|}{|\Omega(E - \tilde{E}_2, V - \tilde{V}, N - \tilde{N}_2)|}, \quad (11.5.1)$$

That ratio can be written, using (6.4.3), (6.4.8), as:

$$\exp(k^{-1}N(s(e - \tilde{e}_1, v - \tilde{v}, 1 - \tilde{n}_1) - s(e - \tilde{e}_2, v - \tilde{v}, 1 - \tilde{n}_2)), \quad (11.5.2)$$

with the lower case letters referring to the variables in capital letters divided by N. Now we can proceed as in the proof of (6.6.15), using $\frac{\partial s(e,v,l)}{\partial l} = -\frac{\mu}{T}$, which follows from (5.6.17), to obtain:

$$\frac{|\Omega(E - E(\mathbf{x}_1), V - \tilde{V}, N - N(\mathbf{x}_1))|}{|\Omega(E - E(\mathbf{x}_2), V - \tilde{V}, N - N(\mathbf{x}_2))|}$$

$$\approx \exp\left(-\frac{1}{kT}(E(\mathbf{x}_1) - E(\mathbf{x}_2)) + \frac{\mu}{kT}(N(\mathbf{x}_1) - N(\mathbf{x}_2))\right), \quad (11.5.3)$$

which justifies (6.6.6).

6.3. We have $\sum_{i=1}^{N} s_i = N_+ - N_- = E$, and we have also $N_+ + N_- = N$. So, $N_+ = \frac{N+E}{2}$, $N_- = \frac{N-E}{2}$ So, $|\Omega(E, N)| = \frac{N!}{N_+! N_-!} = \frac{N!}{(\frac{N+E}{2})!(\frac{N-E}{2})!}$. Thus, we get:

$$s(e) = k \lim_{\substack{N,E \to \infty \\ \frac{E}{N}=e}} \frac{S(E, N)}{N} = -k\left(\frac{1+e}{2}\ln\frac{1+e}{2} + \frac{1-e}{2}\ln\frac{1-e}{2}\right)$$

$$(11.5.4)$$

where $-1 \le e \le 1$.

6.4. We have

$$Z(\beta, N) = \sum_{\mathbf{s}} \exp(-\beta E(\mathbf{s})) = \prod_{i=1}^{N}\left(\sum_{s_i=\pm 1} \exp(-\beta s_i)\right) = (2\cosh\beta)^N$$

$$(11.5.5)$$

and:

$$f(T) = -kT \lim_{N \to \infty} \frac{\ln Z(\beta, N)}{N} = -kT \ln(2\cosh\beta) \quad (11.5.6)$$

6.5. From exercice 6.3, and $\frac{1}{T} = \frac{\partial s(e)}{\partial e}$, we get that $e(T) = -\tanh\beta$, and then we use $f(T) = e(T) - Ts(e(T))$.

6.6. We have $(E + N - 1)!$ permutations of the set of marbles and sticks, but nothing changes to the final distribution of marbles and sticks if we permute the marbles or if we permute the sticks.

So, we get: $\Omega(E, N) = \frac{(E+N-1)!}{E!(N-1)!}$, and

$$s(e) = \lim_{\substack{N,E \to \infty \\ \frac{E}{N}=e}} \frac{S(E, N)}{N} = k\left(\ln(1 + e) + e\ln\left(1 + \frac{1}{e}\right)\right). \quad (11.5.7)$$

6.7. We have:

$$Z(\beta, N) = \sum_{\mathbf{n}} \exp(-\beta E(\mathbf{n})) = \prod_{i=1}^{N} \sum_{n_i \in \mathbb{N}} \exp(-\beta n_i) = \left(\frac{1}{1 - \exp(-\beta)}\right)^N.$$

$$(11.5.8)$$

Thus:

$$f(T) = -kT \lim_{N \to \infty} \frac{\ln Z(\beta, N)}{N} = kT \ln(1 - \exp(-\beta)). \tag{11.5.9}$$

6.8. From exercice 6.6, and $\frac{1}{T} = \frac{\partial s(e)}{\partial e}$, we get that $e(T) = \frac{\exp(-\beta)}{1 - \exp(-\beta)}$, and then we use $f(T) = e(T) - Ts(e(T))$.

6.9. We have:

$$S = k \ln \frac{N!}{n!(N-n)!} \approx k(N \ln N - n \ln n - (N-n) \ln(N-n))$$

For the temperature, we get: $\frac{1}{T} = \frac{\partial S}{\partial E} = \frac{\partial S}{\epsilon \partial n} = \frac{k}{\epsilon}(-\ln n - 1 + \ln(N-n) + 1) = \frac{k}{\epsilon} \ln \frac{N-n}{n}$ and thus, $n(T) = \frac{N}{1 + \exp(\frac{\epsilon}{kT})}$.

6.10. Let n_i be the number of molecules in state $i = 1, 2, 3, 4$. We have: $N = n_1 + n_2 + n_3 + n_4$, $E = (n_2 + n_3 + n_4)\epsilon = (N - n_1)\epsilon$. To compute the entropy of that system, we have to choose n_1 molecules among the N sites and count 3 possibilities for each of the remaining $N - n_1$ molecules. So, $\Omega = \frac{N!}{n_1!(N-n_1)!} 3^{N-n_1}$ and

$$S \approx k(N \ln N + (N - n_1)(\ln 3 - \ln(N - n_1)) - n_1 \ln n_1)$$

For the temperature, we have:

$$\frac{1}{T} = \frac{\partial S}{\partial E} = -\frac{1}{\epsilon} \frac{\partial S}{\partial n_1} = -\frac{k}{\epsilon}(-(\ln 3 - \ln(N - n_1)) - \ln n_1)),$$

so, $\beta\epsilon = \ln \frac{3n_1}{N-n_1}$, and, for the energy, we get: $E(T) = (N - n_1)\epsilon = \frac{3N\epsilon}{3 + \exp(\beta\epsilon)}$. As $T \to 0$, $E \to 0$ and all molecules are para. As $T \to \infty$, $E \to \frac{3N\epsilon}{4}$, all states are equiprobable and we have a fraction $\frac{1}{4}$ for each type of molecules.

6.11. Consider first the integral

$$\mathcal{I}(E, N) = \int_{E(q_i, p_i)_{i=1}^{N} \leq E} \prod_{i=1}^{N} dq_i dp_i, \tag{11.5.10}$$

which gives the volume of a sphere of radius $\sqrt{E}$ in a space of $2N$ dimensions. This equals $\frac{1}{N!}(\pi E)^N$. We have $\Omega(E, N) = \frac{\partial}{\partial E}\mathcal{I}(E, N) = \frac{1}{(N-1)!}\pi^N E^{N-1}$. So, $s(e) \approx k \ln e + C$.

Comparing with the result of exercise 6.6, we have:

$$k\left(\ln(1 + e) + e \ln\left(1 + \frac{1}{e}\right)\right) \approx k \ln e$$

when e is large, since then $\ln(1 + e) \approx \ln e$ and $e \ln(1 + \frac{1}{e}) \approx 1$.

6.12. For the canonical partition function, we have:

$$Z_c(T, N, V) = \frac{1}{N!}(2\pi kT)^{\frac{3N}{2}} \left(\int_\Lambda \exp(-\beta V(\mathbf{q}))d\mathbf{q} \right)^N \equiv \frac{1}{N!}W^N \quad (11.5.11)$$

with $W = (2\pi kT)^{\frac{3}{2}} \left(\int_\Lambda \exp(-\beta V(\mathbf{q}))d\mathbf{q} \right)$. And, for the grand-canonical one:

$$\mathcal{Z}_g(T, \mu, V) = \exp[zW], \quad (11.5.12)$$

with $z = e^{\beta\mu}$.
The pressure is given by, see (6.5.13):

$$PV \approx kT \ln \mathcal{Z}_g(T, V\mu) = kTzW \quad (11.5.13)$$

and the average number of particles is given (taking the derivative with respect to z of the logarithm of the grand-canonical partition function (6.5.8)) by:

$$N = z\frac{\partial}{\partial z} \ln \mathcal{Z}_g(T, \mu, V) = zW, \quad (11.5.14)$$

so that $PV = NkT$ still holds.

6.13. The canonical partition function equals:

$$Z_c = \sum_{\mathbf{n}\in\mathbb{N}^L} \exp\left(-\beta \sum_{i=1}^L \omega_i n_i \right) = \prod_{i=1}^L \sum_{n_i\in\mathbb{N}} \exp(-\beta\omega_i n_i) = \prod_{i=1}^L \frac{1}{1 - \exp(-\beta\omega_i)}. \quad (11.5.15)$$

This gives for the Helmoltz free energy:

$$F = kT \sum_{i=1}^L \ln(1 - \exp(-\beta\omega_i)). \quad (11.5.16)$$

6.14. We have $\omega_M = (\frac{9N}{CV})^{\frac{1}{3}}$. From the solution of exercise 6.13,
$F = CkTV \int_0^{\omega_M} \ln(1 - \exp(-\beta\omega))\omega^2 d\omega$, and
$E = \frac{\partial\beta F}{\partial\beta} = CV \int_0^{\omega_M} \frac{\exp(-\beta\omega)}{1-\exp(-\beta\omega)}\omega^3 d\omega$
Change variables $x = \beta\omega$; we get $E = CV(kT)^4 \int_0^{kTx_M} \frac{\exp(-x)}{1-\exp(-x)}x^3 dx$ with
$kTx_M = \omega_M$.

6.15. We obtain, as in exercise 6.14,

$$E = \frac{V}{\pi^2 c^3} \int_0^\infty \frac{\exp(-\beta\omega)}{1 - \exp(-\beta\omega)}\omega^3 d\omega$$

Change variables $x = \beta \omega$; we get $E = \frac{V}{\pi^2 c^3}(kT)^4 \int_0^\infty \frac{\exp(-x)}{1-\exp(-x)}x^3 dx$, which using the formulas given in the exercise to compute $\int_0^\infty \frac{\exp(-x)}{1-\exp(-x)}x^3 dx = \frac{3!\pi^4}{90} = \frac{\pi^4}{15}$ give

$$E = \frac{V\pi^2 k^4}{15c^3}T^4$$

11.6 Exercises of Chap. 7

7.1. For a uniform distribution on an interval $[a, b]$, we get $S(p) = \int_a^b \log(b-a)$, which is negative for $b - a < 1$.

7.2. Using (11.1.11), (7.2.1), we get, for the Shannon entropy of the multinomial distribution:

$$- \log N! - N \sum_{i=1}^L p_i \log p_i + \sum_{j=1}^L \sum_{n_j=0}^N \frac{N!}{n_j!(N-n_j)!}p_j^{n_j}(1-p_j)^{N-n_j} \log(n_j!),$$

(11.6.1)

where we used $\mathbb{E}(n_i) = Np_i$ where $\mathbb{E}$ is the expectation value with respect to the multinomial distribution to get the second term, and $\sum_{\{n_i;i\neq j\},n_i=0,\ldots,N} \frac{N!}{\prod_{i=1,i\neq j}^L n_i!}$

$\prod_{i=1,i\neq j}^L p_i^{n_i} = \frac{N!}{(N-n_j)!}(1-p_j)^{N-n_j}$ by the multinomial formula:

$$(1-p_j)^{N-n_j} = \left(\sum_{i=0i\neq j}^L p_i\right)^{N-n_j} = \sum_{\{n_i;i\neq j\},n_i=0,\ldots,N} \frac{(N-n_j)!}{\prod_{i=1,i\neq j}^L n_i!}\prod_{i=1,i\neq j}^L p_i^{n_i}$$

to get the third term. For the binomial distribution, take $L = 2$ in (11.6.1).

7.3. Since, by the law of large numbers, $P(i \notin G(N, \epsilon)) \to 0$ as $N \to \infty$, it is enough, for fixed ϵ in the definition of $G(N, \epsilon)$ to choose N large enough.

7.4. After one operation, we have a 1 for symbol 5 and a 0 for symbol 4 and we are left with a set with $K = 4$ and $p(1) = 0.3$, $p(2) = 0.25$, $p(3) = 0.25$, $p(4) = 0.2$. Assign a 1 to 4 (which is the previous 3, before combining 4 and 5, so that we assign a 1 to 3) and a 0 to 3 (which is the previous 2, before combining 4 and 5, so that we assign a 0 to 2). Now combine those two together: we have a probability $0.25 + 0.2 = 0.45$. We thus have $K = 3$, with $p(1) = 0.45$, $p(2) = 0.3$, $p(3) = 0.25$. Assign a 1 to 2 (which is the previous 1, before combining 3 and 4, so that we assign a 1 to the 1 of the previous step, which means that we assign a 1 to 4 and 5, since that 1 resulted from the combination of 4 and 5) and a 0 to 3 (which is the previous 2, before combining 3 and 4, which itself was the previous 1, so that we assign a 0 to 1). We are left with $K = 2$, with $p(1) = 0.55$, $p(2) = 0.45$ and we assign a 0 and a 1 to those two symbols (so we add a 0 to the original symbols 1, 4 and 5 and a 1 to the original symbols 2 and 3) so that we get the code:

1. $C(1) = 00$
2. $C(2) = 10$
3. $C(3) = 11$
4. $C(4) = 010$
5. $C(5) = 011$

7.5. a. For the marginal probabilities P_X and P_Y of the X and Y variables, we get
$P_X = \sum_{Y=1}^{N} P_{XY}$, $P_Y = \sum_{X=1}^{N} P_{XY}$.

  b. $S_{XY} = S_G(P_{XY}) = -\sum_{X,Y=1}^{N} P_{XY} \ln P_{XY}$,
$S_X = S_G(P_X) = -\sum_{X=1}^{N} P_X \ln P_X$,
$S_Y = S_G(P_Y) = -\sum_{Y=1}^{N} P_Y \ln P_Y$.

  c. We have

$$S_{XY} - S_X - S_Y$$

$$= -k\left(\sum_{X,Y=1}^{N} P_{XY} \ln P_{XY} - \sum_{X=1}^{N} P_X \ln P_X - \sum_{Y=1}^{N} P_Y \ln P_Y\right)$$

$$= -k\left(\sum_{X,Y=1}^{N} P_{XY} \ln P_{XY} - \sum_{X=1}^{N}\sum_{Y=1}^{N} P_{XY} \ln P_X - \sum_{Y=1}^{N}\sum_{X=1}^{N} P_{XY} \ln P_Y\right)$$

$$= k\sum_{X,Y=1}^{N} P_{XY} \ln \frac{P_X P_Y}{P_{XY}}$$

$$\tag{11.6.2}$$

The function $x - 1 - \ln x$ is positive with a minimum at $x = 1$, where its value is 0. So, $k\sum_{X,Y=1}^{N} P_{XY} \ln \frac{P_X P_Y}{P_{XY}} \le k\sum_{X,Y=1}^{N} P_{XY}(\frac{P_X P_Y}{P_{XY}} - 1)$, and $\sum_{X,Y=1}^{N} P_{XY}(\frac{P_X P_Y}{P_{XY}} - 1) = \sum_{X,Y=1}^{N} P_X P_Y - P_{XY} = 0$.

  d. We have equality if and only if $\frac{P_X P_Y}{P_{XY}} = 1$, namely if the X and Y variables are independent.

7.6. Shannon's entropy of that distribution is: $S = \frac{1}{2} \log 2 + \frac{1}{4} \log 4 + \frac{1}{4} \log 8 = \frac{1}{2} + \frac{1}{2} + \frac{3}{4} = 1.75$. For the codes:

1. This code is a prefix code and its expected length is 1.75, i.e. it equals Shannon's entropy and its length is thus the shortest possible one for a uniquely decodable code.
2. This code is a prefix code and its expected length is 2.
3. This code has an expected length equal to 1.25, which is less than the Shannon entropy, but is not uniquely decodable code: the sequences $acdbac$ and $cabdca$ both code as 000111000, as it should because of Proposition 7.8.

7.7. 1. $C(1) = 000000$
      2. $C(2) = 01$

3. $C(3) = 0001$
4. $C(4) = 001$
5. $C(5) = 1$
6. $C(6) = 000001$
7. $C(7) = 00001$

11.7 Exercises of Chap. 8

8.1. Eliminating factors that are the same on both sides of (8.7.5), $(2N)!(\frac{1}{2})^N$ and multiplying both sides by $2N(N-n)!(N+n)!$ we get $2N$ on the left hand side of (8.7.5) and

$$\frac{(N+n)!}{(N+n-1)!} + \frac{(N-n)!}{(N-n-1)!} = N+n+N-n = 2N, \qquad (11.7.1)$$

on its right hand side, and (8.7.5) is verified.

8.2. Since the denominators are the same on both sides of (8.7.6), it is enough to check that $P(n(t-1) = n, n(t) = m) = P(n(t+1) = n, n(t) = m)$, with $m = n \pm 1$. For $m = n - 1$, it is enough to verify that $P^*(n-1)\frac{N-n+1}{2N} = P^*(n)\frac{N+n}{2N}$, which can be checked explicitly as in exercise 8.1. The case $m = n + 1$ is similar.

8.3. Since, by (8.7.9), the average time spent in a state n is $\frac{(2N)!}{(N-n)!(N+n)!}(\frac{1}{2})^{2N}$, which is minimal for $n = \pm N$, and since we estimate the return time as the inverse of the time spent in the state n, we get the estimate 2^{2N} for the maximum return time.

8.4. Rewrite (8.7.1) for the fraction $\frac{P(n,t+1)}{G_N(n)}$; we get:

$$\frac{P(n, t+1)}{G_N(n)} = \frac{N-n+1}{2N} \frac{G_N(n-1)}{G_N(n)} \frac{P(n-1,t)}{G_N(n-1)}$$
$$+ \frac{N+n+1}{2N} \frac{G_N(n+1)}{G_N(n)} \frac{P(n+1,t)}{G_N(n+1)},$$

which is of the form $\sum_m c_{m,n} \frac{P(m,t)}{G_N(m)}$ where the sum is over $m = n \pm 1$, and $c_{n-1,n} = \frac{N-n+1}{2N} \frac{G_N(n-1)}{G_N(n)}$, $c_{n+1,n} = \frac{N+n+1}{2N} \frac{G_N(n+1)}{G_N(n)}$. So, $c_{m,n} \geq 0$,

$$\sum_{m=n\pm1} c_{m,n} = \frac{N-n+1}{2N} \frac{G_N(n-1)}{G_N(n)} + \frac{N+m+1}{2N} \frac{G_N(n+1)}{G_N(n)} = 1,$$

and $c_{m,n} G_N(n) = c_{n,m} G_N(m)$.

8.5. Using the variable ξ_t, we have (if we condition on n_t, the conditioning on n_0 is irrelevant since $n(t+1)$ depends on $n(0)$ through n_t):

$$< n(t+1)|n(t) >= n_t + < \xi_t|n(t) >= n_t + \left(1 - \frac{n_t}{N}\right)\frac{n_t}{N} = \left(1 - \frac{2}{N}\right)n_t + 1.$$

Let $\delta_{t+1} =< n(t+1)|n(t) > -\frac{N}{2}$; we get $\delta_{t+1} = (1 - \frac{2}{N})\delta_t$ since $\frac{2}{N}\frac{N}{2} = 1$.
Upon iteration, $\delta_t = (1 - \frac{2}{N})^t\delta_0$, which, in terms of $< n(t)|n(0) >$ is (8.7.13).

8.6. We get:

$$| \Omega((N_+^i(t))_{i=1}^{2^L}) |= \prod_{i=1}^{2^L} \frac{2^{M-L}!}{(2^{M-L} - N_+^i)!N_+^i!}$$

8.7. We may just repeat the proof of Proposition 8.5, for each $i = 1, \ldots, 2^L$. If we set $L = M$, we don't have macroscopic variables any more and N_+^i oscillates with $\eta_i(t)$. We get from exercise 8.6 that the entropy is 0 because then $2^{M-L} = 1$ and $N_+^i = 0$ or 1. N_+^i oscillates also if we set $L = M - m$ for any fixed m.

8.8. The conjugation (4.6.3) between T_b and the shift on symbol sequences, which we will denote here by $\mathbf{c}$, shows that the time evolution shifts sequences to the left. So, the constraint that $\mathbf{c} \in T_b^{-t}C_a$ or $T_b^t\mathbf{c} \in C_a$ imposes on $\mathbf{c}$ is $(c_i)_{i=-l+t}^{l+t} = (a_i)_{i=-l}^l$, and the constraint that $\mathbf{c} \in C_b$ imposes on $\mathbf{c}$ is $(c_i)_{i=-n}^n = (b_i)_{i=-n}^n$. Thus, for t large so that $-l + t > n$, those constraints do not overlap and thus, the size of set of sequences $\mathbf{c}$ satisfying both constraints will be the product of the size $|C_b|$, times the size of C_a, namely $|C_b||C_a| = 2^{-(2l+1)}|C_a|$.

11.8 Exercises of Chap. 9

9.1. We have $< s_A >= \frac{\sum_{s_i, i\in\Lambda} s_A \exp(\beta \sum_{B\subset\Lambda} J_B s_B)}{\sum_{s_i, i\in\Lambda} \exp(\beta \sum_{B\subset\Lambda} J_B s_B)}$. Now apply the formula for the derivative of a ratio of functions.

9.3. Using the transfer matrix (9.4.19), for boundary conditions $s_{-L-1}, s_{L+1} = \pm 1$ we get, for the partition function:

$$Z_{\Lambda,s_{-L-1},s_{L+1}} = < v_{-L-1}T^{2L+2}v_{s_{L+1}} >= \lambda_1^{2L+2} < v_{s_{-L-1}}v_1 >< v_1 v_{s_{L+1}} >$$
$$+ \lambda_2^{2L+2} < v_{-L-1}v_2 >< v_2 v_{s_{L+1}} >,$$

where $v_{s_{-L-1}} = \delta(s - s_{-L-1})$ and similarly for $v_{s_{L+1}}$; v_1, v_2 are the eigenvectors of T corresponding to the eigenvalues λ_1, λ_2. One only needs to check that $< v_{s_{-L-1}}v_1 >< v_1 v_{s_{L+1}} >> 0$ (which holds because the vector v_1 has positive entries) to obtain (9.4.20) in the thermodynamic limit. For the expectation value $< s_0 >$, one gets in the thermodynamic limit, following the same method, $< v_1 S v_1 >$, where S is the diagonal 2×2 matrix with ± 1 on the diagonal.

9.4. For $|A|$ odd, in the expansion (9.5.5) only graphs containing a path from at least one site of A to the boundary give a non-zero contribution. We can then repeat the arguments leading to (9.5.8).

9.5. If we write

$$< s_A >_{\Lambda,\beta;\bar{s}_{\Lambda^c}} - < s_A >_{\Lambda,\beta;\bar{s}'_{\Lambda^c}} = \sum_{s_i,s'_i \in \Lambda} (s_A - s'_A)\mu_{\bar{s}_{\Lambda^c}}(s_\Lambda)\mu_{\bar{s}'_{\Lambda^c}}(s'_\Lambda),$$

and for each pair $< ij >$, write an identity similar to (9.5.10):

$$\exp(\beta J(s_i s_j + s'_i s'_j + 2)) = 1 + (\exp(\beta J(s_i s_j + s'_i s'_j + 2)) - 1)$$
$$\equiv 1 + g(s_i, s_j, s'_i, s'_j) \tag{11.8.1}$$

and then proceed as in the derivation of (9.5.11). We get a sum over graphs and we can observe that, for any graph that does not contain a path going from a site in A to the boundary of Λ, we can use the symmetry between the sum over s_i and s'_i to show that the contribution of this graph vanishes. And, for the remaining graphs, we can bound the sum over paths going from a site in A to the boundary of Λ to show that $< s_A >_{\Lambda,\beta;\bar{s}_{\Lambda^c}} - < s_A >_{\Lambda,\beta;\bar{s}'_{\Lambda^c}}$ goes exponentially to 0 as $\Lambda \uparrow \mathbb{Z}^d$.

To prove (9.5.9), write $2(< s_A s_{B+i} > - < s_A >< s_B >) = \lim_{\Lambda\uparrow\mathbb{Z}^d} \sum_{s_i,s'_i\in\Lambda}$ $(s_A - s'_A)(s_{B+i} - s'_{B+i})\mu(s_\Lambda)\mu(s'_\Lambda)$, where we do not indicate the boundary conditions since they all yield the same result in the thermodynamic limit. Now use again (11.8.1) and do the same expansion as in the previous proof. One can then notice that, in any graph that does not contain a path going from a site in A to a site in B, we can again use the symmetry between the sum over s_i and s'_i to show that the contribution of this graph vanishes. And, for the remaining graphs, we can bound the sum over paths going from a site in A to a site in B to prove the bound (9.5.9).

9.6. We can define contours as connected sets of such "plaquettes". Then there will be a unique contour associated to a configuration with $s_0 = -1$ and formula (9.6.2) will hold, as well as estimate (9.6.4). To control the sum over contours as in (9.6.5), just observe that, for any connected set, there exists a "path" made of adjacent elements of the set (here, plaquettes touching each other) going through each element of that set and visiting each element at most twice. From that, one can deduce that the number of contours with P plaquettes grows at most like C^P for some $C < \infty$ and, inserting this and (9.6.4) in (9.6.5), proves (9.6.1).

9.7. It follows from the fact that $\sinh x$ is analytic and does not vanish near x_0 where $\sinh x_0 = 1$, so that $(1 - (\sinh \beta J)^{-4}) \approx |T - T_c|$, with $\sinh \beta_c J = 1$.

9.8. We get the formula (9.9.7) from (9.3.3) and the fact that $m_0 = 0$ for $T \geq T_c$.

9.9. The transfer matrix is:

$$T = \begin{pmatrix} \exp(\beta J) & 1 & \exp(-\beta J) \\ 1 & 1 & 1 \\ \exp(-\beta J) & 1 & \exp(\beta J) \end{pmatrix}, \tag{11.8.2}$$

Its eigenvalues are

$$\lambda_1, \lambda_2 = \frac{1}{2}(1 + 2\cosh\beta J \pm [(2\cosh\beta J - 1)^2 + 8]^{\frac{1}{2}}),$$

and $\lambda_3 = 2\sinh\beta J$.   We   get   for $f = -kT\ln(\frac{1}{2}(1 + 2\cosh\beta J + [(2\cosh\beta J - 1)^2 + 8]^{\frac{1}{2}}))$, coming from the largest eigenvalue.

9.10. We have

$$\mathcal{Z}_g = \sum_{\rho_i=0,1} z^{\sum_{i=1}^{L}\rho_i}\exp(\beta J\sum_{i=1}^{L}\rho_i\rho_{i+1}) = \mathrm{Trace}\,T^L \qquad (11.8.3)$$

with the matrix T having elements $T(\rho, \rho') = z^{\frac{\rho}{2}}\exp(\beta J\rho\rho')z^{\frac{\rho'}{2}}$. The eigenvalues of T are $\lambda_1, \lambda_2 = \frac{1}{2}(1 + z\exp(\beta J)) \pm [(z\exp(\beta J) - 1)^2 + 4z]^{\frac{1}{2}}$, so that

$$\mathcal{Z}_g = \lambda_1^V + \lambda_2^V.$$

Glossary

We give here some definitions of concepts used in this book and references to where they appear; we do not, in general, include expressions that are titles of sections or subsections.

Adiabatic transformations. Transformations without exchange of heat, see Sect. 5.7.4.

Almost everywhere. A statement is true almost everywhere if it holds except possibly on a set a measure zero.

Asymptotic equipartition. The idea that most sequences of N symbols, each of which appears with a given frequency, have the same probabilities for N large, in the sense that their rate of exponential decay in N is approximately the same, see Proposition 7.6.

Average time τ_A. For a transformation T, τ_A is the average time spent in A by a trajectory of the dynamical system: $\tau_A = \lim_{N \to \infty} \frac{1}{N} \sum_{n=0}^{N-1} \mathbb{1}_A(T^n x)$.

Bayesianism. See subjective probability.

Binomial distribution. For N independent events with two possible outcomes, denoted 0 and 1, with respective probabilities p and $1 - p$, the probability of N_0 results 0 is:

$$\frac{N!}{N_0! N_1!} p^{N_0} (1-p)^{N_1} == \frac{N!}{N_0!(N-N_0)!} p^{N_0} (1-p)^{N-N_0},$$

see (2.3.5).

Boltzmann's entropy. $S_B(\mathbf{x}) = k \ln |\Omega_M(\mathbf{x})|$, where $|\Omega_M(\mathbf{x})|$ is the size of the set of microstates corresponding to a given value of the macrostate: $M = M(\mathbf{x})$, see Sect. 6.4 for the equilibrium situation and Sect. 8.1.5 for the time evolution of Boltzmann's entropy.

Borel sets. Elements of the smallest σ algebra containing all open sets (or, equivalently, all closed sets). See Sects. 2.A.1 and 2.A.2.

Bounded support. A function defined on $\mathbb{R}^n$ has a bounded support if it vanishes outside a bounded subset of $\mathbb{R}^n$.

© Springer Nature Switzerland AG 2022
J. Bricmont, *Making Sense of Statistical Mechanics*, Undergraduate Lecture Notes in Physics, https://doi.org/10.1007/978-3-030-91794-4

Canonical ensemble.   A probability distribution dv_c on $\cup_E \Omega(E, V, N)$, indexed by (T, V, N):

$$dv_c(\mathbf{x}) = \frac{\exp(-\beta E(\mathbf{x}))d\mathbf{x}}{Z_c},$$

with Z_c the canonical partition function, see Sect. 6.6.1.

Canonical partition function.

$$Z_c(T, V, N) = \frac{1}{N!} \int_\Omega \exp(-\beta E(\mathbf{x}))d\mathbf{x}$$

with $\beta = \frac{1}{kT}$ and the symbol $\int \cdot d\mathbf{x}$ means either an integral over $\Lambda^{3N} \times \mathbb{R}^{3N}$ with $d\mathbf{x} = d\mathbf{q}d\mathbf{p}$ or a sum over all discrete states of the system, with N fixed and without the factor $\frac{1}{N!}$, see Sect. 6.5.

Cantor sets.   Special uncountable sets of Lebesgue measure zero, see Appendix 2.A.7. They sometimes appear as chaotic attractors of dynamical systems, and are then called "strange", see Sect. 4.6.2.

Carnot's cycle.   An (idealized) closed curve in the space of equilibrium states whose existence leads to the definition of the thermodynamic entropy, see Sects. 5.4.1 and 5.5.

Central limit theorem.   Loosely speaking, it gives a correction to the law of large numbers:

$$\sum_{i=1}^N a_n \approx N\mathbb{E}(f) + \sqrt{N}X,$$

where X is a Gaussian random variable. For a precise formulation, see Theorem 2.1.

Cesaro convergence.   A sequence $(a_n)_{n=1}^\infty$ converges to a in the Cesaro sense if $\lim_{N\to\infty} \frac{1}{N} \sum_{i=1}^N a_n = a$, see Sect. 4.3.

Chaos.   Synonymous, for a dynamical system, to having the property of sensitive dependence to initial conditions. Quite different from the chaotic hypothesis used in deriving Boltzmann's equation.

Chemical potential.   The variation of energy of a system when one adds one particle of type i to the system, and is denoted μ_i:

$$\frac{\partial E}{\partial N_i} = \mu_i(S, V, N_1, \ldots, N_l),$$

see Sect. 5.6.4.

Clausius entropy.   is defined by the integral of the differential form $dS = \frac{\delta Q}{T}$, see Sect. 5.5.1.

Code.   Given a set of symbols $i = 1, \ldots, K$ a code is a map:

$$C : \{1, \ldots, K\} \to \{0, 1\}^{\mathbb{N}}.$$

from that set to sequences of 0's and 1's, see Sect. 7.7.2.

Coding. In the theory of dynamical systems, an assignment of a sequence of symbols to every trajectory, which records the subsets of a partition of phase space visited by that trajectory, see Sect. 4.6.1.

Conditional probability distribution. See Appendix 2.A.6.

Conditional probability. The conditional probability of an event A given and event B is:

$$P(A \mid B) = \frac{P(A \cap B)}{P(B)},$$

see Sect. 2.2.2.

Conjugations of maps and measures. Two maps $T : \Omega \to \Omega$ and $T' : \Omega' \to \Omega'$ are *conjugated* when they are related by a bijection $\Phi : \Omega \to \Omega'$:

$$T' = \Phi T \Phi^{-1}$$

Given a map $\Phi : \Omega \to \Omega'$, a measure μ on Ω, Σ, and a measure μ^* on (Ω', Σ') with $\Sigma' = \{A \mid \exists B \in \Sigma, \Phi(B) = A\}$ are conjugated if: $\forall A \in \Sigma'$

$$\mu^*(A) = \mu(\Phi^{-1}(A)), \tag{G.1}$$

see Definition 4.1.

Conservation of energy. It is either a theorem of mechanics, see Theorem 3.1, or the first law of thermodynamics, see Sect. 5.3. That law is a consequence of the theorem of mechanics, but can be formulated independently of it.

Correlation. For a measure μ on Ω, Σ and two random variables f, g, the correlation between those variables, which measures their degree of mutual dependence, is defined as: $C(f, g) = \int_\Omega f(x)g(x)d\mu(x) - \int_\Omega f(x)d\mu(x) \int_\Omega g(x)d\mu(x)$. For two sets $A, B \in \Sigma$, with $f = \mathbb{1}_A$, $g = \mathbb{1}_B$, the correlation $C(f, g) = \mu(A \cap B) - \mu(A)\mu(B)$ measures the degree of dependence between the events in A and those in B, see Sect. 4.4.

Correlation functions. In spin systems, correlation functions are expectation values $< s_A >= \mathbb{E}_\mu(s_A)$ for A a finite subset of $\mathbb{Z}^d$ and μ a finite volume measure or a Gibbs measure, see Sect. 9.2.

Correlation length ξ. It characterizes the exponential decay of correlation functions:

$$< s_0 s_i > \approx \exp(-\frac{|i|}{\xi}),$$

see (9.9.4) and (9.5.9).

Cournot's principle. The idea that, if the probability of an event A is very small, given some set of conditions C, then one can be practically certain that the event A will not occur on a single realization of those conditions, see Sect. 2.5.

Critical exponent. Exponent that measures the way certain quantities tend to 0 of ∞ at a critical point, see Sect. 9.9.

Critical point. The point in parameter space at which the difference between phases disappears, see Figs. 9.1 and 9.2; they are discussed in Sect. 9.9.

Cylinder sets. Subsets of a product space $\Omega = \times_{i \in \mathbb{Z}} \Omega_i$ where $\Omega_i, i \in \mathbb{Z}$, is a copy of a set Ω. If Σ is a σ-algebra of subsets of Ω, a cylinder set in Ω is a set of the form:

$$\times_{i < -N} \Omega_i \times A_{-N} \times A_{-N+1} \cdots \times A_N \times_{i > N} \Omega_i$$

where $A_i \in \Sigma$ for $-N \leq i \leq N$ The σ-algebra generated by those sets is called the cylindrical σ-algebra, see Appendix 2.A.2.

Delta measures. Measures that give a measure one to points or countable sets of points or surfaces in $\mathbb{R}^n$, see Appendix 2.A.2.

Deterministic. That notion is defined and discussed in Sect. 4.8, specially by opposition to predictability.

Empirical distribution. If we have N random variables $\mathbf{x} = (x_i)_{i=1}^N$, and a partition of the set of values taken by those variables, $(A_1, \ldots A_n)$, the empirical distribution of those random variables is the set of frequencies $(n_\alpha(\mathbf{x}))_{\alpha=1}^n$ with which those variables belong to the subsets of the partition: $n_\alpha(\mathbf{x}) = \frac{|\{i=1,\ldots,N \,|\, x_i \in A_\alpha\}|}{N}$, see Sect. 2.3.2.

Ensemble. Expression used by physicists as meaning probability distribution on phase space: one considers three such probability distributions, the microcanonical, the canonical and the grand-canonical ones.

Equivalence of ensembles. The three ensembles, the microcanonical, canonical and grand canonical ones, are equivalent in two different senses: first, the microcanonical, canonical and grand canonical partition functions give rise to fundamental relations, the entropy $S = S(E, V, N)$, the free energy $F = F(T, V, N)$ and the grand potential $\Phi = \Phi(T, V, \mu)$ that are Legendre's transforms of one another, see Sect. 6.5.

Moreover, these ensembles, or probability distributions, give the same value to suitable macroscopic quantities, see Sect. 6.6.3, in particular Proposition 6.6.

Ergodic. This is defined in Sect. 4.3 and enters the discussion of the explanation of irreversible behavior in Sect. 8.3.

Expected value. The average with respect to a measure μ of a random variable f: $<f> = \int_\Omega f(x) d\mu(x)$; also denoted $\mathbb{E}(f)$ see Appendix 2.A.3.

First law of thermodynamics. Law of conservation of energy as expressed in thermodynamics, see Sect. 5.3.

Ideal gas. Gas of non interacting molecules, i.e. for which the interaction potential V_{ij} in (3.2.5) equals zero.

Ideal gas law. See (5.2.2) and (5.7.1). Also known as Boyle-Mariotte law.

Indifference Principle. A way to assign rationally probabilities when we have a set possible results for a "random" event, about which we have no reason to think that one of them is more likely to occur than another one. Then, assign to each of them an equal probability. The expression "indifference principle" is due to Keynes; Laplace called it the "principle of insufficient reason", see Sect. 2.2.1. This "principle" is just another expression of our equal ignorance.

Itinerary. Given a partition of the phase space of a dynamical system, the itinerary of an initial point x is the sequence of subsets of that partition visited by the trajectory of x, see Sect. 4.6.1.

Frequentism. See subjective probabilities.

Fundamental relations. A relation of the form $E=E(S, V, N)$, $S=S(E, V, N)$, $F = F(T, V, N)$, or $\Phi = \Phi(T, V, \mu)$ that gives all the thermodynamic information on a system and from which all other relations can be obtained by differentiation, see Sects. 5.6 and 5.7.

Gamma space. Denoted Γ (Γ for gas), it is the full phase space of a system, usually a subset of $\mathbb{R}^n$, with n being proportional to the number of particles and equal to $3N$ or $6N$ in three dimensions, depending on whether we consider only the particles' positions or their positions and velocities.

Gibbs' entropy. Function of a probability measure, see definition 6.5. For the constancy in time of the Gibbs' entropy under the Liouville evolution, see proposition 8.1. For the evolution of the Gibbs' entropy under another time evolution see Sect. 8.8.1.

Gibbs' inequality. For two probability distributions $(p_i)_{i=1}^N$ and $(q_i)_{i=1}^N$,

$$\sum_{i=1}^N q_i \ln p_i \leq \sum_{i=1}^N q_i \ln q_i,$$

see Proposition 7.5.

Gibbs' paradox. Not a real paradox, but refers to the fact that exchanging both the positions and the velocities of two particles does not change the microstate of a system. Hence, when integrating over all microstates, one must divide the usual volume in phase space of N identical particles by $N!$, see (6.4.3), Sects. 6.4 and 8.9.

Gibbs states. They represent the equilibrium states for spin systems on infinite lattices and they are defined by equations satisfied by their conditional probabilities, see (9.2.8) and Sect. 9.2.

Grand canonical ensemble. The probability distribution dv_g on $\cup_{E,N}\Omega$ (E, V, N), indexed by (T, V, μ), and given by:

$$dv_g(\mathbf{x}) = \frac{1}{N!} \frac{\exp(\beta\mu N(\mathbf{x}) - \beta E(\mathbf{x}))d\mathbf{x}}{\mathcal{Z}_g},$$

with $\mathcal{Z}_g$ is the grand canonical partition function. See Sect. 6.6.1.

Grand canonical partition function. See (6.5.8).

Grand potential. In thermodynamics it is the Legendre transform of the Helmhotz free energy with respect to the particle number, see Sect. 5.6.4. In statistical mechanics it is given by formula (6.5.11).

Heat. The heat transferred to a system during a thermodynamic transformation is defined as the difference between the total energy being transferred to the system and the work done on the system, see (5.3.7).

Helmhotz free energy.   In thermodynamics, it is the Legendre transform of the energy with respect to the entropy, see Sect. 5.6.3. In statistical mechanics, it is given by formula (6.5.5).

Identity operator   The operator, denoted Id, defined by $\mathrm{Id}(x) = x$.

Independent random variables.   See (2.A.21).

Indicator function.   The indicator function of a set A, denoted $\mathbb{1}_A$ or $\mathbb{1}(A)$, is defined by $\mathbb{1}_A(x) = 1$, if $x \in A$ and $\mathbb{1}_A(x) = 0$, if $x \notin A$.

Invariant measure.   See Appendix 2.A.5 for the definition, Sect. 3.4.2 for the invariance of the Liouville measure under the Hamiltonian flow, and Chap. 4 for the use of that concept in the theory of dynamical systems.

Irreversible.   In thermodynamics, a process that cannot be run backwards. For macroscopic equations, equations that are not reversible in time, like the diffusion equation (8.1.3), see Sect. 8.1.1.

Kolmogorov-Sinai entropy.   Given a partition of the space of dynamical system, the Kolmogorov-Sinai entropy measures the average amount of information obtained by knowing which element of the partition is visited by consecutive points of a trajectory. See Sect. 4.7.

Lagrange's multiplier.   To find the extrema of a function $f : \mathbb{R}^n \to \mathbb{R}$ under a constraint $g(x) = 0$, $g : \mathbb{R}^n \to \mathbb{R}$, one finds the extrema of the function $F(x, \lambda) = f(x) - \lambda g(x)$, where λ is the Lagrange's multiplier. See Sect. 6.2 for applications of this idea in this book.

Laplace's asymptotic method.   A method that allows us to compute the behavior of integrals of the form $\int_0^\infty \exp(\lambda f(x)) dx$ for λ large and positive, under certain conditions on the function f. See Appendix 6.A.1.

Law of large numbers.   It shows that averages of sums of a large number of independent random variables, all having the same distribution, converge to the expected value of those random variables, see Sect. 2.3. That result is central in this book, see Propositions 6.6 and 6.7.

Lebesgue measure.   The measure that extends the size of intervals or of rectangles to all Borel sets of $\mathbb{R}$ or $\mathbb{R}^n$, see Appendix 2.A.2.

Legendre transform.   It allows us to parametrize a convex function as the enveloppe of a set of straight lines, namely the tangents to the curve $(x, f(x))$ parametrized by their slope p and their intersection with the y axis. See Appendix 5.B.

Liouville's evolution.   It gives the time evolution of probability densities under the Hamiltonian flow, see (3.4.8).

Loschmidt's objection to Boltzmann.   An objection to Boltzmann's ideas based on the reversibility of the microscopic equations of motion proven in Sect. 3.5. This objection is discussed in Sect. 8.1.3.

Macrostates.   They are defined in general by a map from the phase space of a system into $\mathbb{R}^L$:

$$M : \Omega \to \mathbb{R}^L,$$

where $L << N$, see Sect. 6.1.

Magnetization. For spin systems, the average value $< s_i >$ of a single spin. For an extremal translation invariant Gibbs state μ, for which a form of the law of large numbers holds, that average value coincides with $\lim_{\Lambda \uparrow \mathbb{Z}^d} \frac{1}{|\Lambda|} \sum_{i \in \Lambda} s_i \, \mu$ almost everywhere, see (9.2.13) and Sect. 9.2.

Marginal distribution. It is defined in Appendix 2.A.6. One shows that the canonical and grand-canonical distributions are marginal distributions of the microcanonical one in Sect. 6.6.3.1.

Maximum entropy principle. In thermodynamics, it characterizes the equilibrium states as corresponding to values of the variables that maximizes the Clausius entropy of the system, given the external constraints imposed on the latter; this implies that when systems with different temperatures, pressures and chemical potentials are put into contact, those quantities evolve as expected, see Sect. 5.6.5. A totally different principle of maximum entropy, the one of the maximization of the Shannon entropy, characterizes the most spread out or the least biased probability distribution, see Chap. 7.

Mean field theory. A simple approximation which is useful in order to study, among others, lattice spin systems, see Sects. 9.3 and 9.9.2.

Measurable sets and functions. In this book, measurable sets are either Borel sets (in $\mathbb{R}^n$) or elements of the cylindrical σ algebra (on product spaces); a function $F : \Omega \to \mathbb{R}^n$ is measurable if

$$\forall A \in \mathcal{B}(\mathbb{R}^n), \quad F^{-1}(A) \in \Sigma,$$

where $\mathcal{B}(\mathbb{R}^n)$ are the Borel sets in $\mathbb{R}^n$, see Appendix 2.A.3.

Microcanonical ensemble; It is given by the normalized Liouville measure on a constant energy surface, see (6.6.1).

Microcanonical partition function. $\frac{|\Omega(E,V,N)|}{N!}$, see Sect. 6.6.1.

Microstate. The microstate of a classical mechanical system is a pair $\mathbf{x} = (\mathbf{q}, \mathbf{p})$, with $\mathbf{q} = (\vec{q}_1, \vec{q}_2, \ldots, \vec{q}_N) \in \mathbb{R}^{3N}, \mathbf{p} = (\vec{p}_1, \vec{p}_2, \ldots, \vec{p}_N) \in \mathbb{R}^{3N}$, and its time evolution is given by Hamilton's equations (3.3.3, 3.3.4, 3.3.10).

Mixing. It is defined in Sect. 4.4; for its (alleged) relevance to the approach to equilibrium, see Sect. 8.3.2.

Molecular chaos. A simplifying hypothesis that enters into the derivation of Boltzmann's equation, see Sect. 8.8. A similar, but simpler hypothesis is made in the context of Kac ring model, see (8.7.19).

Multinomial distribution. It gives the probability of having, after N draws from an urn containing balls of L different colors, with p_i being the fraction of the balls of color $i = 1, \ldots, L$, a sequence $(n_i)_{i=1}^L$ of balls of color $i = 1, \ldots, L$. It equals:

$$\frac{N!}{\prod_{i=1}^{L} n_i!} \prod_{i=1}^{L} p_i^{n_i},$$

see exercise [2.12].

Mu space   (μ for molecule). It is a subset of the "real" space, i.e. of $\mathbb{R}^3$ or of $\mathbb{R}^6$, depending on whether we consider only the set of possible positions of the particles or of their possible positions and velocities.

Objective probabilities.   They can either refer to the probabilities of intrinsically random events (in a non-deterministic universe) or, as advocated by von Mises, to idealized frequencies in infinite sequences of repetitions of the "same" random event, under certain conditions that are supposed to guarantee that those sequences are "really random", see Sect. 2.4.

Past hypothesis.   The idea that, in order to fully account for the increase of entropy, we must assume that the universe was in a lower entropy state in the past, which implies that it was in very low entropy state at the time of the Big Bang, see Sect. 8.5.

Poincaré's objections to Boltzmann.   There were several of them, some based on Zermelo's and Loschmidt's objections but also one based on a misunderstanding of the notion of statistical equilibrium and discussed in Sect. 8.2.2.

Poincaré's recurrence.   The fact that, for a dynamical system on a space with a finite measure, the set of points starting from a set of positive measure (no matter how small) that do not revisit that set infinitely often is of measure zero, see Sect. 4.2 for a precise statement and its proof and Sect. 8.2.1 for a discussion of the implications of that theorem in statistical mechanics.

Predictability.   See Sect. 4.8 for a definition of this notion in contrast with the one of determinism.

Prefix code.   A code such that no codeword is a prefix of another codeword, meaning that one does not have codewords c and d and a tail string t of the codeword d so that $ct = d$, see Sect. 7.7.2. A prefix code is always uniquely decodable.

Pressure.   It is defined in thermodynamics by (5.6.3), (5.6.5), (5.6.22) and in statistical mechanics by (6.5.13).

Probability measure.   A measure on a space Ω with $\mu(\Omega) = 1$.

Product space and measure.   In this book, the product spaces that we consider are infinite Cartesian products of copies Ω_i, $i \in \mathbb{Z}$, of a set Ω: $\Omega = \times_{i \in \mathbb{Z}} \Omega_i$ or $\Omega = \times_{i \in \mathbb{N}} \Omega_i$. They are endowed with the cylindrical σ-algebra, namely the σ-algebra generated by the cylinder sets. If μ is a probability measure on Ω, one defines the corresponding product measure on Ω by the following formula for the measure of cylinder sets:

$$\mu(\times_{i<-N} \Omega_i \times A_{-N} \times A_{-N+1} \cdots \times A_N \times_{i>N} \Omega_i) = \prod_{i=-N}^{i=N} \mu(A_i),$$

see Appendix 2.A.2.

Propensity.   A property, introduced mostly by Karl Popper, that is supposed to explain why objects behaving "randomly" behave as they do, see Sect. 2.4.

Random variable.   A measurable function from a measure space Ω to $\mathbb{R}$ or $\mathbb{R}^n$.

Reductionism.   The idea that scientific explanations ultimately point to the laws governing the elementary constituents of the universe, see Chap. 10.

Reversible.   In mechanics, a property of the microscopic equations of motion, see Sect. 3.5, and, in thermodynamics, a transformation in the space of equilibrium states that can be run backwards.

Second law of thermodynamics.   That law takes different forms, first as a statement about the impossibility of certain transformations (either the one of Lord Kelvin or of Clausius), see Sect. 5.4 or as the fact that the thermodynamic entropy increases or stays constant in an isolated system, see Sect. 5.5.2. In statistical mechanics, it is expressed by the increase of Boltzmann's entropy, see Sect. 8.1.

Sensitive dependence on initial conditions.   A dynamical system exhibits such a dependence when the distance between trajectories increases exponentially in time, see definition 4.11 and Sect. 4.5.

Size of a set.   It is denoted $|E|$: for a finite set E, its number of elements or, for a subset of $\mathbb{R}^n$, its Lebesgue measure.

Shannon's entropy.   See Sect. 7.2 for its definition and Chap. 7 for its relevance and applications.

Shift map.   It is defined on a product space by Remark 2.10; a special case is considered in Chap. 4 and is defined by (4.1.12).

Specific heat.   The quantity of heat required to raise the temperature of one gram of a substance by one Celsius degree. In Ising systems it is given (up to a factor) by (9.9.2) and (9.9.3). For its mean field critical exponent, see (9.9.7). At the end of the nineteenth century, the behavior of the specific heats of solids was observed to disagree with the classical predictions and that was one of the sources of quantum mechanics, see Sect. 7.8.2.

Spin variables.   Classical variables associated to each site of a lattice $\mathbb{Z}^d$; they take values ± 1 in the simplest case (Ising models), see Sect. 9.2 but can take values in $\mathbb{R}$, see Sect. 9.7.4 or on a unit sphere in $\mathbb{R}^n$, see Sect. 9.7.5.

Stirling's formula.   It gives the asymptotic behavior of $N!$ for N large: $N! \approx N^N \exp(-N)\sqrt{2\pi N}$, see Appendix 6.A.2.

Subjective probabilities.   The classical view of probabilities, going back to Bayes and Laplace and advocated by Jaynes, for which probabilities refer to a form of judgment and not to objective facts in the world. These judgments can be either totally subjective (only avoiding Dutch bets) or following rules of rationality like the indifference principle or the maximization of the Shannon entropy, given the information we have on the system; see Sect. 2.2.

Temperature.   In thermodynamics, it is defined first by phenomenological relations like (5.2.2); then, once the notion of thermodynamic entropy is introduced, via the Carnot cycle that uses the phenomenological notion of temperature, it is defined by relations of the form $\frac{\partial S}{\partial E} = \frac{1}{T}$, see Sect. 5.6.2. In statistical mechanics, the temperature is defined by $\beta = \frac{1}{kT}$, where β enters the definition of the canonical partition function and of the canonical distribution, see Sects. 6.5 and 6.6.1; the identification between those two notions of temperature is done in Sect. 6.7.

Thermodynamic entropy.   See Clausius entropy.

Transfer matrix.   A matrix that allows us to reduce the solution of one dimensional spin systems to a calculation of its eigenvalues and eigenvectors, Sect. 9.4. It can also be applied to some higher dimensional spin systems, for example the two-dimensional Ising model, see [249].

Typical. One says that a property is typical if it occurs with a probability close to one, see Sects. 2.3.2, 6.3 and 6.8.

Uniquely decodable code. A code is uniquely decodable if $\forall i, j \in \{1, \ldots, K\}$, $i \neq j$, $C(i) \neq C(j)$, see Sect. 7.7.2. All prefix codes are uniquely decodable.

Variance. For a random variable f, its variance is:

$$\text{Var}(f) \equiv \mathbb{E}[(f - \mathbb{E}(f))^2] = \mathbb{E}(f^2) - \mathbb{E}(f)^2 \geq 0.$$

Work. In thermodynamics, the change of energy of a system when a force is exerted either by it or on it see Sect. 5.3.1. For its statistical mechanical definition, see Sect. 6.7, in particular (6.7.2), (6.7.4).

Zermelo's objection to Boltzmann's ideas. It was based on the Poincaré recurrence theorem, see Sect. 8.2.1 for a discussion.

Zeroth law of thermodynamics. It states that there exists equilibrium states for systems that are isolated long enough and which are characterized by equations of the form (5.2.1).

Suggested Reading

Obviously, we could give thousands of references, but we will give only a few, namely those that have helped us most in the writing of this book.

Chapter 2. In [183] Jaynes gives an detailed and lively defense of the Bayesian approach with many applications of Bayesian reasoning; see also Laplace [209]. For a defense of the frequentist viewpoint, see von Mises [324].

Chapter 3. The book of classical mechanics on which this chapter is based is Arnol'd [15].

Chapter 4. For nice mathematical introductions, see Hirsch, Smale and Devaney [172], Lanford [207], Sinai [296], or Walters [327], and, for more physical approaches, see Dorfman [109], Gaspard [142] or Strogatz [306]. For more detailed expositions see Collet and Eckmann [85, 86], Cornfeld et al. [87] and Katok and Hasselblatt [186].

Chapter 5. For a nice introduction see Fermi [124] and for a detailed exposition, see Callen [66].

Chapter 6. Amit and Verbin [12], Callen [66], Chandler [76] and Thompson [308] are standard textbooks.

For more discussion of foundational issues, see lectures by Tumulka [314], lectures by Zanghì [333], the collective volumes of Bricmont et al. [56] and of Allori [10], and the review article by Goldstein et al. [155]. See also Gallavotti [141] for an overview of the statistical mechanics formalism.

For an extension of this formalism to the quantum situation, see [155, 314].

Chapter 7. We followed Jaynes [183] for the part on entropy; see also, e.g. Bais and Doyne Farmer [17], Ben-Naim [24] or Grandy [159] for a defense of the information theoretic approach. We followed MacKay [234] for the part on information theory. See also Khinchin [192].

Chapter 8. The book by Kac [185] discusses approach to equilibrium in simple models; see Spohn [304] and Cercignani, Illner and Pulvirenti [73] for

© Springer Nature Switzerland AG 2022

J. Bricmont, *Making Sense of Statistical Mechanics*, Undergraduate Lecture Notes in Physics, https://doi.org/10.1007/978-3-030-91794-4

mathematical non equilibrium statistical mechanics, and see the suggested readings of Chap. 6 as well as articles by Goldstein and Lebowitz [149–151, 154, 218, 225] for foundational issues.

Chapter 9. Thompson [308] is a good introduction. For more advanced books, see Simon [292, 293] and Sinai [297]. Friedli and Velenik [131] give a thorough mathematical, but pedagogical, discussion of phase transitions in lattice systems.

Others For the history of statistical physics, see Brush [61, 62]; those works contain most of the classical texts. See Cercignani [74] for the life and work of Boltzmann, and Uffink [316] and references therein for an overview (noting however that the general "philosophy" of this latter author is different from ours). For a comprehensive (but not recent) discussion of approaches to the foundations of statistical mechanics, see Sklar [298].

References

1. P. Adriaans, J. van Benthem, *Handbook of Philosophy of Information*, Elsevier, Amsterdam (2008)
2. M. Aizenman, S. Goldstein, J.L. Lebowitz, Conditional equilibrium and the equivalence of micro-canonical and grandcanonical ensembles in the thermodynamic limit. Comm. Math. Phys. **62**, 279–302 (1978)
3. M. Aizenman, Translation invariance and instability of phase coexistence in the two-dimensional Ising system. Comm. Math. Phys. **73**, 83–94 (1980)
4. M. Aizenman, Geometric analysis of ϕ^4 fields and Ising models. I, II. Comm. Math. Phys. **86**, 1–48 (1982)
5. M. Aizenman, R. Graham, On the renormalized coupling constant and the susceptibility in ϕ_4^4 field theory and the Ising model in four dimensions. Nuclear Phys. B **225**, 261–288 (1983)
6. M. Aizenman, R. Fernández, On the critical behavior of the magnetization in high-dimensional Ising models. J. Stat. Phys. **44**, 393–454 (1986)
7. M. Aizenman, D.J. Barsky, R. Fernández, The phase transition in a general class of Ising-type models is sharp. J. Stat. Phys. **47**, 343–374 (1987)
8. M. Aizenman, J. T. Chayes, L. Chayes, C.M. Newman, Discontinuity of the magnetization in one-dimensional $\frac{1}{|x-y|^2}$ Ising and Potts models. J. Stat. Phys. **50**, 1–40 (1988)
9. D. Albert, *Time and Chance*, Harvard University Press, Cambridge (2000)
10. V. Allori (ed.), *Statistical Mechanics and Scientific Explanation: Determinism, Indeterminism and Laws of Nature*, World Scientific, Singapore (2020)
11. V. Allori, Some reflections on the statistical postulate: typicality, probability and explanation between deterministic and indeterministic theories. In [10], p. 65–111
12. D.J. Amit, Y. Verbin, *Statistical Physics. An Introductory Course*, World Scientific, Singapore (1999)
13. P.W. Anderson, More is different. Science **177**, 393–396, 04 1977
14. V.I. Arnol'd, A. Avez, *Ergodic Problems of Classical Mechanics*, Benjamin, New York (1968)
15. V.I. Arnol'd, *Mathematical Methods of Classical Mechanics*, translated by K. Vogtmann and A. Weinstein, 2nd ed., Springer-Verlag, New York (1989)
16. V.I. Arnol'd, V.V. Kozlov, A.I. Neishtadt, *Mathematical Aspects of Classical and Celestial Mechanics*, translated from the Russian by E. Khukhro, 3rd ed., Springer-Verlag, Berlin, Heidelberg (2006)
17. F.A. Bais, J. Doyne Farmer, The physics of information, in [1], p. 617–691
18. M. Baldovin, L. Caprini, A. Vulpiani, Irreversibility and typicality: A simple analytical result for the Ehrenfest model. Physica A **524**, 422–429 (2019)

© Springer Nature Switzerland AG 2022
J. Bricmont, *Making Sense of Statistical Mechanics*, Undergraduate Lecture Notes in Physics, https://doi.org/10.1007/978-3-030-91794-4

19. H. Barnum, C.M. Caves, C.A. Fuchs, R. Schack, D.J. Driebe, W. G. Hoover, H. Posch, B. L. Holian, R. Peierls, J.L. Lebowitz, Is Boltzmann entropy time's arrow's archer? Letters in Physics Today, **47**, issue 11 (November 1994), pages 11 and 13.
20. V. Bauchau, Universal Darwinism. Nature **361**, 489 (1993)
21. E.T. Bell, *The Development of Mathematics*, 2nd edition, McGraw-Hill, New York (1945)
22. J.S. Bell, *Speakable and Unspeakable in Quantum Mechanics. Collected Papers on Quantum Philosophy*, 2nd edn, with an introduction by Alain Aspect, Cambridge University Press, Cambridge (2004)
23. Y. Ben-Menahem, M. Hemmo (eds), *Probability in Physics*, Springer, Berlin, Heidelberg (2012)
24. A. Ben-Naim, *Entropy, the Truth, the Whole Truth, and Nothing But the Truth*, World Scientific, Singapore (2017)
25. C.H. Bennett, The thermodynamics of computation. A review. International Journal of Theoretical Physics **21**, 905–940 (1982)
26. J. Bernoulli, *The Art of Conjecturing, together with Letter to a Friend on Sets in Court Tennis* (English translation), translated by E. Sylla, Johns Hopkins Univ Press, Baltimore (2005) [1713]
27. J. Bertrand, *Calcul des probabilités*, Gauthier-Villars, Paris (1889)
28. P. Billingsley, *Ergodic Theory and Information*, J. Wiley, New York (1965)
29. K. Binder, Ising model, Encyclopedia of Mathematics. http://encyclopediaofmath.org/index.phptitle=Ising_modeloldid=50835 (2020)
30. D. Bohm, A suggested interpretation of the quantum theory in terms of "hidden variables", Parts 1 and 2. Physical Review **89**, 166–193 (1952)
31. D. Bohm, B.J. Hiley, *The Undivided Universe*, Routledge, London (1993)
32. L. Boltzmann, Weitere Studien über das Wärmegleichgewicht unter Gasmolekülen. Sitzungsberichte der Akademie der Wissenschaften Wien **66**, 275–370 (1872). English translation: Further studies on the thermal equilibrium of gas molecules, in [62], p. 262–349
33. L. Boltzmann, Über die Beziehung eines allgemeine mechanischen Satzes zum zweiten Hauptsatze der Wärmetheorie. Sitzungsberichte der Akademie der Wissenschaften Wien **75**, 67–73 (1877). English translation: On the relation of a general mechanical theorem to the second law of thermodynamics, in [62], p. 362–367
34. L. Boltzmann, Vorlesungen über Gastheorie. 2 vols., Barth, Leipzig (1896, 1898). English translation by S.G. Brush, *Lectures on Gas Theory*, Cambridge University Press, London (1964)
35. L. Boltzmann, Entgegnung auf die wärme–theoretischen Betrachtungen des Hrn. E. Zermelo. Annalen der Physik **57**, 773–784 (1896). English translation: Reply to Zermelo's remarks on the theory of heat, in [62], p. 392–402
36. L. Boltzmann, Zu Hrn. Zermelo's Abhandlung Über die mechanische Erklärung irreversibler Vorgange. Annalen der Physik **60**, 392-398 (1897) English translation: On Zermelo's Paper "On the mechanical explanation of irreversible processes", in [62], p. 412–419
37. L. Boltzmann, *Populäre Schriften*, Barth, Leipzig (1905)
38. L. Boltzmann, *Theoretical Physics and Philosophical Problems. Selected Writings*, B. McGuinness (ed.), Reidel, Dordrecht (1974)
39. E. Borel, *Les Probabilités et la Vie*, Presses Universitaires de France, Paris (1943). English translation by M. Baudin, *Probabilities and Life*, Dover, New York (1962)
40. E. Borel, *La Mécanique Statistique et l'Irréversibilité*, in *Œuvres*, Tome 3, p.1697, éd. CNRS, Paris (1972)
41. C. Borgs, J.Z. Imbrie, A unified approach to phase diagrams in field theory and statistical mechanics. Commun. Math. Phys. **123**, 305–328 (1989)
42. C. Borgs, R. Waxler, First order phase transitions in unbounded spin systems: construction of the phase diagram. Commun. Math. Phys. **126**, 291–324 (1989)
43. M. Born, *Natural Philosophy of Cause and Chance*, Clarendon Press, Oxford (1949)
44. A. Bovier, M. Zahradník. The low-temperature phase of Kac-Ising models. J. Statist. Phys. **87**, 311–332 (1997)

45. R. Bowen, *Equilibrium States and the Ergodic Theory of Anosov Diffeomorphisms*, Lecture Notes in Mathematics, Vol. 470, Springer-Verlag, Berlin-New York (1975)

46. R. Bowen, D. Ruelle, The ergodic theory of Axiom A flows. Invent. Math. **29**, 181–202 (1975)

47. J. Bricmont, J.L. Lebowitz, C.E. Pfister, On the equivalence of boundary conditions. J. Stat. Phys. **21**, 573–582 (1979)

48. J. Bricmont, J.L. Lebowitz, C.E. Pfister, periodic Gibbs states of ferromagnetic spin systems, J. Stat. Phys. **24**, 269–277 (1981)

49. J. Bricmont, T. Kuroda, J. Lebowitz, First order phase transitions in lattice and continuum systems: extension of Pirogov-Sinai theory. Commun. Math. Phys. **101**, 501–538 (1985)

50. J. Bricmont, J. Slawny. First order phase transitions and perturbation theory. In: *Statistical mechanics and field theory: mathematical aspects (Groningen, 1985)*, Lecture Notes in Phys., Vol. 257, p. 10–51, Springer, Berlin (1986)

51. J. Bricmont, A. Kupiainen, Lower critical dimensions for the random field Ising model. Physical Review Letters **59**, 1829–1832 (1987); Phase transition in the 3d random field Ising model. Commun. Math. Phys. **116**, 539–572 (1988)

52. J. Bricmont and J. Slawny. Phase transitions in systems with a finite number of dominant ground states. J. Statist. Phys. **54**, 89–161 (1989)

53. J. Bricmont, Science of chaos, or chaos in science? Physicalia Magazine, **17**, 159–208 (1995)

54. J. Bricmont, A. Kupiainen,, High-temperature expansions and dynamical systems. Commun. Math. Phys. **178**, 703–732 (1996)

55. J. Bricmont, Bayes, Boltzmann and Bohm: Probabilities in Physics. In: [56], p. 3–21

56. J. Bricmont, D. Dürr, M.C.Galavotti, G. Ghirardi, F. Petruccione, N. Zanghì (eds), *Chance in Physics. Foundations and Perspective*, Springer, Berlin, New York (2001)

57. J. Bricmont, *Making Sense of Quantum Mechanics*, Springer, Berlin (2016)

58. J. Bricmont, Probabilistic explanations and the derivation of macroscopic laws. In: [10], p. 31–64

59. L. Brillouin, Maxwell's demon cannot operate: information and entropy I. In: [227], p. 120–123

60. S.G. Brush, History of the Lenz-Ising model. Reviews of Modern Physics **39**, 883–893 (1967)

61. S.G. Brush, *The Kind of Motion we Call Heat: a History of the Kinetic Theory of Gases in the 19th Century*, North-Holland, Amsterdam (1976)

62. S.G. Brush, *The Kinetic Theory Of Gases. An Anthology of Classic Papers with Historical Commentary*, Nancy S Hall (ed.), Imperial College Press, London (2003)

63. D.C. Brydges, A short course on cluster expansions. In: *Critical Phenomena, Random Systems, Gauge Theories. Les Houches Session XLIII*, K. Osterwalder, R. Stora (eds), p. 139–183, Elsevier, Amsterdam (1984)

64. L. Bunimovich, Dynamical Billiards. Scholarpedia, **2(8)**, 1813 (2007)

65. E. Caglioti, N. Chernov, J.L. Lebowitz, Stability of solutions of hydrodynamic equations describing the scaling limit of a massive piston in an ideal gas. Nonlinearity **17**, 897–923 (2004)

66. H.B.Callen, *Thermodynamics and an Introduction to Thermostatistics*, 2d edition, Wiley and sons, New York (1985)

67. C. Callender, Reducing thermodynamics to statistical mechanics: the case of entropy. Journal of Philosophy **96**, 348–373 (1999)

68. F. Camia, C. Garban, C.M. Newman, The Ising magnetization exponent on $\mathbb{Z}^2$ is $\frac{1}{15}$. Probab. Theory and Related Fields **160**, 175–187 (2014)

69. E. Carlen, M. Carvalho, Entropy production estimates for Boltzmann equation with physically realistic collision kernels. J. Stat. Phys. **74**, 743–782 (1994)

70. S.M. Carroll, J. Chen: Spontaneous inflation and the origin of the arrow of time, http://arxiv.org/abs/hep-th/0410270 (2004)

71. M. Cassandro, E. Presutti, Phase transitions in Ising systems with long but finite range interactions. Markov Process and Related Fields **2**, 241–262 (1996)

72. C. Cercignani, *The Boltzmann Equation and Its Applications*, Springer-Verlag, Berlin, New York (1988)

73. C. Cercignani, R. Illner, and M. Pulvirenti, *The Mathematical Theory of Dilute Gases*, Springer-Verlag, Berlin (1994)
74. C. Cercignani, *Ludwig Boltzmann, the Man who Trusted Atoms, with a foreword by Roger Penrose*, Oxford University Press, Oxford (1998)
75. L. Cerino, F. Cecconi, M. Cencini, A. Vulpiani, The role of the number of degrees of freedom and chaos in macroscopic irreversibility. Physica A **442**, 486–497 (2016)
76. D. Chandler, *Introduction to Modern Statistical Mechanics*, Oxford University Press, Oxford (1987)
77. S. Chandrasekhar, Stochastic problems in physics and astronomy. Rev. Mod. Phys., **15** 1–89 (1943)
78. S. Chibarro, L. Rondoni, A. Vulpiani, *Reductionism, Emergence and Levels of Reality. The Importance of Being Borderline*, Springer, Cham (2014)
79. R. Clausius, Über verschiedene für die Anwendung bequeme Formen der Hauptgleichungen der mechanischen Wärmetheorie. Annalen der Physik und Chemie **125**, 353–400 (1865). English translation by J. Tyndall, On Several Convenient Forms of the Fundamental Equations of the Mechanical Theory of Heat. In: R. Clausius, *The Mechanical Theory of Heat, with its Applications to the Steam Engine and to Physical Properties of Bodies*, p. 327–365, van Voorst, London (1867)
80. N. Chernov, J.L. Lebowitz, Ya. Sinai, Dynamics of a massive piston in an ideal gas. Russian Mathematical Surveys **57**, 1045–1125 (2002)
81. N. Chernov, J.L. Lebowitz, Dynamics of a massive piston in an ideal gas: oscillatory motion and approach to equilibrium J. Stat. Phys. **109**, 507–527 (2002)
82. N. Chernov, J.L. Lebowitz, Ya. Sinai, Scaling dynamics of a massive piston in a cube filled with ideal gas: exact results. J. Stat. Phys. **109**, 529–548 (2002)
83. L. Chierchia, J. N. Mather, Kolmogorov-Arnold-Moser theory. Scholarpedia **5**(9), 2123 (2010)
84. J. Cohen, I. Stewart, *The Collapse of Chaos*, Penguin Books, New York, 1994.
85. P. Collet, J-P. Eckmann, *Iterated Maps of the Interval as Dynamical Systems*, Birkhaüüser, Basel (1980)
86. P. Collet, J-P. Eckmann, *Concepts and Results in Chaotic Dynamics: A Short Course*, Springer, Berlin, Heideberg (2006)
87. I.P. Cornfeld, S.V. Fomin, Ya-G. Sinai, *Ergodic Theory*, Springer, New York (1982)
88. A. A. Cournot, *Exposition de la théorie des chances et des probabilités*, Hachette, Paris (1843)
89. P. Coveney, The Second Law of thermodynamics: entropy, irreversibility and dynamics. Nature **333**, 409–415 (1988)
90. R.T.Cox, Probability, frequency and reasonable expectation, Am. J. Phys. **14**, 1–13 (1946)
91. M.A.B. Deakin, The wine/water paradox: background, provenance and proposed resolutions. The Australian Mathematical Society Gazette **33**, 200–205 (2006)
92. S. de Bièvre, P.E. Parris, A rigourous demonstration of the validity of Boltzmann's scenario for the spatial homogenization of a freely expanding gas and the equilibration of the Kac ring. J. Stat. Phys. **168**, 772–793 (2017)
93. B. de Finetti, La prévision: ses lois logiques, ses sources subjectives. Annales de l'Institut Henri Poincaré **7** 1–68 (1937). English translation: Foresight. Its Logical Laws, Its Subjective Sources, in: *Studies in Subjective Probability*, H. E. Kyburg, Jr. and H. E. Smokler (eds), p. 55–118, Robert E. Krieger Publishing Company, Malabar (Fl) (1980).
94. B. de Finetti, *Theory of Probability: A Critical Introductory Treatment*, Wiley, New York (2017)
95. R. De la Llave, A tutorial on KAM theory. Proceedings of Symposia in Pure Mathematics **69**, 175–296 (2001)
96. K. Denbigh, How subjective is entropy? In: [226], p. 109–115
97. L. Desvillettes, C. Villani, On the trend to global equilibrium for spatially inhomogeneous kinetic systems: the Boltzmann equation. Inventiones Mathematicae **159**, 245–316 (2005)
98. R. Devaney, *An Introduction To Chaotic Dynamical Systems*, 2d edition, CRC Press, Boca Raton (Fl) (2003)

99. P. Diaconis, S. Holmes, R. Montgomery, Dynamical bias in the coin toss. SIAM Review **49**, 211–235 (2007)

100. E.L. Dinaburg, Ya.G. Sinai, An analysis of ANNNI model by Peierls contour method. Commun. Math. Phys. **98**, 119–144 (1985)

101. E.L. Dinaburg, A.E. Mazel, Ya.G. Sinai, The ANNNI model and contour models with interactions. Sov. Sci. Rev. C Math./Phys., **6** 113–168 (1987)

102. R.L. Dobrushin. Existence of a phase transition in the two-dimensional and three-dimensional Ising models. Soviet Physics Dokl. **10**, 111–113 (1965)

103. R.L. Dobrushin. Description of a random field by means of conditional probabilities and conditions for its regularity. Teor. Verojatnost. i Primenen **13**, 201–229 (1968)

104. R.L. Dobrushin. Gibbs states describing a coexistence of phases for the three-dimensional Ising model. Th. Prob. and its Appl. **17**, 582–600 (1972)

105. R.L. Dobrushin, S.B. Shlosman. Absence of breakdown of continuous symmetry in two-dimensional models of statistical physics. Comm. Math. Phys. **42**, 31–40 (1975)

106. R.L. Dobrushin, S.B. Shlosman, Completely analytical Gibbs fields. In: *Statistical Physics and Dynamical Systems*, J. Fritz, A. Jaffe, and D. Szasz (eds), p. 347–370; p. 371–403, Birkenhauser, New York (1985)

107. R.L. Dobrushin, M. Zahradník, Phase diagrams for continuous spin models. Extention of Pirogov-Sinai theory. In: *Mathematical Problems of Statistical Mechanics and Dynamics*, R.L. Dobrushin (ed.), p. 1–123, Reidel, Dordrecht (1986)

108. R.L. Dobrushin, S.B. Shlosman, Completely analytical interactions: constructive description. J. Statist. Phys. **46**, 983–1014 (1987)

109. J.R. Dorfman, *An Introduction to Chaos in Nonequilibrium Statistical Mechanics*, Cambridge University Press, Cambridge (1999)

110. W. Driessler, L. Landau, J. Fernando Perez, Estimates of critical lengths and critical temperatures for classical and quantum lattice systems. J. Stat. Phys. **20**, 123–162 (1979)

111. N. Dunford, J.T. Schwartz, *Linear Operators; Part 1: General Theory*, Interscience, New York (1958)

112. D. Dürr, S. Goldstein, N. Zanghí, Quantum equilibrium and the origin of absolute uncertainty. J. Stat. Phys. **67**, 843–907 (1992)

113. D. Dürr, S. Teufel, *Bohmian Mechanics. The Physics and Mathematics of Quantum Theory*, Springer, Berlin, Heidelberg (2009)

114. D. Dürr, S. Goldstein, N. Zanghì, *Quantum Physics Without Quantum Philosophy*, Springer, Berlin, Heidelberg (2012)

115. D. Dürr, W. Struyve, Typicality in the foundations of statistical physics and Born's rule. In: V.Allori, A. Bassi, D. Dürr, N. Zanghì (eds), Do Wave Functions Jump? Perspectives of the Work of GianCarlo Ghirardi, p. 35–43, Springer, Cham (2021)

116. F. J. Dyson, Existence of a phase-transition in a one-dimensional Ising ferromagnet. Comm. Math. Phys., **12**, 91–107 (1969)

117. J. Earman, *A Primer on Determinism*, Reidel, Dordrecht (1986)

118. J. Earman, M. Rédei, Why ergodic theory does not explain the success of equilibrium statistical mechanics, British Journal for the Philosophy of Science **47**, 63-78 (1996)

119. J. Earman, J.D. Norton, Exorcist XIV: The wrath of Maxwell's demon. Part I: From Maxwell to Szilard. Studies in the History and Philosophy of Modern Physics, **29**, 435–471(1998); Part II: From Szilard to Landauer and beyond. Studies in the History and Philosophy of Modern Physics, **30**, 1–40 (1999)

120. J-P. Eckmann, D. Ruelle, Ergodic theory of chaos and strange attractors. Rev. Mod. Phys. **57**, 617–656 (1985)

121. P. Ehrenfest, T. Ehrenfest, Begriffliche Grundlagen der statistischen Auffassung in der Mechanik. In: Enzyklopädie der Mathematischen Wissenschaften, Vol. IV-4, Art. 32. English translation: *The Conceptual Foundations of the Statistical Approach in Mechanics*, translated by M. J. Moravesik, Cornell Univ. Press, Ithaca (1959)

122. A. Einstein, Autobiographical Notes. In: P. A. Schilpp (ed.), *Albert Einstein, Philosopher-Scientist*, The Library of Living Philosophers, Evanston (Ill.) (1949)

123. A. Erdelyi, *Asymptotic Expansions*, Dover, New York (2010)
124. E. Fermi, *Thermodynamics*, Dover, New York (1936)
125. E. Fermi, J. Pasta, S. Ulam, Studies of Nonlinear Problems, Document LA-1940, Los Alamos National Laboratory (1955)
126. R. Fernández, J. Fröhlich, A.D. Sokal, *Random Walks, Critical Phenomena, and Triviality in Quantum Field Theory*, Texts and Monographs in Physics, Springer-Verlag, Berlin (1992)
127. R. Fernández, Contour ensembles and the description of Gibbsian probability distributions at low temperature. 21o Colóquio Brasileiro de Matemática. [21st Brazilian Mathematics Colloquium]. Instituto de Matemática Pura e Aplicada (IMPA), Rio de Janeiro (1997)
128. R. Feynman, *The Character of Physical Law*, MIT Press, Cambridge (1967)
129. R. Feynman, R.B. Leighton, M. Sands, *The Feynman Lectures on Physics*, Volume 1, Addison-Wesley, Reading (Mass.) (1963)
130. M.E. Fisher, The free energy of a macroscopic system. Archive for Rational Mechanics and Analysis **17**, 377–410 (1964)
131. S. Friedli, Y. Velenik, *Statistical Mechanics of Lattice Systems: A Concrete Mathematical Introduction*, Cambridge University Press, Cambridge (2017)
132. R. Frigg, Typicality and the approach to equilibrium in Boltzmannian statistical mechanics. Philosophy of Science **76**, 997–1008 (2009)
133. R. Frigg, Why typicality does not explain the approach to equilibrium. In: Suárez, M. (ed.), *Probabilities, Causes and Propensities in Physics*, M. Suarez (ed.) p. 77–93, Synthese Library, Springer, Dordrecht (2011)
134. R. Frigg, C. Werndl, Demystifying typicality. Philosophy of Science **79**, 917–929 (2012)
135. R. Frigg, C. Werndl, Explaining thermodynamic-like behaviour in terms of epsilon-ergodicity. Philosophy of Science **78**, 628–652 (2013)
136. J. Fröhlich, B. Simon, T. Spencer, Infrared bounds, phase transitions and continuous symmetry breaking. Comm.Math. Phys. **50**, 79–95 (1976)
137. J. Fröhlich, T. Spencer, The Kosterlitz-Thouless transition in two-dimensional abelian spin systems and the Coulomb gas. Comm. Math. Phys. **81**, 527–602 (1981)
138. J. Fröhlich, T. Spencer, The phase transition in the one- dimensional Ising model with $\frac{1}{r^2}$ interaction energy. Comm. Math. Phys., **84**, 87–101 (1982)
139. G. Gallavotti, S. Miracle-Sole, Correlation functions of a lattice system. Commun. Math. Phys. **7**, 274–288 (1968)
140. G. Gallavotti, Ergodicity, ensembles, irreversibility in Boltzmann and beyond. J.Stat. Phys. **78**,1571–1589 (1995)
141. G. Gallavotti, *Statistical Mechanics. A Short Treatise*, Texts and Monographs in Physics. Springer-Verlag, Berlin (1999)
142. P. Gaspard, *Chaos, Scattering and Statistical Mechanics*, Cambridge University Press, Cambridge (2005)
143. K. Gawedzki, R. Kotecký, A. Kupiainen, Coarse-graining approach to first-order phase transitions. J. Stat. Phys. **47**, 701–724 (1987)
144. M. Gell-Mann, *The Quark and The Jaguar*, Little, Brown, London (1994)
145. H.-O. Georgii, *Gibbs Measures and Phase Transitions*, volume 9 of de Gruyter Studies in Mathematics , 2d edition, Walter de Gruyter and Co., Berlin (2011)
146. J.W. Gibbs. *Elementary Principles in Statistical Mechanics: Developed with Especial Reference to the Rational Foundation of Thermodynamics*, Dover, New York, (1960) [1902]
147. J. Glimm, A. Jaffe, T. Spencer, Phase transitions for ϕ_2^4 quantum fields. Commun. Math. Phys **45**, 203–216 (1975)
148. J. Glimm and A. Jaffe, *Quantum Physics: A Functional Integral Point of View*, Springer, Berlin (1987)
149. S. Goldstein, Boltzmann's approach to statistical mechanics. In [56] p. 39–54
150. S. Goldstein, J.L. Lebowitz, On the (Boltzmann) entropy of nonequilibrium systems. Physica D **193**, 53–66 (2004)
151. S. Goldstein, Typicality and notions of probability in physics. In [23] p. 59–71

152. S. Goldstein, Bohmian Mechanics, *The Stanford Encyclopedia of Philosophy*, Edward N. Zalta (ed.) (Spring 2013 Edition), available at: plato.stanford.edu/archives/spr2013/entries/qm-bohm/

153. S. Goldstein, R. Tumulka, N. Zanghì: Is the hypothesis about a low entropy initial state of the universe necessary for explaining the arrow of time? Physical Review D **94**: 023520 (2016)

154. S. Goldstein, Individualist and ensemblist approaches to the foundations of statistical mechanics. The Monist **102**, 439–457 (2019)

155. S. Goldstein, J.L. Lebowitz, R. Tumulka, N. Zanghì, Gibbs and Boltzmann entropy in classical and quantum mechanics. In: [10] p. 519–581

156. S. Goldstein, W. Struyve, R. Tumulka: The Bohmian approach to the problems of cosmological quantum fluctuations. In: Guide to the Philosophy of Cosmology, A. Ijjas and B. Loewer (eds), Oxford University Press, Oxford (2020)

157. G.A. Gottwald, M. Oliver, Boltzmann's dilemma: an introduction to statistical mechanics via the Kac ring. SIAM Rev. **51**, 613–635 (2009)

158. H. Grad: Principles of the Kinetic Theory of Gases. In: *Handbuch der Physik*, S. Flügge (ed.), Vol. 12, p. 205–294 Springer, Berlin (1958)

159. W.T. Grandy, *Foundations of Statistical Mechanics Vol. I: Equilibrium Theory*, Reidel, Dordrecht (1987)

160. R.B. Griffiths, Peierls proof of spontaneous magnetization in a two-dimensional Ising ferromagnet. Phys. Rev. A **136**, 437–439 (1964)

161. R.B. Griffiths, Correlations in Ising ferromagnets I. J. Math. Phys. **8**, 478–483, Correlations in Ising ferromagnets II. J. Math. Phys. **8**, 484–489 (1967)

162. R.B. Griffiths, Rigorous results and theorems. In: C. Domb and M.S. Green (ed.) *Phase Transitions and Critical Phenomena. Vol. 1*, Academic Press, New York (1972)

163. C. Gruber, A. Lesne, The adiabatic piston. In: *The Encyclopedia of Mathematical Physics*, J.-P. Françoise, G. L. Naber, T.S. Tsun (eds), p. 160–173, Elsevier, Amsterdam (2006)

164. J. Hadamard, Les surfaces à courbures opposées et leurs lignes géodésiques. J. Math.pures et appl. **4**, 27–74 (1898)

165. J. Hadamard, Les principes du calcul des probabilités. Revue de Métaphysique et de Morale **29**, 289–293 (1922)

166. A. Hájek, Interpretations of Probability, The Stanford Encyclopedia of Philosophy (Fall 2019 Edition), Edward N. Zalta (ed.), available at: https://plato.stanford.edu/archives/fall2019/entries/probability-interpret/.

167. P.M. Harman, *The Natural Philosophy of James Clerk Maxwell*, Cambridge University Press, Cambridge (1998)

168. W. Heisenberg, *The Physicist's Conception of Nature*, Hutchinson, London (1958)

169. C. Hempel, The function of general laws in history. Journal of Philosophy **39**, 35–48 (1942)

170. C. Hempel, P. Oppenheim, Studies in the logic of explanation. Philosophy of Science, **15**, 135–175 (1948)

171. T. Hill, *Statistical Mechanics*, McGraw-Hill, New York (1956)

172. M. Hirsch, S. Smale, R.L. Devaney, *Differential Equations, Dynamical Systems, and an Introduction to Chaos*, Academic Press, New York (2012)

173. W. Holsztynski, J. Slawny, Phase transitions in ferromagnetic spin systems at low temperatures. Commun. Math. Phys. **66** 147–166 (1979)

174. F. Hoyle, The Asymmetry of Time, Third annual lecture to the research student's association, Camberra (1962)

175. J. Imbrie, Phase diagrams and cluster expansions for low temperature $P(\phi)_2$ models. 1. The phase diagram. Commun. Math.Phys. **82**, 261–304 (1981). 2. The Schwinger functions. Commun. Math. Phys. **82**, 305–343 (1981)

176. E. Ising, Beitrag zur Theorie des Ferromagnetismus. Z. Phys. **31**, 253–258 (1925)

177. E.T. Jaynes, Gibbs vs Boltzmann entropies. Amer. J. of Phys. **33**, 391–398 (1965)

178. E.T. Jaynes, Violation of Boltzmann's H-theorem in real gases. Phys. Rev. **A4**, 747–750 (1971)

179. E.T. Jaynes, The well-posed problem. Foundations of Physics **3** , 477–493 (1973)

180. E.T. Jaynes, *Papers on Probability, Statistics and Statistical Physics*, R. D. Rosencrantz (ed.), Reidel, Dordrecht (1983)
181. E.T. Jaynes, Clearing up mysteries-the original goal. In: *Maximum Entropy and Bayesian Methods*, J. Skilling (ed.), Kluwer, Dordrecht (1989)
182. E.T. Jaynes, The Gibbs paradox. In: *Maximum Entropy and Bayesian Methods*, G. Erikson, P. Neudorfer, C. R. Smith (eds) Kluwer Acad. Pub., Dordrecht (1991)
183. E.T. Jaynes, *Probability Theory: the Logic of Science*, Cambridge University Press, Cambridge (2003)
184. M. Kac, Foundations of kinetic theory. In: *Proceedings of the Third Berkeley Symposium on Mathematical Statistics and Probability*, vol. III, J. Neyman (ed.), p. 171–197, University of California Press, Berkeley (1956)
185. M. Kac, *Probability and Related Topics in the Physical Sciences*, Interscience Pub., New York (1959)
186. A. Katok, B. Hasselblatt, *Introduction to the Modern Theory of Dynamical Systems. With a Supplementary Chapter by Katok and Leonardo Mendoza*, Encyclopedia of Mathematics and its Applications, Cambridge University Press, Cambridge (1995)
187. J. B. Keller, The probability of heads. American Mathematical Monthly **93**, 191–197 (1986)
188. D.G. Kelley, S. Sherman, General Griffiths' inequalities on correlations in Ising ferromagnets. J. Math. Phys. **9**, 466–484 (1969)
189. J. Kemeny, Fair bets and inductive probabilities. Journal of Symbolic Logic, **20**, 263–273 (1955)
190. J.M. Keynes, *A Treatise of Probability*, MacMillan, London (1921)
191. A.I. Khinchin, *Mathematical Foundations of Statistical Mechanics*, translated by G. Gamow, Dover, New York (1949)
192. A.I. Khinchin, *Mathematical Foundations of Information Theory*, translated by R. A. Silverman and M. D. Friedman, Dover, New York (1957)
193. R. Kindermann, J.L. Snell, *Markov Random Fields and Their Applications*, volume 1 of Contemporary Mathematics, American Mathematical Society, Providence (R.I.) (1980)
194. M.J. Klein, Entropy and the Ehrenfest urn model. Physica **22**, 569–575 (1956)
195. S. Kobe, Ernst Ising–physicist and teacher. J. Stat. Phys. **88**, 991–995 (1997)
196. A.N. Kolmogorov, *Foundations of Probability*, Chelsea Publishing Company, New York (1950) [1933]
197. A.N. Kolmogorov, S. V. Fomin, *Elements of the Theory of Functions and Functional Analysis*, Dover, New York (1999)
198. R. Kotecký, D. Preiss, An inductive approach to Pirogov-Sinai theory. Proc. Winter School on Abstract Analysis, Suppl. ai Rend. del Mat. di Palermo (1983)
199. R. Kotecký and D. Preiss, Cluster expansion for abstract polymer models. Comm. Math. Phys. **103**, 491–498 (1986)
200. N.S. Krylov, *Works on the Foundations of Statistical Mechanics*, Princeton University Press, Princeton (1979)
201. L.D. Landau, E.M. Lifshitz, *Statistical Physics. Course of Theoretical Physics, Volume 5*, Butterworth-Heinemann, Oxford (1980)
202. R. Landauer, Irreversibility and heat generation in the computing process. IBM Journal of Research and Development **5**, 183–191 (1961)
203. P.T. Landsberg, *The Enigma of Time*, Adam Hilger, Bristol (1982)
204. O.E. Lanford, III, D. Ruelle, Observables at infinity and states with short range correlations in statistical mechanics. Comm. Math. Phys. **13**, 194–215 (1969)
205. O.E. Lanford, Entropy and equilibrium states in classical statistical mechanics. In: *Statistical Mechanics and Mathematical Problems*, A. Lenard (ed.), Lecture Notes in Physics **20**, p. 1–113, Springer-Verlag, Berlin (1973)
206. O.E. Lanford, Time evolution of large classical systems, In: *Dynamical Systems, Theory and Applications: Battelle Seattle 1974 Rencontres*, Jürgen Moser (ed.) Lecture Notes in Physics vol. 38, p. 1–111, Springer-Verlag, Berlin (1975)

207. O.E. Lanford, Introduction to the modern theory of dynamical systems. In: *Chaotic Behaviour of Dynamical Systems*, G. Iooss, R.H.G. Helleman, B. Stora (eds), p. 5–51, North-Holland, Amsterdam (1983)
208. O.E. Lanford, On a derivation of the Boltzmann equation. Astérisque **40**, 117–137, 1976. Reprinted in: *Nonequilibrium Phenomena 1: The Boltzmann Equation*, J.L. Lebowitz, E. W. Montroll (eds), North-Holland, Amsterdam (1983)
209. P.S. Laplace, *Essai philosophique sur les probabilités*, C. Bourgeois, Paris 1986, text of the fifth edition (1825). English translation: *A Philosophical Essay on Probabilities*, translated by F. W. Truscott and F. L. Emory, Dover, New York (1951)
210. D.A. Lavis , The concept of probability in statistical mechanics. In: *Frontiers of Fundamental Physics 4*, B.G. Sidarth and M.V. Altaisky (eds) p. 293–308, Kluwer, Dordrecht (2001)
211. D.A. Lavis, Boltzmann, Gibbs and the concept of equilibrium. Philosophy of Science **75**, 682–696 (2008)
212. D. Lazarovici, P. Reichert, Typicality, irreversibility and the status of macroscopic laws. Erkenntnis **80**, 689–716 (2015)
213. D. Lazarovici, On Boltzmann versus Gibbs and the equilibrium in statistical mechanics. Philosophy of Science **86**, 785–793 (2019)
214. D. Lazarovici, P. Reichert, Arrow of time without a past hypothesis, in [10] p. 343–386
215. D. Lazarovici, P. Reichert, Against (epsilon-) ergodicity (Unpublished)
216. J.L. Lebowitz, O. Penrose, Rigorous treatment of the van der Waals Maxwell theory of the liquid-vapor transition. J. Mat. Phys. **7**, 98–113 (1966)
217. J.L. Lebowitz, Macroscopic laws, microscopic dynamics, time's arrow and Boltzmann's entropy. Physica **A194**, 1–27 (1993)
218. J.L. Lebowitz, Boltzmann's entropy and time's arrow. *Physics Today*, Sept. 1993, **46**, 32–38; for replies in Physics Today, see [19]
219. J.L. Lebowitz, Microscopic reversibility and macroscopic behavior: physical explanations and mathematical derivations, Turkish Journal of Physics, **19**, 1–20 (1995). Also in: *25 Years of Non-Equilibrium Statistical Mechanics*, Proceedings, Sitges Conference, Barcelona, Spain, 1994, Lecture Notes in Physics, J.J. Brey, J. Marro, J.M. Rubi, M. SanMiguel (eds), Springer-Verlag, Berlin, Heidelberg (1995)
220. J.L. Lebowitz, A. Mazel, Improved Peierls argument for high dimensional Ising models. J. Stat. Phys. **90**, 1051–1059 (1998)
221. J.L. Lebowitz, Statistical mechanics: a selective review of two central issues. Reviews of Modern Physics **71**, 346–357 (1999)
222. J.L. Lebowitz, A. Mazel, E. Presutti, Liquid-vapor phase transitions for systems with finite range interactions. J. Stat. Phys. **94**, 955–1025 (1999)
223. J.L. Lebowitz, Microscopic Origins of irreversible macroscopic behavior. Physica **A 263**, 516–527 (1999)
224. J.L. Lebowitz, C. Maes, Entropy – a dialogue. In: *Entropy*, A. Greven, G. Keller and G. Warnecke (eds), p. 269–276. Princeton University Press, Princeton and Oxford (2003)
225. J.L. Lebowitz: From time-symmetric microscopic dynamics to time-asymmetric macroscopic behavior: an overview. In: *Boltzmann's Legacy*, G. Gallavotti , W. L. Reiter, J. Yngvason (eds) p. 63–88, European Mathematical Society, Zürich (2008)
226. H.S. Leff, A. Rex (eds), *Maxwell's Demon: Entropy, Classical and Quantum Information, Computing*, Adam Hilger, Bristol (1990)
227. H.S. Leff, A. Rex (eds), *Maxwell's Demon 2: Entropy, Classical and Quantum Information, Computing*, Institute of Physics Publishing, Bristol and Philadelphia (2003)
228. J. Leinaass, J. Myrheim, On the theory of identical particles. Il Nuovo Cimento **37B**, 1–23 (1977)
229. W. Lenz, Beiträge zum Verständnis der magnetischen Eigenschaften in festen Körpern. Physikalische Zeitschrift **21**, 613–615 (1920)
230. P. Lévy, *Calcul des probabilités*, Gauthier-Villars, Paris (1925)
231. J. Lighthill, The recently recognized failure of predictability in Newtonian dynamics. Proc. Roy. Soc. **A 407**, 35–50 (1986)

232. J. Loschmidt, Über den Zustand des Wärmegleichgewichtes eines Systems von Körpern mit Rücksicht auf die Schwerkraft. Sitzungberichte, K. Akademie der Wissenschaften in Wien, Math.-Naturwiss. Kl., 73: 128-42 (1876)
233. S.K. Ma, *Statistical Mechanics*, World Scientific, Singapore (1985)
234. D.J.C. MacKay, *Information Theory, Inference, and Learning Algorithms*, Cambridge University Press, Cambridge (2003)
235. V.A. Malyshev, R.A. Minlos, *Gibbs Random Fields. Cluster Expansions*, Kluwer, Dordrecht (1991)
236. T. Maudlin, The grammar of typicality, in [10] p. 231–251
237. J.C. Maxwell, *Theory of Heat*, Longmans, Green and Co., London (1871)
238. J.C. Maxwell, Tait's "Thermodynamics". Nature **17**, 257–259 (1878)
239. J.C. Maxwell, *Scientific Letters and Papers*, Vol. 2, 1862–1873, P. M Harman (ed.), Cambridge University Press, Cambridge (1995)
240. A. Mazel, I. Stuhl, Y. Suhov, High-density hard-core model on $\mathbb{Z}^2$ and norm equations in ring $Z[\sqrt{3}]$. arXiv:1909.11648v2 (2019)
241. J. Miekisz, Low-temperature equilibrium states of ferromagnetic lattice systems Journal of Physics A **21**, 1679–1688 (1988)
242. T. Norsen, *Foundations of Quantum Mechanics: An Exploration of the Physical Meaning of Quantum Theory*, Springer International Publishing, Cham, Switzerland (2017)
243. J.D. Norton, Eaters of the Lotus: Landauer's principle and the return of Maxwell's demon. Studies in History and Philosophy of Modern Physics **36**, 375–411 (2005)
244. J.D. Norton, Waiting for Landauer. Studies in History and Philosophy of Modern Physics **42**, 184–198 (2011)
245. J.D. Norton, Author's reply to 'Landauer defended'. Studies in History and Philosophy of Modern Physics **44**, 272 (2013)
246. J.D. Norton, All shook up: fluctuations, Maxwell's demon and the thermodynamics of computation. Entropy **15**, 4432–4483 (2013)
247. J.D. Norton, Maxwell's demon does not compute. In: *Physical Perspectives on Computation, Computational Perspectives on Physics*, M. E. Cuffaro and S. C. Fletcher (eds) p. 240–256, Cambridge University Press, Cambridge (2018)
248. J.D. Norton, The worst thought experiment. In: *The Routledge Companion to Thought Experiments*, M. T. Stuart, J. R. Brown, and Y.Fehige (eds), p. 454–468, Routledge, London (2018)
249. L. Onsager, Crystal statistics. I. A two-dimensional model with an order-disorder transition. Phys. Rev. (2), **65**, 117–149 (1944)
250. W. Pauli, Wahrscheinlichkeit und Physik. Dialectica, **8**, 112–124 (1954). English translation in: W. Pauli *Writings on Physics and Philosophy*, Charles P Enz and Karl von Meyenn (eds), translated by R. Schlapp, Springer-Verlag, Berlin, Heidelberg (1994)
251. R. Peierls, On Ising's model of ferromagnetism. Proc. Camb. Phil. Soc. **32**, 477–481 (1936)
252. C.S. Peirce, The fixation of belief. Popular Science Monthly **12**, 1–15 (1877)
253. C.S. Peirce, Notes on the doctrine of chances. In: *Essays in the Philosophy of Science* (The American Heritage Series) p. 74–84, Bobbs-Merrill, Indianapolis and New York (1957)
254. A. Pelissetto E. Vicari, Critical phenomena and renormalization-group theory. Physics Reports **368**, 549–727 (2002)
255. R. Penrose, *The Emperor's New Mind*, Oxford University Press, New York (1990)
256. R. Penrose, On the Second Law of thermodynamics. J. Stat. Phys. **77**, 217–221 (1994)
257. S.A Pirogov, Ya.G. Sinai, Phase diagrams of classical lattice systems. Theor. and Math. Phys. **25**, 358–369, **26**, 1185–1192 (1975)
258. I. Pitowsky, Typicality and the role of Lebesgue measure in statistical mechanics. In: [23] p. 41-58
259. M. Plancherel, Beweis des Unmöglichkeit ergodischer mechanischer Systeme. Annalen der Physik **4**, 1061–1063 (1913). English translation: Proof of the impossibility of ergodic mechanical systems, in [62], p. 521–523
260. H. Poincaré, Sur les tentatives d'explication mécanique des principes de la thermodynamique. Comptes rendus de l'Académie des Sciences **108**, 550–553 (1889)

261. H. Poincaré, Sur le problème des trois corps et les équations de la dynamique. Acta Math. **13**, 1–270 (1890). The recurrence theorem is on pp. 67–72. For that theorem in English, see [62] p. 368–376
262. H. Poincaré Le Mécanisme et l'expérience. Revue de métaphysique et de morale **1**, 534–537 (1893). English translation: Mechanism and Experience in [62] p. 377–381
263. H. Poincaré, *Science et Méthode*, Flammarion, Paris, 1909
264. H. Poincaré, *The Foundations of Science: Science and Hypothesis, The Value of Science, Science and Method*, translated by G.B. Halsted, The Science Press, New York (1913)
265. K.R. Popper, *Quantum Theory and the Schism in Physics*, Rowman & Littlefield, Totowa (N.J.) (1956)
266. K.R. Popper, *The Open Universe. An Argument for Indeterminism.*, Rowman & Littlefield, Totowa (N.J.) (1956)
267. K.R. Popper, *The Logic of Scientific Discovery* , Hutchinson, London (1958)
268. K.R. Popper, Autobiography. In: *The Philosophy of Karl Popper*, P. A. Schilpp (ed), p. 3–181, Open Court, La Salle (Ill.) (1974)
269. E. Presutti, *Scaling Limits in Statistical Mechanics and Microstructures in Continuum Mechanics*, Theoretical and Mathematical Physics, Springer, Berlin (2009)
270. I. Prigogine, I. Stengers, *La Nouvelle Alliance. Métamorphoses de la science*, Gallimard, Paris (1979)
271. I. Prigogine, Un siècle d'espoir. In: *Temps et Devenir, Colloque de Cerisy, à partir de l'oeuvre d'Ilya Prigogine*; J. P. Brans, I. Stengers, P. Vincke (eds), Patino, Genève (1983)
272. I. Prigogine, I. Stengers, *Entre le temps et l'éternité*, Fayard, Paris (1988)
273. I. Prigogine, *Les Lois du Chaos*, Flammarion, Paris (1994)
274. M. Reed, B. Simon, *Methods of Modern Mathematical Physics I: Functional Analysis*, Academic Press, New York (1972)
275. P. Reichert, Essentially ergodic behaviour. British Journal for the Philosophy of Science, https://www.journals.uchicago.edu/doi/10.1093/bjps/axaa007
276. L.E. Reichl, *The Transition to Chaos in Conservative Classical Systems: Quantum Manifestations*, Springer, New York (1992)
277. A. Rosenthal, Beweis des Unmöglichkeit ergodischer Gassyteme. Ann. Phys. **42**, 796–806 (1913). English translation: Proof of the impossibility of ergodic gas systems, in [62], p. 513–520
278. H.L. Royden, *Real Analysis*, Macmillan, New York (1968)
279. W. Rudin, *Real and Complex Analysis*, McGraw-Hill, London (1970)
280. D. Ruelle, *Statistical Mechanics: Rigorous Results*, Benjamin, London (1969)
281. D. Ruelle, A measure associated with axiom-A attractors. Amer. J. Math. **98**, 619–654 (1976)
282. D. Ruelle, *Thermodynamic Formalism*, Encyclopedia of Mathematics and Its Applications No 5, Addison Wesley, New York (1978)
283. B. Russell, On the notion of cause with application to the free-will problem. Chap. 8 in: B. Russell, *Our Knowledge of the External World: As a Field for Scientific Method in Philosophy*, The Open Court Pub. Co., Chicago (1914)
284. N. Sator, N. Pavloff, *Physique Statistique*, Vuibert, Paris (2016)
285. E. Schrödinger, *What is Life? And Other Scientific Essays*, Doubleday Anchor Books, New York (1965)
286. E. Schrödinger, *What is Life? : The Physical Aspect of the Living Cell with Mind and Matter and Autobiographical Sketches*, Cambridge University Press, Cambridge (1992)
287. J.T. Schwartz, The pernicious influence of mathematics on science. In: *Logic, Methodology and Philosophy of Science*, E. Nagel, P. Suppes, A. Tarski (eds), Stanford University Press, Stanford (1962)
288. E. Seneta, *Non-negative Matrices and Markov Chains*, Springer-Verlag, New York (1981)
289. G. Shafer, V. Vovk, The sources of Kolmogorov's *Grundbegriffe*. Statistical Science **21**, 70–98 (2006)
290. G. Shafer: Why did Cournot's principle disappear? Talk at École des Hautes Études en Sciences Sociales, Paris. May 19, 2006, http://www.glennshafer.com/assets/downloads/disappear.pdf

291. C.E. Shannon, A mathematical theory of communication. Bell System Technical Journal **27**, 379–423 (1948)
292. B. Simon, *The Statistical Mechanics of Lattice Gases. Vol. I*, Princeton University Press, Princeton, (1993)
293. B. Simon, *Phase Transitions in the Theory of Lattice Gases*, Cambridge University Press, in preparation.
294. Ya.G. Sinai, On the notion of entropy of a dynamical system. Dokl. Akad. Nauk SSSR, **124** no. 4, 768–771 (1959)
295. Ya.G. Sinai, Gibbs measures in ergodic theory. Russian Mathematical Surveys **27**, 21–69 (1972)
296. Ya.G. Sinai, *Introduction to Ergodic Theory*, translated from the Russian by V. Sheffer, Princeton University Press, Princeton (1976)
297. Ya.G. Sinai, *Theory of Phase Transitions: Rigorous Results*, translated from the Russian by J. Fritz, A. Krámli, P.Major and D. Szász, Pergamon Press, Oxford (1982)
298. L. Sklar, *Physics and Chance. Philosophical Issues in the Foundations of Statistical Mechanics*, Cambridge University Press, Cambridge (1993)
299. J. Slawny, Low temperature properties of classical lattice systems: Phase transitions and phase diagrams. In: *Phase Transitions and Critical Phenomena*, vol. 11, p. 127–205, C. Domb, J.L. Lebowitz (eds), Academic Press, New York (1987)
300. M. Smoluchowski, Experimentell nachweisbare, der üblichen Thermodynamik widersprechende Molekularphänomene. Physikalische Zeitschrift **13**, 1069–1080 (1912)
301. E. Sober, *Philosophy of Biology*, Westview Press, Boulder (CO) (2000)
302. A.D. Sokal. A rigorous inequality for the specific heat of an Ising or ϕ^4 ferromagnet. Phys. Lett. A **71**, 451–453 (1979)
303. A.D. Sokal, J. Bricmont: *Fashionable Nonsense: Postmodern Intellectuals' Abuse of Science*, Picador, New York (1998)
304. H. Spohn, *Large Scale Dynamics of Interacting Particles*, Springer, Berlin (1991)
305. D. Stove *Popper and after. Four Modern Irrationalists*, Pergamon Press, Oxford (1982)
306. S.H. Strogatz, *Nonlinear Dynamics and Chaos (with applications to Physics, Biology, Chemistry and Engineering)*, Addison-Wesley, New York (1994)
307. L. Szilard, On the decrease of entropy in a thermodynamic system by the intervention of intelligent beings. In: *The Collected Works of Leo Szilard: Scientific Papers*, p.120–129, MIT Press, Cambridge (1972)
308. C.J. Thompson, *Classical Equilibrium Statistical Mechanics*, Oxford University Press, Oxford (1988)
309. W. Thomson, Kinetic theory of the dissipation of energy. Nature **9**, 441–444 (1874)
310. R.C. Tolman, *The Principles of Statistical Mechanics*, Oxford University Press, London (1938)
311. M. Tribus, *Boelter Anniversary Volume*, McGraw Hill, New York (1963)
312. M. Tribus, E.C. McIrvine, Energy and information. Scientific American **225**, No. 3, 179–190 (1971)
313. R. Tumulka, Understanding Bohmian mechanics – A dialogue. American Journal of Physics **72**, 1220–1226 (2004)
314. R. Tumulka, *Lecture Notes on Mathematical Statistical Physics* (2019) available on https://www.math.uni-tuebingen.de/de/forschung/maphy/lehre/ss-2019/statisticalphysics/dateien/lecture-notes.pdf
315. R. Tumulka: The problem of Boltzmann brains and how Bohmian mechanics helps solve it. To appear in the *Proceedings of the 15th Marcel Grossmann Meeting on General Relativity* (Rome 2018), E. Battistelli, R. T. Jantzen, R. Ruffini (eds) World Scientific, Singapore (2021). Available on: https://arxiv.org/abs/1812.01909
316. J. Uffink, Compendium of the foundations of classical statistical physics. In: *Handbook for the Philosophy of Physics*, J. Butterfield, J. Earman (eds), p. 923-1047, Elsevier, Amsterdam (2007)
317. H. van Beijeren, Interface sharpness in the Ising system. Comm. Math. Phys. **40**, 1–6 (1975)

318. A.C.D. van Enter, R. Fernández, A.D. Sokal, Regularity properties and pathologies of position-space renormalization-group transformations: scope and limitations of Gibbsian theory, J. Statist. Phys. **72**, 879–1167 (1993)

319. R. van Riel, R. Van Gulick, Scientific Reduction, The Stanford Encyclopedia of Philosophy (Spring 2019 Edition), Edward N. Zalta (ed.), available at: https://plato.stanford.edu/archives/spr2019/entries/scientific-reduction/

320. S. Vauclair, *Eléments de Physique Statistique*, Interéditions, Paris (1993)

321. C. Villani, Convergence to equilibrium: entropy production and hypocoercivity. Harold Grad lecture, delivered in the 24th Rarefied Gas Dynamics conference, Bari (July 2004)

322. S. Vineberg, Dutch Book Arguments, The Stanford Encyclopedia of Philosophy (Spring 2016 Edition), Edward N. Zalta (ed.), available at: https://plato.stanford.edu/archives/spr2016/entries/dutch-book/

323. S.B. Volchan, Probability as typicality. Studies in History and Philosophy of Modern Physics **38**, 801–814 (2007)

324. R. Von Mises, *Probability, Statistics and Truth*, 2d revised English edition, Allen and Unwin, London (1957)

325. J. von Neumann, *Mathematische Grundlagen der Quantenmechanik*, Springer Verlag, New York-Heidelberg-Berlin 1932; English translation by R.T. Beyer, *Mathematical Foundations of Quantum Mechanics*, Princeton University Press, Princeton, N.J., 1955

326. D. Wallace, The necessity of Gibbsian statistical mechanics, in [10], p. 583–616

327. P. Walters, *An Introduction to Ergodic Theory*, Springer, New York (1982)

328. S. Weinberg, *Dreams of a Final Theory*, Vintage, London (1993)

329. I. Wilhelm, Typical: a theory of typicality and typicality explanation. The British Journal for the Philosophy of Science, to appear.

330. J. Woodward, L. Ross, "Scientific Explanation", The Stanford Encyclopedia of Philosophy (Summer 2021 Edition), Edward N. Zalta (ed.), available at: https://plato.stanford.edu/archives/sum2021/entries/scientific-explanation/

331. C.N. Yang. The spontaneous magnetization of a two-dimensional Ising model. Physical Rev. (2), **85**, 808–816 (1952)

332. M. Zahradník, An alternate version of Pirogov-Sinai theory. Commun. Math. Phys. **93**, 559–581 (1984)

333. N. Zanghì, Lecture notes on Statistical Physics, available on http://www.ge.infn.it/~zanghi/.

334. E. Zermelo, Uber einen Satz der Dynamik und die mechanische Wärmetheorie. Annalen der Physik **57**, 485–494 (1896). English translation: On a Theorem of Dynamics and the Mechanical Theory of Heat, in [62], p. 382–391

335. E. Zermelo Über mechanische Erklärungen irreversibler Vorgänge. Annalen der Physik, **59**, 793–801 (1896). English translation: On the Mechanical Explanation of Irreversible Processes, in [62], p. 403–411

Index

© Springer Nature Switzerland AG 2022
J. Bricmont, *Making Sense of Statistical Mechanics*, Undergraduate Lecture Notes in Physics, https://doi.org/10.1007/978-3-030-91794-4